Urban Informatics

Urban Informatics: Using Big Data to Understand and Serve Communities introduces the reader to the tools of data management, analysis, and manipulation using R statistical software. Designed for undergraduate and above level courses, this book is an ideal onramp for the study of urban informatics and how to translate novel data sets into new insights and practical tools.

The book follows a unique pedagogical approach developed by the author to enable students to build skills by pursuing projects that inspire and motivate them. Each chapter has an Exploratory Data Assignment that prompts readers to practice their new skills on a data set of their choice. These assignments guide readers through the process of becoming familiar with the contents of a novel data set and communicating meaningful insights from the data to others.

CHAPMAN & HALL/CRC DATA SCIENCE SERIES

Reflecting the interdisciplinary nature of the field, this book series brings together researchers, practitioners, and instructors from statistics, computer science, machine learning, and analytics. The series will publish cutting-edge research, industry applications, and textbooks in data science.

The inclusion of concrete examples, applications, and methods is highly encouraged. The scope of the series includes titles in the areas of machine learning, pattern recognition, predictive analytics, business analytics, Big Data, visualization, programming, software, learning analytics, data wrangling, interactive graphics, and reproducible research.

Published Titles

Statistical Foundations of Data Science
Jianqing Fan, Runze Li, Cun-Hui Zhang, and Hui Zou

A Tour of Data Science: Learn R and Python in Parallel
Nailong Zhang

Explanatory Model Analysis
Explore, Explain, and, Examine Predictive Models
Przemyslaw Biecek, Tomasz Burzykowski

An Introduction to IoT Analytics
Harry G. Perros

Data Analytics
A Small Data Approach
Shuai Huang and Houtao Deng

Public Policy Analytics
Code and Context for Data Science in Government
Ken Steif

Supervised Machine Learning for Text Analysis in R
Emil Hvitfeldt and Julia Silge

Massive Graph Analytics
Edited by David Bader

Data Science
An Introduction
Tiffany-Anne Timbers, Trevor Campbell and Melissa Lee

Tree-Based Methods
A Practical Introduction with Applications in R
Brandon M. Greenwell

Urban Informatics
Using Big Data to Understand and Serve Communities
Daniel T. O'Brien

For more information about this series, please visit: https://www.routledge.com/Chapman--HallCRC-Data-Science-Series/book-series/CHDSS

Urban Informatics

Using Big Data to Understand and Serve Communities

Daniel T. O'Brien

CRC Press is an imprint of the
Taylor & Francis Group, an **informa** business
A CHAPMAN & HALL BOOK

First edition published 2023
by CRC Press
6000 Broken Sound Parkway NW, Suite 300, Boca Raton, FL 33487-2742

and by CRC Press
4 Park Square, Milton Park, Abingdon, Oxon, OX14 4RN

CRC Press is an imprint of Taylor & Francis Group, LLC

Library of Congress Cataloging-in-Publication Data

Names: O'Brien, Daniel T., 1983- author.
Title: Urban informatics : using big data to understand and serve communities / Daniel T. O'Brien.
Description: First edition. | Boca Raton : CRC Press, [2023] | Series: Chapman & Hall/CRC data science series | Includes bibliographical references and index.
Identifiers: LCCN 2022011262 (print) | LCCN 2022011263 (ebook) | ISBN 9781032264592 (pbk) | ISBN 9781032274683 (hbk) | ISBN 9781003292951 (ebk)
Subjects: LCSH: City planning. | Big data.
Classification: LCC HT166 .O333 2023 (print) | LCC HT166 (ebook) | DDC 307.1/216--dc23/eng/20220323
LC record available at https://lccn.loc.gov/2022011262
LC ebook record available at https://lccn.loc.gov/2022011263

ISBN: 978-1-032-27468-3 (hbk)
ISBN: 978-1-032-26459-2 (pbk)
ISBN: 978-1-003-29295-1 (ebk)

DOI: 10.1201/9781003292951

Typeset in LM Roman
by KnowledgeWorks Global Ltd.

Contents

Preface

They say the "age of smart cities" is upon us. This is often equated with science fiction-y technologies that are under development, such as autonomous vehicles and algorithms that can predict crimes before they happen. But there are more immediate opportunities to transform cities and communities thanks to the vast proliferation of digital data. Novel data resources, including administrative records, social media platforms, and sensor technologies, offer original insights that can help us to refine, improve, and reimagine the ways communities work and the services and products that can support them.

As the name suggests, this book teaches the fundamental skills of *urban informatics*, or the use of data and technology to better understand and serve communities. This includes the technical skills of accessing, manipulating, analyzing, and visualizing complex, messy, and "big" data sets using R, as well as the ability to interpret and make sense of them. Come be a part of the smart cities revolution and help transform the communities of the 21st century.

This book is designed to support learners who would like to leverage data science *for the purpose* of having public impact. As with any textbook, many readers will use this as a reference, dipping in and out to learn or refresh skills as needed. Others may work through it linearly as a full curriculum, either as part of a formal course or a self-directed venture. In either case, the book takes an experiential learning approach, including the following features:

* **Urban Informatics integrates technical and conceptual skills**, guiding students to make informed decisions about the interpretation of data and their analysis and visualization. Within this there is an especial emphasis on unpacking questions of equity. * More than just another statistics textbook, **the technical curriculum consists of both data management and analytics,** including both as needed to become acquainted with and reveal the content of a new data set. * **All content is contextualized in real-world applications relevant to community concerns.** * **Real-world worked examples, all drawn from greater Boston, MA,** are made possible through public data sets from the Boston Area Research Initiative's (https://www.bostonarearesearchinitiative.net) Boston Data Portal (https://cssh.northeastern.edu/bari/boston-data-portal/). * **Each chapter features traditional problem sets and an Exploratory Data Assignment that prompts students to practice their new skills on a data set of their choice.** This alternative set of assignments guides students through the process of becoming familiar with the contents of a novel data set and communicating meaningful insights from the data to others. * **Unit-level assignments, including Community Experiences** that prompt students to evaluate the assumptions they have made about their data against real-world information.

Please enjoy, and welcome to the burgeoning world of urban informatics!

Acknowledgments

In one sense, this book was a sprint, completed in a single semester. In another, it has been a long-term project beginning in Summer 2014 when I first prepared the curriculum for the course "Big Data for Cities," which has evolved into this text. I see it as only appropriate to thank everyone along the way.

I was hired by Northeastern University in 2014—a decision made primarily by David Lazer, who was leading a search for computational social scientists, Joan Fitzgerald, who was then Director of the School of Public Policy and Urban Affairs, and Uta Poiger, Dean of the College of Social Sciences and Humanities—ostensibly to deepen the bench of faculty for their newly established Masters of Science in Urban Informatics. I was asked to design and teach Big Data for Cities (BD4C) in my first semester as the introductory course to the program. With advice from my new colleagues James Connolly, Dietmar Offenhuber, and Nick Beauchamp, as well as the new Director of both the School and Program Matthias Ruth, I set out to design a course that wasn't quite a statistics course or a data management course, but something in between that gave students the true experience of "discovering" and working through a new data set. The first year went fine. The second year was better. The third year I developed it as an online course. By the fourth we were filling up sections in both Fall and Spring semesters (though I was only teaching the former).

I have to thank Saina Sheini Mehrab Zadeh my (recently graduated) PhD student and long-time teaching assistant for the course. She took the course in 2017 and has TA-ed it every year since, also teaching it herself in three semesters. There is no one who has had more influence on the evolution of the course. I am grateful to another (recently graduated) PhD student, Talia Kaufmann, who TA-ed the course before Saina did. I also thank Geoff Boeing (now at University of Southern California), Curt Savoie, and Connor MacKay (both previously of the City of Boston and Commonwealth of Massachusetts), who have also taught the course. Last, I appreciate the influence that all of the students who have taken the course have had on the refinement of the curriculum through their experience, though I couldn't possibly list them all by name.

A major element of this book is the use of real-world datasets curated by the Boston Area Research Initiative in our Boston Data Portal (BDP). These datasets were made possible by the hard work of many postdocs and research assistants, including: Mehrnaz Amiri, David Brade, Edgar Castro, Qiliang Chen, Alexandra Ciomek, Bidisha Das, Justin de Benedictis-Kessner, Chelsea Farrell, Sage Gibbons, Forrest Hangen, Laiyang Ke, Sam Levy, Barrett Montgomery, Petros Papadopoulos, Josiah Parry, Will Pfeffer, Nolan Phillips, Alina Ristea, Michael Shields, Xin Shu, Riley Tucker, Shunan You, and Michael Zoorob. This also was only possible thanks to a long-term partnership with the City of Boston's Department of Innovation and Technology, driven by previous Chief Information Officers (Bill Oates, Jascha Franklin-Hodge, Alexandra Lawrence), Chief Data Officers (Andrew Therriault, Stephanie Costa Leabo), and their team members (including the aforementioned Curt Savoie and Connor MacKay), as well as the collaborative leadership of the members of the Mayor's Office of New Urban Mechanics (Kris Carter, Nigel Jacob, Kimberly Lucas, and Chris Osgood). The BDP has been funded by the National Science Foundation's Resource Implementations for Data Intensive Research (RIDIR) program (award no. 1637124 for those curious), the John D. and Catherine T. MacArthur Foundation, and the Herman and Frida L. Miller Foundation.

In June 2021 I found myself at a crossroads. About to embark on sabbatical, I wanted to convert that opportunity into something that would have impact. Over that summer I engaged in a variety of conversations with friends and mentors that led me to the decision to write this book, including with John Wihbey and Alisa Lincoln of Northeastern University, Luis Bettencourt of University of Chicago, Ken Steif (then of University of Pennsylvania), Ben Levine (then Executive Director of the MetroLab Network), and Chris Winship and Rob Sampson (co-founding directors of BARI at Harvard University). The consensus was that there was need for an urban informatics textbook and that I had the opportunity to write it. I also received valuable advice from Brandon Welsh (Northeastern University) and Russell Schutt (UMass Boston), two other colleagues and mentors who had written textbooks and helped me to understand the industry.

The writing of the book itself owes the greatest debt to Northeastern University's Interdisciplinary Sabbatical program. By becoming a member of the College of Engineering I was able to have the full year to work on multiple projects, enabling me to truly dedicate the time needed to the textbook. This was granted by Provost David Madigan and his team following the strong collective endorsement of Chair of Civil and Environmental Engineering Jerry Hajjar, Dean of Engineering Gregory Abowd, Director of Public Policy and Urban Affairs Jennie Stephens, and Dean of Social Sciences and Humanities Uta Poiger. During the semester, I am grateful to the members of the BARI team—many of whom are mentioned above in the paragraph about the BDP—for providing advice on structure and content and otherwise listening to me prattle on about the work. This thanks also goes to faculty affiliated with BARI, including Nigel Jacob, Kimberly Lucas, and Moira Zellner. All of them not only gave good advice but were too kind to tell me I was getting annoying, which I appreciate.

Last, thank you to my family. My parents, Bonnie and Bill, and siblings, Tania and Liam, for listening and offering input when helpful. My in-laws for the same, especially my ever-supportive mother-in-law Kathy and my sister-in-law Jiordan, who writes books of a very different sort and loves to compare notes on the process. My older son, Beckett, bounced on my knee as a newborn when I developed Big Data for Cities for the first time. My younger son, Sebastian, has also lived alongside the development of the course and this book for his whole life. They are my youngest "math students" (and I can literally hear Beckett teaching Sebastian arithmetic as I write this).

And, of course, thank you to my wife, Leslie, for listening, supporting, and always helping me believe that the final product would be exactly what it needed to be.

1

Introduction

At the end of August 2011, Hurricane Irene struck the East Coast of the United States, battering the country from North Carolina's Outer Banks to Vermont. To give a small example of Irene's power and the havoc she wreaked, there were as many as 1,045 downed trees and limbs in Boston, Massachusetts, alone, one of which is pictured in Figure 1.1. This is the same amount as the city would see in a typical year.

Wait. How do we know this? It turns out that the City of Boston has a 311 system by which people can report issues to public agencies. These 1,045 "tree emergencies" were received by the 311 system and relayed to the Parks & Recreation Department for fixing. The system does so by generating a database cataloging every report that it receives and then sending each to the appropriate agency. On the day Irene struck, 311 also received reports about downed streetlights and signs and requests for highway maintenance, which were forwarded to the Public Works Department and the Department of Transportation, respectively.

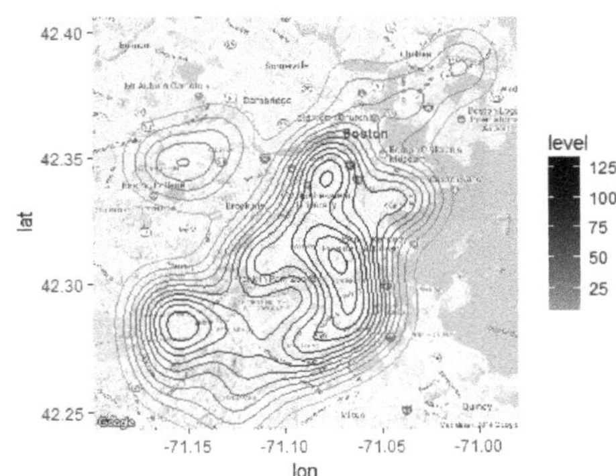

FIGURE 1.1
One of the many *trees* knocked down by Hurricane Irene in Boston, this one in the South End neighborhood (left) and a density plot of which parts of the city had more and fewer tree emergencies following the hurricane, as reported to 311 (right). (Credit: Boston.com; Author)

The power of a database is more than just being able to relay information to agencies, however. In retrospect, it becomes a digital record of the damage that Irene wrought on the city and its infrastructure, one that we can learn a lot from. By making maps, like the one on the right side of Figure 1.1, we see that tree emergencies came from across the city, though were more common in some places, including coastal areas exposed to the strongest winds. We can even watch in this video[1] (also captured in Figure 1.2) how different that day truly was from the rest of the year.

[1] https://vimeo.com/41535798

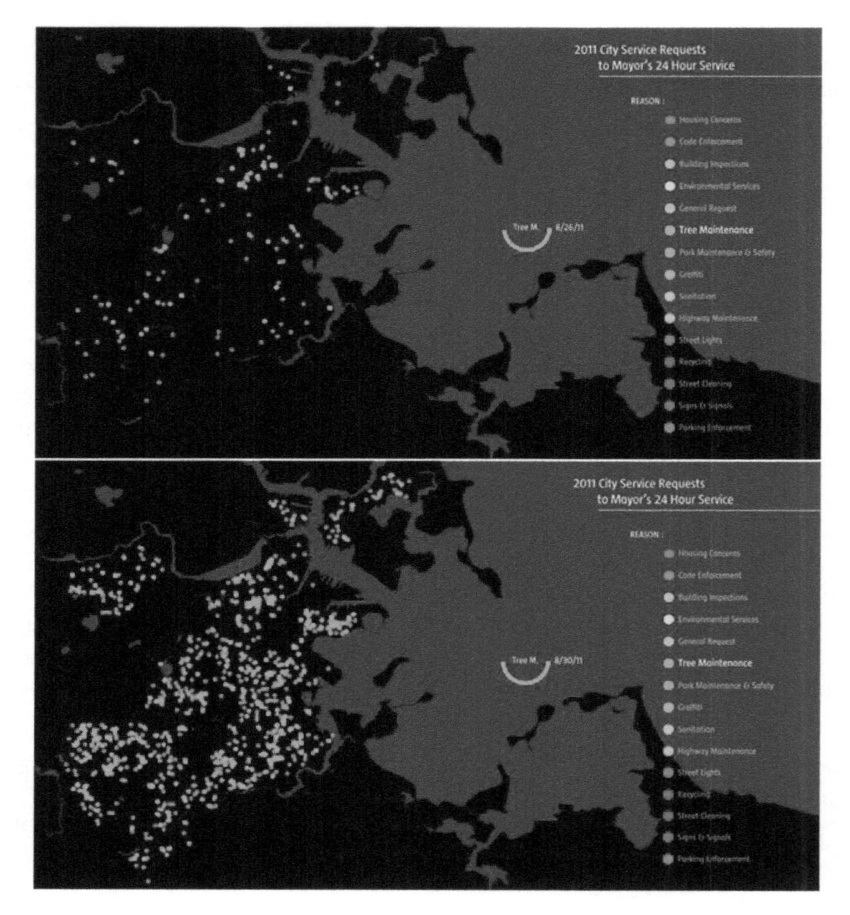

FIGURE 1.2
Though the City of Boston's 311 system typically receives only a handful of tree emergency reports daily (top), Hurricane Irene was no average day, generating 1,045 such reports (bottom). You can also watch the [full video](https://vimeo.com/41535798) comparing this day to the rest of the year. (Credit: Author)

Hurricane Irene is just one (rather dramatic) example of how 311 systems enlist constituents in the maintenance of public spaces and infrastructure. Everyday examples include people reporting potholes, cracks in the sidewalk, and graffiti tags. While more mundane, these are no less important to upkeeping the city and its many communities. And we can learn from these records, too. What types of challenges does each community face? How often? Can we better understand and even anticipate their needs? Can we evaluate and improve the effectiveness of the government agencies responsible for addressing these needs? All of these lessons can and have been learned by cities through the data generated by 311 systems.

For me, 311 is the iconic illustration of urban informatics. *Urban informatics* is a relatively young discipline, coming of age with the vast proliferation of digital tools for collecting, organizing, and analyzing data over the last decade. We might define it as *the use of digital technology and data to better understand and serve communities.* Whereas 311 can help us better understand communities and their needs, there is a broad range of other systems and databases—from education records to crime reports to Yelp reviews to Craigslist postings to library card transactions to traffic cameras—that can help us to answer questions and design innovative solutions. This is the promise and potential of urban informatics. Importantly, despite the word "urban" in the name, the tools associated with urban informatics can be of value to any community, be it in a city, a suburb, or the country, as long as there are data to work with.

1.1 This Book: The Practice of Urban Informatics

The goal of this book is to teach the practice of urban informatics. First and most obviously this comprises data science skills, including how we access data, manipulate it to expose the desired information, analyze it to answer questions, and visualize and otherwise communicate our insights to others. Those skills are the core curriculum of the book. But we want to do more than just play with data. Often, the data used to serve communities are "naturally occurring" in some way, generated as the byproduct of some system, like 311. As such, we are forced to determine how to interpret them. What do these data *mean*? How do we *understand* the information in them? These skills of interpretation are crucial for guiding our use of data science tools to generate valuable insights and impacts and help us frame the data science we will do.

Last, you may have already noticed a philosophical angle. In my opinion, data and technology give us a special opportunity to have impact, to hear and respond to local needs, and to reimagine 21st century communities. I think that this civic bent is central to the field, and it will be visible throughout the book, helping to frame the rest of the material. To be clear, this does not mean the book is meant exclusively for those looking to go into public service. There are a multitude of ways to serve communities, including through public service, non-profits, the private sector, or even (as I have chosen to do) through academic research. The skills in this book will be useful for anyone with any of those goals.

That said, before we start playing with data, the rest of this chapter describes the field of urban informatics in a bit more detail. Also, at the end of this chapter you will find more on the layout of the content of the book and how it is organized so that you can best utilize it to gain the skills that you need.

1.2 The Themes of Urban Informatics

As with any field, we might describe urban informatics in terms of three questions: (1) What inspired it?, (2) How does it work?, (3) What are its products? First, the inspiration has been the emergence of novel data that could be used in original ways. But it also includes new technologies that generate some of these data, especially those that are designed to "sense" conditions and events at precise times and places. Second, urban informatics is rather distinctive in being driven by widespread data-sharing and collaboration across institutions, including public agencies, non-profits, private corporations, and colleges and universities. We might call this the *civic data ecosystem* of a community. Third, the field has generated two closely related products: innovations in policy, practice, and services pertaining to communities; and a "new urban science" focused on developing a deeper understanding of

FIGURE 1.3
Many daily activities generate digital records, including commercial transactions, like credit card swipes, administrative processes, like vehicle inspections, and social media platforms, like Twitter. (Credit: touchbistro.com, fleetio.com, rss.app)

those communities. This makes for five themes of urban informatics: novel data, sensing and crowdsourcing, the civic data ecosystem, policy innovations, and a new urban science. Let's unpack each of these a bit further.

1.3 Novel Digital Data: "Big" Data or Something More?

In our digital age, computers are everywhere. And where there are computers, there are data. Every process and every transaction results in a data record. Some of the data sets are things that previously existed only on paper but are now supported by computer systems, including some of the things pictured in Figure 1.3. Think credit card purchases, utility bills (and, thus, energy and water usage), annual vehicle inspections, the marriage registry, restaurant inspections, and, of course, requests for public service through 311 systems. Others are more novel, arising from the introduction of computers themselves, especially social media. Facebook, Craigslist, Twitter, Yelp, Airbnb, newspaper comment boards, and cell-phone GPS pings all constitute "databases" of the activities of users. These lists are far from exhaustive, but they give you a sense of the diverse range of data generated by the individuals and institutions of the city. All told, they offer an unprecedented wealth of information on the behaviors, movements, social interactions, commerce and industry, and physical and environmental conditions of communities.

Many of the data sets I have named are included under the term "big data." To be certain, "big data" is not a rigorous term, nor does it have a clear definition. It is really a catch all for the many new types of data that have been proliferating in recent years, a good number of which happen to be considerably larger than the kinds of data we used to have on communities. That said, the census is pretty large and has been since China's Han Dynasty was the first to count all of its subjects (58 million, in fact) in 2 CE. Thus, it would be good for us to unpack what actually makes these novel data different apart from their size, and what those characteristics mean for the kinds of things we can do with them.

1.3.1 What Are "Big Data"?

A colloquial meme has summarized the distinctiveness of "big data" in "3 Vs": *volume*, or "big"-ness; *velocity*, or the fact that many of these data update often, sometimes in real time; and *variety*, or the breadth of content. These features are often different from traditional data from surveys, observations, and experiments, which are typically more limited in size and content and are rarely updated. But are these the things that really make modern data special? Rob Kitchin and Gavin McArdle sought to answer this question by comparing the characteristics of various novel data sources (Kitchin and McArdle, 2016). They concluded that volume is a by-product of two other characteristics. First, velocity, with its regular updates, contributes to an ever-expanding data resource. Second, they highlighted *exhaustivity*, or the intent to include all cases—that is, all posts on a social media platform, or all 911 calls, or all student records in a school district—drives size even further. They determined that variety did little to add to size.

If size were the only thing that distinguished modern digital data, however, they would not be very interesting. It is more important to understand how they enable new and different analyses and innovations. There are four things that bear noting:

1. *(Novel) Variety*: We have already noted variety as a special characteristic of modern data. Even if variety is not primarily responsible for size, it is very important to the potential of the data. Most of these data are "naturally occurring," harvested as the by-product of some other process. This is true for 311 and other administrative data sets as well as internet-generated data. In many cases, this results in a new view on some component of behavior and society that was previously more difficult to access directly, vastly expanding the questions we can ask and answer.
2. *Relationality*: Exhaustivity requires that data be indexed, with each element having a unique serial code. Often, this indexing occurs at multiple levels. For example, a 311 record has a unique record ID but also references the address at which services were required and the street and neighborhood where that address is located. In this way, indexing creates *relationality*, meaning data sets that reference the same unit of analysis can be merged. Such mergers further amplify variety and the number of questions that might be asked about a given unit of analysis.
3. *Flexibility of Records*: Whereas traditional data sources usually describe the characteristics of some unit of interest, such as a person, street, or neighborhood, many modern data come in the form of records. The records of a single data set might reference a particular unit once, twice, fifty times, or not at all, creating additional detail by which one can describe all units in the population. As we will see, this is not a scripted process, and an analyst must make multiple decisions to generate the desired measure. To illustrate, the average census tract in Boston

generates about 1,000 requests for service per year via 311, which might be tabulated to understand the needs of a community. But how? Should all types of requests be included? Or only those that reference a particular community need? If the latter, how many different types of needs could we quantify? These same considerations are necessary for any archive of records.

4. *Automation*: Closely related to velocity is the fact that many of these data sets are generated by automated computer systems. Events occur, information comes in, the system updates the database, and that database is immediately available. This does not necessarily affect the nature of the questions that can be asked and answered, but it does change when they can be conducted and what they can be used to build. To the former, if an important event occurs on Tuesday, we can examine it on Wednesday—possibly sooner. To the latter, if the data are automated, the insights they generate can be incorporated into other systems. For instance, a spike in 311 cases in a particular community could be caught by a computer and brought to the attention of administrators. All told, whether we use the term "big data" or not, modern digital data are distinctive and they enable a new set of analyses and innovations that previous data did not. This is owed not so much to their size, but to the combination of variety, relationality, flexibility, and automation.

1.4 "Sensing" the Pulse of Communities

In addition to the data, it is worth noting the rapid growth of technologies for sensing the conditions and events of communities. Take, for example, the Array of Things (AoT), a network of sensor nodes developed by the University of Chicago's Urban Center for Computation and Data (see Figure 1.4). Each node contains sensors that track air pollution, temperature, light intensity, precipitation, noise levels, and physical vibrations. They even process camera footage to estimate the volume of pedestrian, bicycle, car, and truck traffic. The vision of AoT is to deploy these nodes throughout communities to be a "fitness tracker" of sorts, constantly noting conditions and supporting systems for responding to them.

Sometimes, however, the "sensors" are not what we would typically expect. Thanks to smartphones, with their multitude of apps and tools, people can be sensors, too. *Citizen sensing* is when members of the public are a vehicle for observing and recording events and conditions. We can think of it as a form of *crowdsourcing*, a term that has entered common parlance through efforts like Wikipedia. In this case, instead of the "crowd" collectively contributing to knowledge, they are helping to create a real-time snapshot of the local landscape. Sometimes, participation in citizen sensing can be passive, such as when cell phone records register the location and activity of a user every time she engages with a cell tower. In other cases, the "citizens" are actively engaged in the data collection. For example, Sarah Williams of MIT worked with bus drivers in Nairobi's informal transit system who carried GPS trackers in order to better identify their "routes". Likewise, 311 systems enable residents to act as the "eyes and ears of city," observing and reporting issues they encounter in their daily movements.

Sensors are rarely deployed individually. Instead, many of them work together as part of a broader network. This reflects a distinctive approach to measurement in which many narrow observations are combined to build a comprehensive, or composite, view of the world. For

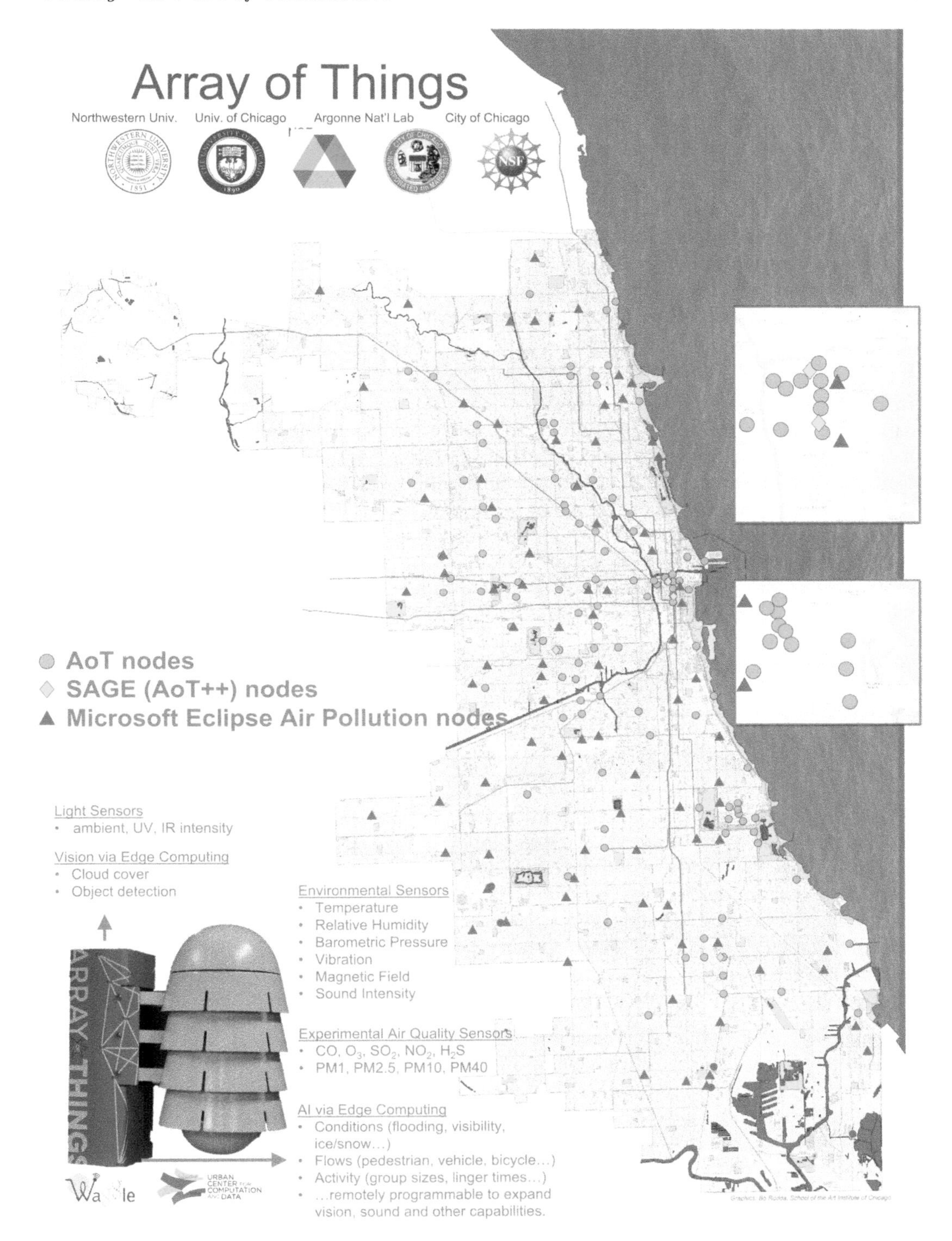

FIGURE 1.4
The Array of Things project has installed sensor "nodes" across Chicago, IL, to capture various conditions locally to gain a composite view of temperature, humidity, sound, light, and air quality, among other things, throughout the city. (Credit: http://arrayofthings.github.io)

instance, networks like AoT are deployed far and wide (see Figure 1.4 for an illustration). Each node observes only a small slice of the world, but together they track the entirety of the city. The same is true for citizen sensing. Wherever the people go, sensors go. Something similar can be said for the administrative records we described in Section 1.3, as it is the whole corpus of records that gives us a full view of communities.

Capturing the characteristics of communities in this composite fashion is not an entirely novel concept, of course. For example, in the 1990s the Project on Human Development in Chicago Neighborhoods surveyed thousands of Chicagoans about their neighborhoods to describe the variation in physical and social conditions across the city. Sensor networks and citizen sensing efforts, however, take this concept to a whole new level, often with many, many more records distributed through space and time. In this way, we can observe the *pulse of the city*, or the daily rhythms and long-term trends of the places, people, and institutions that constitute an urban area, in ways that we never have before.

In Figure 1.5 we see multiple examples of how 311 reports can reveal the pulse of various aspects of maintenance in neighborhoods. Looking at the frequency of pothole requests by months, we see a sharp spike in March, which is just after the damage of snow, ice, and salt from the winter becomes fully apparent. Or if we look at requests by day of the week, we see that people are much more active on weekdays, likely through the movements associated with work and school. If we assess the proportion of snow requests by month, we see that they become the prominent concern of Bostonians during February, the month with the most snow. Importantly, we have already seen how a major event like a hurricane can disrupt the natural rhythm of society, generating events that stand out in the data.

1.5 Civic Data Ecosystem

"Open data" has become a buzz term in recent years. It typically refers to data that a government agency has made publicly available, though it can come from any organization willing to share data with few or no restrictions. The open data trend has arisen in part from a push for government transparency but also from excitement about the valuable insights and tools that might be produced if we put these data in the hands of a broader community of analysts and "hackers." As a result, many cities have passed "open data ordinances" that require departments to publish their data in machine-readable formats (i.e., spreadsheets that can be analyzed) and built "open data portals" where the public can access these data.

Though the narrative has often been around the open data themselves, more attention should be paid to the community of individuals and institutions who stand ready to translate those data into insights and innovations. Without them, the open data would do little more than sit on hard drives and servers, not having very much impact at all. Boston, Massachusetts, the city where I live and work, serves as a good illustration of how such a community can operate.

Boston is renowned for its many colleges and universities and its thriving tech sector, including both industry behemoths, like Microsoft and Google, and hundreds of start-ups. Local city government has also been at the forefront of data- and technology-driven innovation, led by the Mayor's Office of New Urban Mechanics, a unique initiative that acts as a research and development lab that experiments with ways to improve city services and infrastructure. There is a vibrant Code for America brigade, consisting of volunteer "hackers," developers,

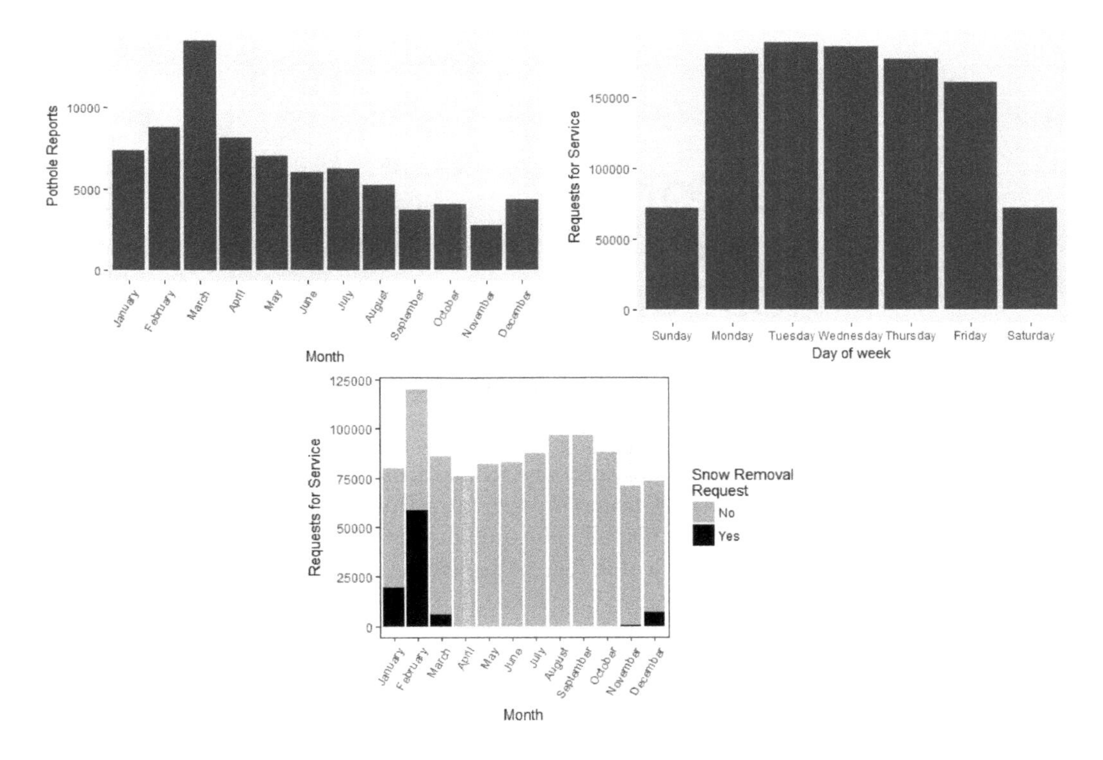

FIGURE 1.5
Examples of how 311 records capture the pulse of the city, including the monthly patterns of pothole requests (top left), the frequency of reports across days of the week (top right), and the proportion of snow removal requests by month (bottom). (Credit: Author)

and analysts. Further, non-profits and community activists have become involved as well. Advocates for housing, education, public safety, and environmental justice regularly ask how data and technology can best be used to support local residents—and question when it appears to be bypassing or even harming those same communities.

The engine underlying urban informatics in Boston and elsewhere is not just the presence of many organizations prepared to work with data. It is the way they work together, sharing data, co-designing questions and identifying challenges, and collaboratively generating solutions. The sum of these overlapping partnerships across a region is a network that is greater than the sum of its parts. This might be referred to as the *civic data ecosystem.*

For over a decade now, I have had the privilege of building and leading an interuniversity consortium at Northeastern and Harvard Universities called the Boston Area Research Initiative (BARI; see also Figure 1.6). BARI's mission is to convene greater Boston's civic data ecosystem to imagine how data and technology will reshape communities in the 21st century. The consortium focuses especially on opportunities to advance core societal values—things like equity, justice, democracy, resilience, and sustainability. One of my favorite moments every year is our annual conference, where hundreds of representatives from local public agencies, private corporations, non-profits, colleges and universities gather together to share the work they have done over the previous year and start new collaborations.

Thus, open data is only a small component of urban informatics. It is made possible by local institutions building sustained partnerships that center on the use of data and technology

FIGURE 1.6
The Boston Area Research Initiative (left) is a consortium of researchers, policymakers, practitioners, and community leaders from across greater Boston committed to supporting a thriving civic data ecosystem in the region. The annual conference (right) is an opportunity for members to share data-driven research and policy efforts and develop new collaborations. (Credit: Author)

to advance communities. Without collaboration, everyone is simply analyzing their own data for their own isolated purposes, creating a collection of narrow insights that do not necessarily intersect. Together they can develop questions and answers that have greater impact. It is in acknowledgment of this spirit of collective effort and public-minded data science that I refer to this not only as the data ecosystem, but the *civic* data ecosystem of the city.

BARI is not the only center that seeks to enable and support its local civic data ecosystem. Others include the Center for Urban Science and Progress at New York University and its partnerships with New York City, Metro21 at Carnegie-Mellon University and its collaborations with Pittsburgh and Allegheny County, the 21st Century Cities Initiative at Johns Hopkins University and Baltimore, and the non-profit Envision Charlotte in North Carolina.

Of course, not every city (or county or town) is as blessed with the density of technology-oriented institutions as these examples. Nonetheless, they all generate data. They all have smart people who can identify challenges and questions facing their communities. And they all have the opportunity to build collaborative relationships that can pursue those challenges and questions. Not every one of them is going to be using artificial intelligence, installing block-by-block sensor systems, or experimenting with autonomous vehicles, but that is not the point. As the skills of analysis, visualization, and interpretation become increasingly accessible (for example, through this book!), all cities, counties, and towns can leverage data for impact. That is all urban informatics aims to be.

1.6 Policy Innovations: Changing How the City Works

There is an adage that you can know a tree by the fruits that it bears. Similarly, you can better know a field or industry by the products it generates. The most visible of these are the innovations in policy, practice, services, and infrastructure that shape daily life for community residents. These come in many different forms. My favorite illustration of this comes from the MetroLab Network, an international consortium of over 30 partnerships between local government and universities focused on research and innovation, which writes a column titled Innovation of the Month published by *GovTech* magazine.

By browsing the Innovation of the Month columns[2], one can see the breadth of impact that urban informatics has already had (see also Figure 1.7). Sensors are advancing environmental justice, as when New York City has tracked flooding street-by-street during storm surges, or when Kansas City measured air quality in COVID hotbeds. "Predictive analytics" enables interventions before crises occurs, as in a Pittsburgh-Carnegie Mellon collaboration to predict fires, or a Fairfax County-George Mason University effort to support at-risk youth. There are the ever-popular "behavioral nudges," like a project by University of Chicago to redesign court summons to increase appearance rates. And, similar to 311's effort to crowdsource public issues, Philadelphia and the University of Pennsylvania crowdsourced community opinions to inform decisions to restore or replace historic housing.

Each of these projects highlights something important about the word "transformation." Transformation has the trappings of science fiction, evoking visions of the flying cars in *The Jetsons*, Tom Cruise solving crimes before they happen in *Minority Report*, and robots managing all daily operations in Isaac Asimov novels. Transformation does not occur overnight, however. It must be built through a multitude of modest, incremental advances on everyday problems—like addressing floods and air pollution, anticipating events like fires or high school dropouts, or simply getting people to show up to court. Each of these can be seen as an isolated innovation, but together they turn society into a smarter, and, hopefully, better place. In other words, transformation is the long-term product of rather mundane advances.

To what end is urban informatics gradually transforming society? In his book on smart cities, Anthony Townsend argued that all of our innovations are just modern solutions to the same problems we have always faced—sanitation, transportation, infrastructure maintenance, education, public safety (Townsend, 2013). This is absolutely true on a surface level, but it ignores the civic values that are often guiding the design and implementation of these solutions. This is apparent from looking closer at 311 systems.

311 systems use technology to make the maintenance of city services more efficient and effective. At their heart, though, they reflect a collaboration between government services and the residents and communities that they serve. It is a democratizing force that makes those services more accessible and responsive, while also encouraging constituents to participate directly in the governance of their own city. 311 is the epitome of *coproduction*, an approach to public administration that directly involves the public in the governance process. It also has been an icon for the closely related trend of *civic tech*, or the effort to use data and technology to enable people to collectively contribute to society, through government services or otherwise. Other examples include the Nextdoor platform, which connects neighbors to each other; participatory budgeting, in which community members have increased input

[2] https://metrolabnetwork.org/projects/innovation-of-the-month/

FIGURE 1.7
MetroLab Network's Innovation of the Month column captures the breadth of impact that urban informatics might have, including crowdsourcing community opinions to guide decisions around historic housing in Philadelphia (top left), localized sensors to manage street-by-street flooding in New York City (top right), and a data-driven system to predict fires in Pittsburgh (bottom). (Credit: MetroLab Network)

into the use of public funds; and even apps like Waze, through which drivers alert others to hazards on the road.

This is all to say that urban informatics promises to transform communities. In some ways, the end point of this transformation is unknown and probably unknowable. It will be the product of many iterative advances that seek to address everyday challenges. As these advances build upon and learn from each other, however, they begin to take a distinctive form, a literal embodiment of the civic mission that inspired and guided the work. Being that "civicness" often translates into concerns for one's neighbors and the overall well-being of society, we see many projects that also seek to advance related values, including racial and socioeconomic equity, sustainability and resilience in the face of climate change, and social and environmental justice in a rapidly changing society. These are the kinds of things that practitioners of urban informatics tend to build.

1.7 The New Urban Science: In Search of a Paradigm

In the early 20th century, the field of sociology was just taking shape, and no group of scholars was more influential than the Department of Sociology at the University of Chicago. The so-called Chicago School of Sociology was captivated by the emergence of the industrial city and the ways in which it was transforming society. They concentrated especially on how our historic reliance on close relationships among family, friends, and neighbors for most information and activities had been replaced by societal institutions, like newspapers, schools, and other public services. These insights on what they referred to as the *social organization* of a community gave rise to ideas that continue to frame how we think about the dynamics of neighborhoods, including aspects of crime, public health, education, economic activity, and public advocacy. We might call this the original urban science.

Cities might not be all that "new" anymore, but academic scholars continue to be drawn to them. This is especially true now as data and technology not only provide novel lenses for understanding communities but also instigate transformations that themselves merit closer study. And they attract scholars of all disciplines. It is often quipped that cities are a stage upon which all aspects of human society are played out, in which case urban science can support just about any question or topic from across the social sciences. The opportunity is greater than that, though. To stretch the metaphor, engineers want to study the stage itself and to develop guidelines for building a better one. Environmental scientists have found that the same phenomena they tend to study in nature, such as soil, air, and water conditions and weather patterns, also vary across neighborhoods in meaningful ways. Physicists and chemists are discovering that many of their theories about physical structure can be extended to the construction of cities.

An interdisciplinary urban science is the second set of products generated by urban informatics. But what does that mean? What value does it create? A lot of it is driven by the new data and methods that are now available. As we will see throughout this book, geographical information systems have unlocked the power of spatial data; network science has made sense of connectivity between individuals and communities; and rich data derived from administrative records, social media posts, and sensor systems expose the pulse of the city to study in ways that have never before been possible. Much as the policy innovations we are seeing are just the newest approaches to classic challenges, a new urban science is re-asking the same questions that were asked by the Chicago School of Sociology and their contemporaries, just with new tools and information.

Of course, it is not as simple as that. Iteratively asking new questions can lead to more comprehensive transformation. For example, network science allows us to empirically examine classic ideas regarding the social organization of a neighborhood. But there are currently no good theories for the network of communication or mobility for a city of a million people—something that can now be studied through cellphone records. Likewise, coproduction was a new idea in the 1970s that responded to and catalyzed new institutions like local crime watches and parent-teacher organizations. The scholars who originally wrote about it probably never envisioned a world in which reporting public issues and other public contributions were just a tap on a cellphone away. As such, civic tech has enabled a complete reevaluation of this area of scholarship.

Additionally, there has been an increasing awareness in academia of the power of research to have positive impact. To wit, anything that a scientist learns about a community should be of interest to those who serve that community, and anything that policymakers care about should be important enough to merit scientific study. Consistent with this ideal, all of the examples of policy innovations listed in Section 1.6 include academic partners. By bringing advanced methods, frameworks for critical thinking, and an insatiable appetite for new questions and challenges to the table, academics can partner with local leaders, both in government and otherwise, to demonstrate the mutual reinforcement between research, policy, and practice. And, in the process, they are advancing a new urban science that deepens our fundamental understanding of society.

1.8 This Book: Learning Objectives and Structure

1.8.1 Learning Objectives

I hope that to this point I have not only informed you about the field of urban informatics but also excited you to become involved yourself—or at least intrigued enough to continue reading this book. Just as transformation can be the gradual product of many mundane advances, the chapters that follow will impart the tools and skills at the foundation of urban informatics. This includes both the technical and the conceptual.

The technical skills will center on accessing, curating, manipulating, analyzing, and visualizing data, especially those that are naturally occurring. There are many software packages available for working with data, and this book focuses on one called R. This is for a few reasons. First, R is freeware, meaning that anyone with a computer can use it without requiring an expensive license. Second, thanks to a rich user community, there are add-on packages that can execute pretty much any analysis or data manipulation that exists, and certainly everything you will need for this book and beyond. R does require coding, which is a great benefit as it gives you as the analyst complete control over what you are doing. That said, do not be daunted if you have not coded before as we will work from the ground up.

On the conceptual side, data cannot speak for themselves. As an analyst, you must understand and interpret them and give them meaning for your audience. Likewise, the products of analysis and visualization are not pre-ordained. The analyst must make dozens or even hundreds of decisions along the way, all reflecting a specific use of the data. This is not to say that research is completely subjective or that people can "tell any story they want" with data. To the contrary, it requires discipline and awareness regarding the strengths and weaknesses of the data, which questions they can and cannot answer, and the purpose of the analyst. Throughout the book we will engage with these considerations as they relate to the technical skills we are developing.

1.8.2 Organization of the Book

The book is broken up into four units with learning objectives that build upon each other. Each unit contains multiple chapters that interweave technical skills with the corresponding conceptual ones, generally centered on an illustrative real-world example. The first unit, *Information*, concentrates on the skills needed to access, interpret, and represent the basic content of data. The second unit, *Measurement*, walks through the steps needed to use records to describe meaningful units of analysis—individuals, streets, neighborhoods, etc. Again, the tools of interpretation and decision-making are crucial in determining whether the measurements we create are precisely what we want and that they mean what we think they mean. The third unit, *Discovery*, introduces statistical analysis and more formal ways of representing relationships between variables. Here as well, the analyst must think critically about how to design models and communicate results based on the meaning of the data and the measures derived from them. The fourth and final unit, *The Other Tools*, contains supplementary chapters on the applications of other techniques and technologies that have become popular in urban informatics, including network science, machine learning, block chain, and 5G.

Every chapter concludes with a series of exercises for practicing the skills that you have learned. These are designed to engage both technical and conceptual skills. The exercises come in two different forms. Depending how you (or the instructor of the course you are taking) want to use this book, you may concentrate on either or both. Each is equally effective at developing the desired skills. (1) *Problem Sets* consist of traditional prompts that have "correct" answers. This approach will be especially useful to students who do not have access to data sets of their own. (2) *Exploratory Assignments* consist of a single, multipart prompt that requires the student to flexibly apply multiple technical and conceptual skills to a data set of their choosing. This approach to learning the content is particularly useful for students who have access to data, either through the course they are taking, their work, or otherwise, as they are designed to build on each other to carry students through the process of finding, making sense of, manipulating, and analyzing a single data set. Also, the exploratory assignment in Chapter 2 invites students to find an open data portal and download a data set they find interesting, so any student who wants to can pursue this option.

Each unit also contains two additional assignments. *Community Experience Assignments* encourage students to interact with the communities represented in their data, either in-person or through online content, in order to better examine the assumptions they have made when analyzing and interpreting. These assignments can be useful for introducing a community-engaged or service-learning spirit to a course using this text. *Unit Final Assignments* give students an opportunity to synthesize the skills of the unit in a single deliverable that represents information, measurement, or discovery derived from their data set. Each of these assignment types works best with the Exploratory Assignments but can also be applied to any of the data sets used in the chapters.

1.8.3 Worked Examples in this Book

Each chapter in this book presents a set of technical and conceptual skills through a worked example. These examples will always come from real-world data describing Boston, in the northeast corner of the United States. This consistency holds a number of benefits. It is always more relatable to learn skills through data sets that have real-world meaning and implication, rather than the simplistic, "canned" data sets that methods textbooks often utilize. Further, unlike those canned data sets, the real-world data force us to tangle with messiness, which is itself an important skill for the practice of urban informatics. Additionally, as noted, I have spent many years studying Boston and supporting others to do the same through my involvement with the Boston Area Research Initiative. These activities have generated a wide range of data sets that can be leveraged to illustrate all nature of methodological tool. Last, this approach gives you, the reader, an opportunity to see how an in-depth engagement with multiple data sets can illuminate the places, processes, and dynamics of a single city. My hope is that many of you will use the skills and exercises in this book to do the same for a community that is important to you.

Of course, there are also downsides to this approach, the greatest one being for those readers who do not identify with Boston, or any similar city. For example, if you are from the global South, your communities may have very different structures and relationships. I encourage you to consider how these insights translate and how they do not. You are also welcome to follow along with the steps of the worked example with whatever data set you like, as long as you are confident that you are able to evaluate whether your results look like they "should."

1.9 Exercises

1.9.1 Problem Set

1. Define the following terms and explain their relevance to urban informatics:
   a. Crowdsourcing
   b. Civic data ecosystem
   c. Coproduction
   d. Civic tech
2. What are the features of "big" data that most affect the ways that they can impact research and innovation? How do these differ from the characteristics of "big" data highlighted in popular discourse?
3. Are "big" data and naturally occurring data the same thing? How or how not?
4. List the five themes of urban informatics. Which of the five do you think is the most distinctive or important? You are free to select more than one.
5. Browse the winners of MetroLab Network's Innovation of the Month[3]. For each theme of urban informatics, identify at least one example that you think embodies that theme. Justify your selections.

1.9.2 Exploratory Data Assignment

Select a city, maybe your own or one nearby—or one on the other side of the world that you find interesting. Use internet resources to learn more about the civic data ecosystem there and write a 1-2 page memo answering the following questions:

1. Who are the main institutions and actors? What do they each appear to contribute?
2. How do you see the five themes of urban informatics at work there?
3. Do one or two of the five themes of urban informatics stand out as being more central to this civic data ecosystem than some of the other themes?

[3]https://metrolabnetwork.org/projects/innovation-of-the-month/

Unit I

Information

2

Welcome to R

Toronto, Ontario has an open data portal. As with such portals in many other cities, the goal was to put the data in the hands of people who might do something with them. The data available there are quite diverse in their content, enabling analyses on all nature of question. For those interested in homelessness, there are daily occupancy numbers for shelters; for those interested in housing, Airbnb rental registrations; for those interested in climate change, a survey on Torontonians' climate perceptions; for transportation enthusiasts, the locations of bus stops, train stops, and the sidewalk network; for political watchers, election results by ward; for those interested in local economics, all licenses for businesses operating in the city. And, of course, 311 requests are available for download, documenting where and when Torontonians have identified graffiti, potholes, street light outages, and other local needs.

While the breadth and organization of the data on Toronto's open data portal are noteworthy, what really sets it apart is something additional the City did to facilitate use of the data. They worked with Sharla Gelfand, a local developer, to build a custom package for R statistical software—or 'R,' for short—that facilitates browsing and downloading the data, called `opendatatoronto` (Gelfand, 2020) (also see Figure 2.1) . The practical upshot is that statisticians, analysts, visualizers, hackers, and researchers who use R—which is one of the most popular softwares for accessing, managing, and analyzing data—can access data from Toronto's Open Data Portal from within the program, without having to go through the tedious process of manually visiting the web site, browsing data sets, downloading data, and then loading them into R. How convenient! So much so that Toronto has not been alone in this idea. Numerous metro transit systems, including Washington, D.C., have done the same for their data. A developer in Vancouver, British Columbia, built one called, fittingly, `VancouvR` (von Bergmann, 2021) . And Socrata , the biggest vendor of municipal open data portals has one called `RSocrata` (Devlin et al., 2021) that generalizes to all data portals they have built.

This anecdote illustrates just how central R has become to urban informatics and data science in general, which is owed to a few distinctive characteristics. R is free to download and install. It has a very approachable and flexible programming language. Further, developers can use that language to build new "packages"—what you might think of as "add-ons" or "plug-ins"—that offer new capabilities not available in the original software, which is exactly what the City of Toronto and Sharla Gelfand did.

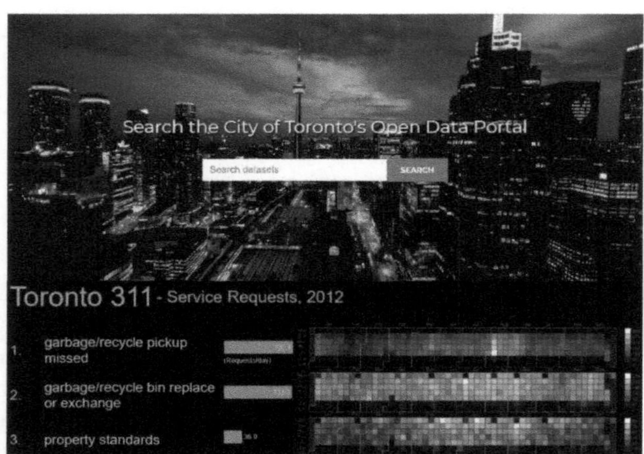

FIGURE 2.1
The City of Toronto's Open Data Portal (top left) is accompanied by an R package designed to directly access its contents, opendatatoronto (right), and also has a gallery featuring ways in which people have used the data, including this visualization of the most common 311 reports by day across the year. (Credit: https://open.toronto.ca/, https://www.sharlagelfand.com/project/opendatatoronto/, http://neoformix.com/Projects/Toronto311/)

2.1 Worked Example and Learning Objectives

Just like Toronto, Boston , Massachusetts, has a large, well-populated open data portal. It is available at data.boston.gov. We will visit it and use it as inspiration to become familiar with R and its capabilities. We will:

- Install R and its preferred interface, RStudio;
- Use R as a calculator;
- Create and manipulate "data objects", including data sets;
- Access and install packages.

2.2 Getting Set Up with R

2.2.1 What is R?

R is an opensource, freeware program for working with data, especially focused on the production of professional quality analyses and visualizations. Freeware means that R is available for download at no cost for Windows, Mac OS, UNIX, and LINUX through the Comprehensive R Archive Network, or CRAN. Open source means that the underlying code is public and accessible to anyone who might want to work with it. This code is based in a programming language called S, designed at Bell Laboratories (formerly AT&T, now Lucent Technologies) by John Chambers and colleagues specifically for the purpose of executing statistical analyses. R as a standalone program was developed by Robert Gentleman and Ross Ihaka (whose initials account for the name R) starting in 1995 at the Statistics Department of the University of Auckland.

You might be asking yourself, how does a freeware program, which by definition has no profit model, become the most popular and possibly the most versatile tool for working with data? The answer lies at the intersection of being both freeware and opensource. Given its accessibility, R has grown a large community of users, including some of the world's leading statisticians and methodologists. These users have contributed in many ways over the years, including debugging and suggesting improvements for R and developing and publishing thousands of packages that expand its capabilities. When statisticians develop a new tool for analysis or visualization, they often build it for R first, giving other R users access to cutting-edge techniques well before they have been incorporated into for-purchase softwares, like SPSS, SAS, or Stata. In addition, the R Foundation provides the funds necessary for basic maintenance of the software and CRAN, which is the public access point for the program and packages.

2.2.2 Installing R

CRAN is an archive and network. The archive contains code and documentation for R and packages. Like any public archive, it must be hosted on a server somewhere. This is made possible by a network of "mirrors," or web sites and servers that store identical, up-to-date versions of the ever-updating archive. These are located around the world and are often, but not always, hosted by universities. Let us visit the mirror directory: https://cran.r-project.org/mirrors.html.

When you get here, click on the link for a mirror that is close to you. Coincidentally, being in Boston, Massachusetts, the closest to me is at the University of Toronto. Once you click on this, you should reach the following screen:

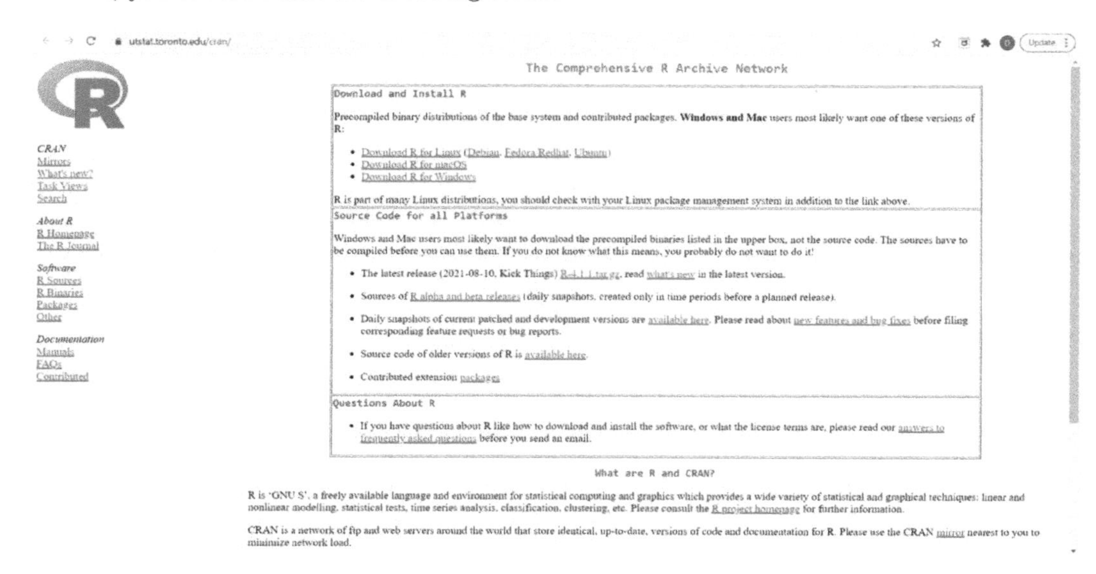

2.2.3 The R Interface

Navigate to where R was installed on your computer and open the program. Depending on how you responded to the prompts in the installer, you will probably find it in a folder titled R. You will see something that looks a lot like this:

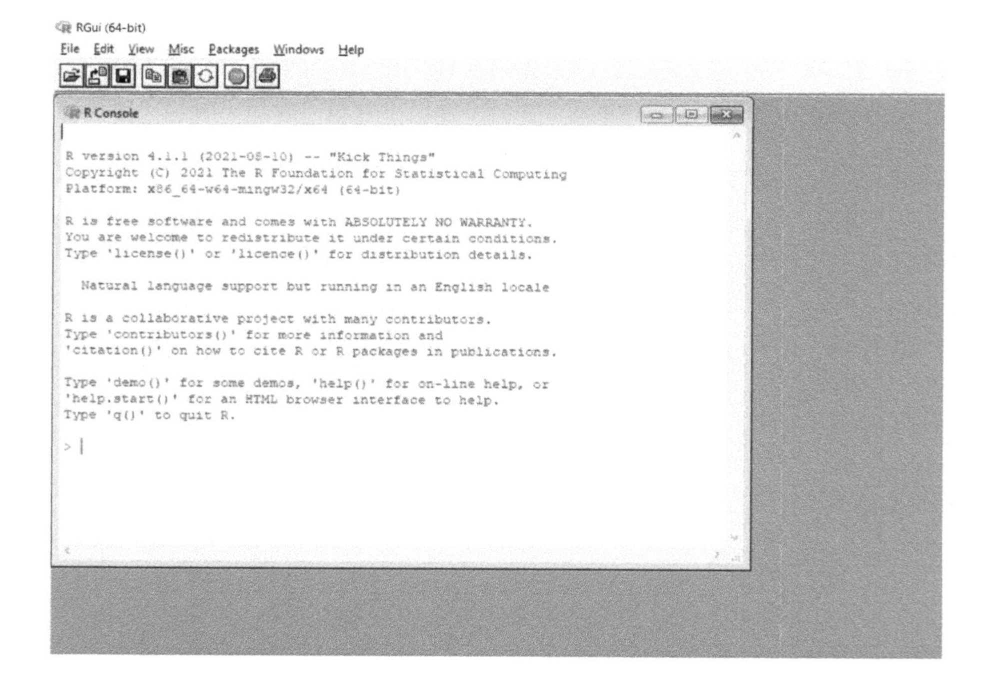

This is the R Console. You can enter code at the prompt, and R will execute that code. You can use code to import data, analyze it, and generate visualizations. There are also a few drop-down menus that facilitate certain actions by point-and-click. The default RGui (general user interface) is a bit spartan, however, with few tools available to even keep track of what you have done and created. We could really use something that is more user-friendly that combines the flexibility and control that we have through coding with some other features that make that work more efficient and its products more accessible. Luckily, such a tool exists.

2.2.4 Installing RStudio

Rather than work in the default RGui, many analysts work instead in RStudio. RStudio is known as an *Integrated Development Environment* (or IDE) . An IDE is software that coordinates numerous tools required to write and test software. Though it might not feel like it, every time you conduct an analysis or create graph while working with this book, you will be writing (and testing) software! An IDE is simply an interface that enhances this process, allowing you to keep track of the code you have written, the products it has generated, and the tools you are using. RStudio is not the only one built for R, but it is the best and most popular at this time.

Let us download RStudio here: https://www.rstudio.com/products/rstudio/download/. You will note that RStudio is not freeware in all cases, but there is a free version that is completely sufficient for our (and most people's) purposes. Click DOWNLOAD under RStudio Desktop and then select the version to corresponds to your operating system.

2.2.5 The RStudio Interface

Once you have installed RStudio, navigate to it on your computer and open it. You should see something that looks like this:

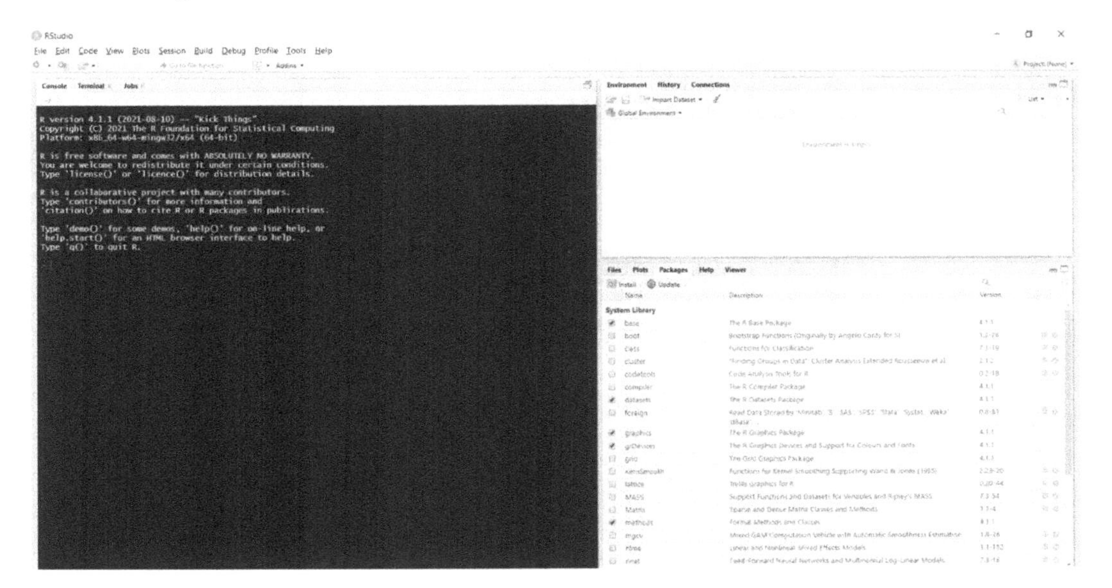

There are three components visible here.

1. On the left side of the screen you have the Console, which, as in the RGui, is where you can submit code directly to be executed.
2. On the top right is the Environment. This is where data objects (including data sets) and some of their basic details will be listed. Behind this there are tabs for your History, or the code you have previously entered, and Connections to external sources. We will not use these often in this book.
3. On the bottom right, there is a list of packages that you have installed, some of which have checks next to them, which means that they are currently enabled or "turned on." This is also where Plots and Help will appear, as well as where you can navigate your File Directory to view the contents of your computer while in R.

We will learn more about how and when we use each of these tabs in the coming sections, but for now it is useful to know that they are there. Also, RStudio is highly customizable, so you can resize and move all of these tabs around and play with the appearance through the Tools menu, selecting Global Options.

2.3 Creating a Project in RStudio

An appealing feature of RStudio is the ability to create *projects*. A project is a combination of files that are all related to each other. Typically, analysts use projects to keep their work organized. Each project references a specific file directory on your computer where it stores all of its products (unless instructed to do otherwise, though we will get to that).

Create a new project in RStudio by clicking the File menu and then clicking "New Project..."

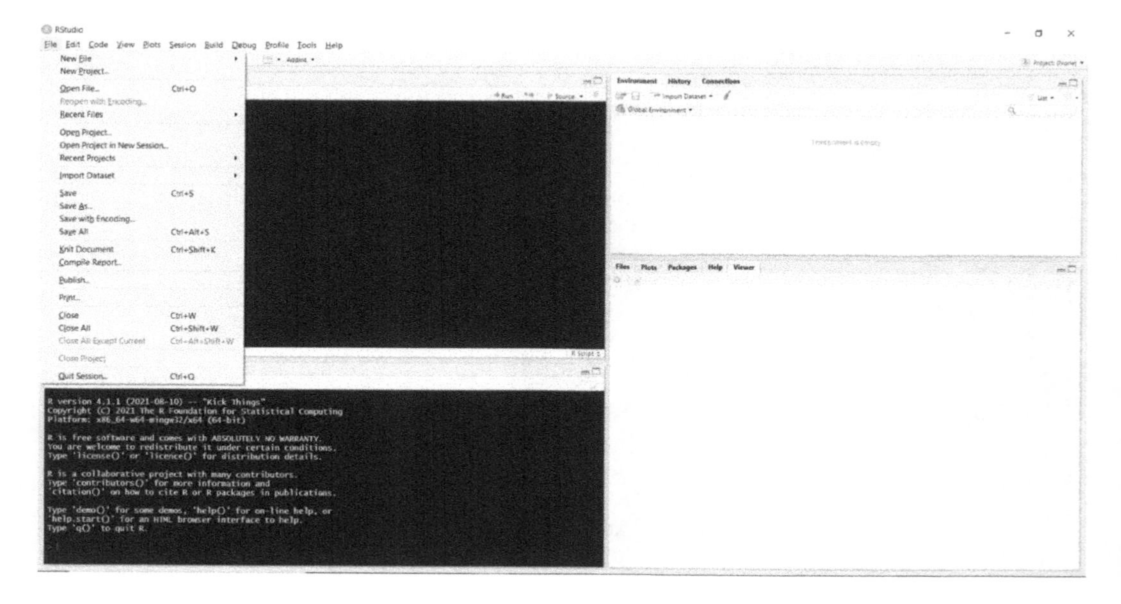

You will then be presented with this interface, where you want to select New Directory.

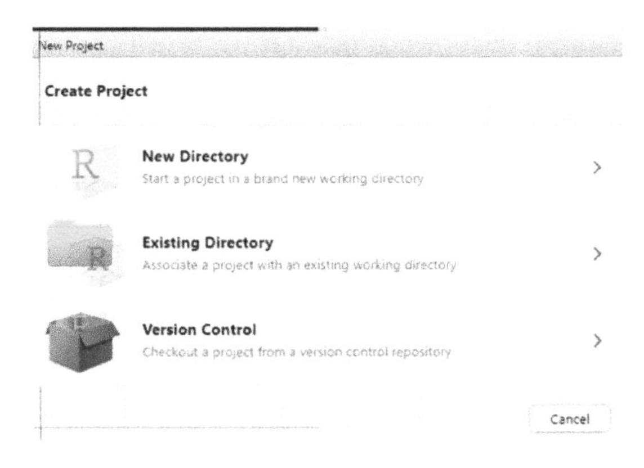

Then select New Project.

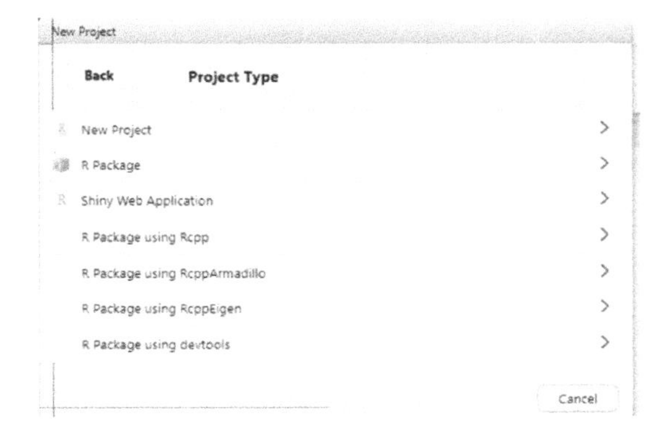

You will then need to specify a directory, as I have done here.

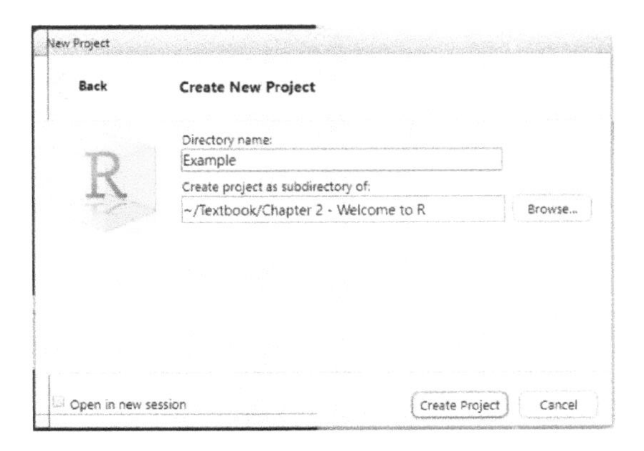

Note that you are creating a new file folder at this time that will then be the home for any files generated by the project. The folder name will also be the name of the project. (The process is slightly different when using an existing directory, but if you can do this successfully, you should have no trouble with the other.)

When you click Create, you will note that your RStudio resets. The Files tab will be on top on the bottom right now, and it will be viewing the contents of the new folder you have created. If you put files in this folder or generate products while using R, they will become visible here.

2.3.1 Creating a New Syntax

Before we get started really exploring R's capabilities, we should have a syntax file open. As noted, we can enter commands through the Console and R will execute them. The challenge here is that this will occur line-by-line, and if we make any mistakes it is hard to go back and correct them. Instead, we want the flexibility to write multiple lines of code and to submit them to the Console in bunches. To do this, we write our code in syntax files.

Create a syntax file by clicking this button in the upper-left-hand corner, beneath the File menu:

And then selecting R Script:

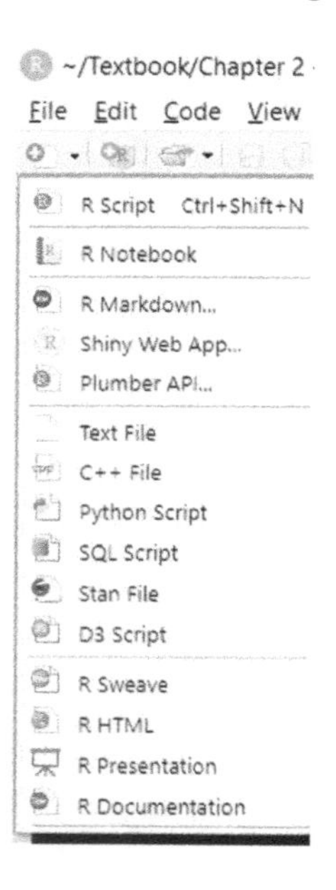

Your screen should then look something like this:

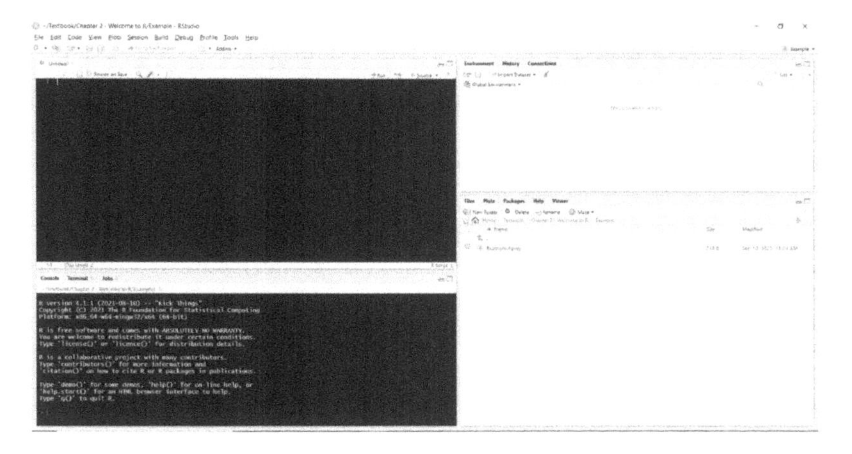

This script file is now our canvas. When we are ready to send lines of code to the console, we can select lines of code and click ctrl+Enter or tap the Run button in the top right of the script window.

Run

If we want to run one line of code, we can put the cursor on that line and hit ctrl+Enter, but be careful—if you do that with the Run button, it will run the entire script, which you might not want.

2.4 How to Work with R

2.4.1 Coding in R

I have mentioned a few times now that work in R is done through code. That is, an analyst must write commands that can be executed by R. This might seem daunting if you have never coded before, especially when one hearkens back to the point-and-click simplicity of Excel. Sure, there are challenges to coding. One has to know what terms the computer will recognize to do a particular job. The code has to be precise, because any typo will lead the computer to do the wrong thing or, more likely, produce an error message. This is especially important here because **R is case-sensitive**; meaning it interprets upper- and lower-case letters as distinct from each other. Analysts who want to do more complex tasks have to fit multiple pieces of code together to achieve that task, which further necessitates close attention to detail and often multiple tests (and failures) that require debugging.

Despite its challenges—and possibly because of them—code is also extremely useful and gives analysts complete control over their work. First, code is often a more direct pathway to accomplish a given task. For instance, in Excel it might take a half-dozen clicks to create a graph. In R, a graph can be generated in one line of code. Second, code allows the analyst to precisely specify and customize details as they go. The same graph in Excel will start with defaults—the names of the axes, the number of tick marks, even the colors of the lines or bars—that then need to be modified manually. In R, the analyst can incorporate these details into their code at the front-end. Third, code is replicable, which holds benefits for the analyst and their colleagues. Let us suppose you have written a line of code for the exact graph that you want, but you want to make this graph repeatedly for 10 variables. All you need to do is copy and paste the code, changing only the variable name. Once the initial work is done, the replication takes minimal effort. Additionally, you can share that code, plus anything you did before or after, with colleagues who can then replicate the work, refine it, or add to it.

Last, though I may be biased here, once you get the hang of it, coding is fun! You get to identify a product or outcome that you want to reach, determine the steps for getting there, and then build and execute those steps. Writing a successful sequence of code is an immensely gratifying process and not all that different from doing a crossword puzzle or solving a brainteaser.

R code is capable of supporting a lot of tasks, but at its foundation it is constructed to do two things: be a calculator and manage data. Let's learn more about each.

2.4.2 R as Calculator

Many of R's functions are built upon the underlying ability to operate as a calculator. Try entering the following line of code into your Console, or practice putting it in your Script and running it from there:

```
6+5
```

```
## [1] 11
```

R supports all traditional operations, like exponents, for example. Try:

```
8^2
```

```
## [1] 64
```

R is more like a scientific calculator than a basic calculator in which you have to enter one operation at a time. R can accommodate equations of any length. In doing so, it follows the order of operations rules of PEMDAS, or: solve what is inside **P**arentheses first, then calculate **E**xponents, then **M**ultiplication and **D**ivision, then **A**ddition and **S**ubtraction.

For example (see if you can check R's math):

```
75/25*3+15/(8-3/3+16/64*2-1264*0)-1
```

```
## [1] 10
```

This would be a good time to take a break and try a few arithmetic equations of your own, just to make sure you are comfortable.

TABLE 2.1
Some arithmetic functions built into R.

Function	What It Calculates
abs(x)	Absolute value of x
cos(x)	Cosine of x
exp(x)	e^x
log(x)	Natural logarithm of x
sin(x)	Sine of x
sqrt(x)	Square root of x
tan(x)	Tangent of x

2.4.3 Functions

This is our first opportunity to engage with *functions* in R. Functions are built-in commands that instruct R to complete certain calculations or tasks. Generally, functions are followed by parentheses, and the function is performed on the contents of the parentheses. In this book, functions will be formatted as R code inline.

R has many functions, including all of the ones you would expect from a calculator, including square root (`sqrt`) , absolute value (`abs`) , and trigonometric functions (e.g., `sin` and `cos`) . Others are listed in Table 2.1.

Let's try a few examples.

```
sqrt(100)
```

```
## [1] 10
```

```
abs(-25)
```

```
## [1] 25
```

We can even incorporate them into more complex equations:

```
sin(0)*sqrt(169)+abs(35)-cos(0)
```

```
## [1] 34
```

Before moving on, try breaking that one down into its pieces to see why that was the answer. Also, this would be a good time to take a break and try a few arithmetic equations of your own, just to make sure you are comfortable.

2.5 R as Data Management Software

R, of course, is more than just a calculator. It is an environment within which data can be managed and manipulated. As such, it is what we call an *object-oriented programming language* . That is, it recognizes multiple classes of objects that contain data in various forms and structures. Each class of object has characteristic features that allow R and its functions to describe and modify it. This may seem a bit abstract at this point, but what it means for our purposes is that: (1) we can store data in objects that have names; (2) functions can be applied to these objects just as if they were the numbers in the arithmetic equations we calculated in the previous section; and (3) nearly every analysis we conduct in this book will generate additional objects, meaning we can apply functions to *them* as well. Like functions, objects in this book will also be formatted as R code inline.

Let's walk through the most common object classes for storing data.

2.5.1 Variables

A variable is a single value stored in an object. It might seem silly to store a single value in an object, but there are cases in which it could come in handy. For example, let us say you are looking at Boston's open data portal and want to calculate certain records per 1,000 residents. In 2020, the population of Boston was 675,647 people. That's a big number to have to enter repeatedly, though. What we can do instead is store it in an object that we will call `bos_pop`. See Table 2.2 for more on naming objects.

TABLE 2.2
Rules about naming objects in R.

Rules for naming objects.
1) Must begin with a letter or dot (.).
2) Can only contain letters, numbers, underscores (_), or dots (.).
3) A variety of words used for programming cannot be used, including if, NA, TRUE, FALSE, else, function, and others.
4) It is good practice to give objects names that help you and others to recognize what they mean when they are embedded in code.
5) Keep in mind that R is case-sensitive, so you will always have to reference the object exactly as you wrote it. Also, it is possible to have multiple objects with the same name but different capitalizations, but that would be highly confusing.

We can create a variable in two ways. The more common is arrow notation:

```
bos_pop<-675647
bos_pop
```

```
## [1] 675647
```

By putting a ‘<-‘ in between the name that we want to give the variable and value we want to attribute to it, we instruct R to create that variable and give it that value. Note that R does not print the variable after this step. Instead, if we want to confirm that R did this, we need to tell it to do so by typing in the name of the variable and submitting it as a separate line of code. Also, your Environment should now have the object `bos_pop` in it. Note that for variables, the Environment also tells us the value. Last, R uses commas for specific purposes of separating information, which is why we enter numbers greater than 1,000 without them.

The second and less common way to do this is with an equals sign:

```
bos_pop=675647
bos_pop
```

```
## [1] 675647
```

Though the arrow might seem odd or uncomfortable at first, it is the more standard practice and eventually becomes natural. You might even think of it as the arrow "storing" the value in the variable. Now, to use our variable. Boston received 251,374 requests for service through 311 in 2020. How many is this per 1,000 residents?

```
251374/bos_pop*1000
```

```
## [1] 372.0493
```

Meanwhile, the City approved 37,460 building permits. How many is this per 1,000 residents?

```
36351/bos_pop*1000
```

```
## [1] 53.80176
```

And there were 70,894 crimes reported, making for

```
70894/bos_pop*1000
```

```
## [1] 104.9276
```

per 1,000 residents.

Suppose we had wanted to simplify our code here, we might have constructed a second variable based on the first

```
bos_pop_thou<-bos_pop/1000
bos_pop_thou
```

```
## [1] 675.647
```

Here we have illustrated a numerical variable that can be incorporated into arithmetic equations. It is, of course, possible to store characters (i.e., letters and other characters) in a variable as well. We will do more with this in the next section (Section 2.5.2) on vectors. In addition, variables can store dates and logical statements (`TRUE`/`FALSE`).

2.5.2 Vectors

A vector is a collection of values of the same type (i.e., numeric, character, etc.). For example, we might want to combine our counts of 311 requests, building permits, and major crimes in a single place. We can do this with `c()`, which stands for combine, with commas between each value:

```
bos_events<-c(251374, 36351, 70894)
bos_events
```

```
## [1] 251374  36351  70894
```

Note that `bos_events` is now visible in your environment as

```
Values
  bos_events            num [1:3] 251374 36351 70894
```

Communicating that it is a numerical (`num`) vector with 3 elements (`[1:3]`). (In R, putting a colon in between 2 numbers indicates a range. In fact, you can use the same convention when writing code. Try the following piece of code and see what happens: `range_vector<-1:3`.)

Vectors can make certain tasks more efficient because calculations can be applied across all of their elements. For instance:

```
bos_events/bos_pop_thou
```

```
## [1] 372.04931  53.80176 104.92757
```

We have now divided all three of our counts by the population of Boston in thousands, calculating the number of events per thousand for each in one line of code. This is illustrative of vector arithmetic in general, with the operation being applied to each of the elements in the vector and the output being a new vector of the same length. We could also store this as a new object if we so choose:

```
bos_events_percap<-bos_events/bos_pop_thou
```

Last, it is possible to have vectors that store character data. Before moving on, we can create one

```
bos_event_types<-c('311','Permits','Major Crimes')
```

For character data, we put ' ' or " " around each value and then commas between them outside the single or double quotation marks.

Also, I have stopped printing the content of newly made objects at this point, but you are welcome to check them on your own to confirm you entered the code correctly.

2.5.3 Data Frames

Possibly the most common and useful class of object in R—and certainly the one we will use most often in this book—is the *data frame*. A data frame is what you might typically envision as a spreadsheet in Excel: a set of rows, or *observations*, and columns, or *variables*. In other words, each observation has a value for each variable.

In R, we can think of the columns as a series of vectors. In fact, we can construct a data frame by combining vectors with the `data.frame` function:

```
boston<-data.frame(bos_event_types,bos_events,bos_events_percap)
```

Data frames are often too large to view easily in the console, so we can use the `View()` function:

```
View(boston)
```

This will produce the following view in a new tab alongside your syntax:

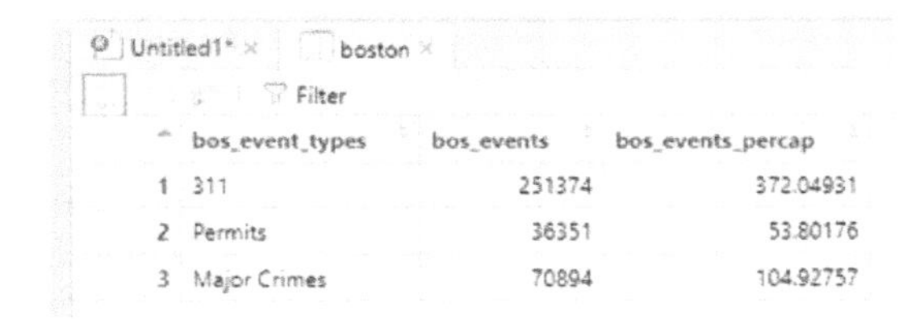

Untitled1* boston

Filter

	bos_event_types	bos_events	bos_events_percap
1	311	251374	372.04931
2	Permits	36351	53.80176
3	Major Crimes	70894	104.92757

Thanks to the way we constructed our vectors, the information here is cleanly aligned, with event types matching their frequency and rate per 1,000 residents. If, however, we had a fourth row for, say, `Votes Cast`, but had not entered events or calculated events per capita, we would have blanks in the fourth observation. R does not do blanks, though. Instead, the value NA would be filled in, as you see here.

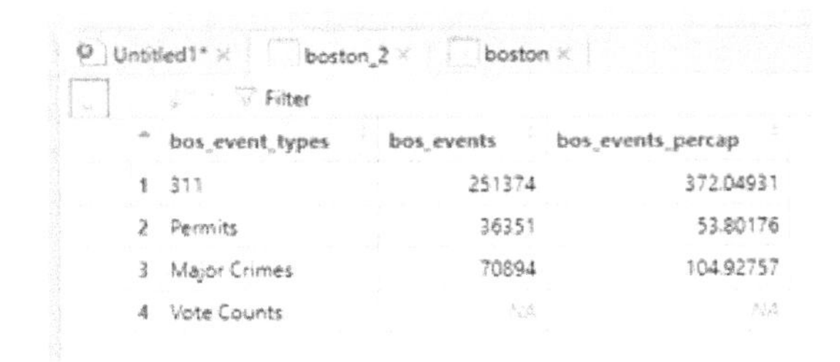

Untitled1* boston_2 boston

Filter

	bos_event_types	bos_events	bos_events_percap
1	311	251374	372.04931
2	Permits	36351	53.80176
3	Major Crimes	70894	104.92757
4	Vote Counts	NA	NA

This is true in any R object: blanks are filled in with the value NA.

2.5.4 Other Object Classes

Variables, vectors, and data frames constitute the most basic and logical ways to organize information and they are the types of objects we will most often work with in this book. Nonetheless, there are various other classes of object that can be useful.

Lists. A list is a container for an any number of objects of any type. They may be useful in complex processes in which you want to hold a set of disparate information together for easy reference.

Matrices. Many of you may have encountered matrices in algebra classes over the years. Matrices are similar to data frames in that they have rows and columns. However, there are two major differences. First, all elements must be of the same type (e.g., numeric or character). Second, R does not treat the columns as independent vectors that can be separately named and examined.

Function-specific objects. Some functions, especially those used in statistical analyses, generate new objects. For instance, the function for running regressions, which we will learn about in Chapter 12, generates objects of the class `lm` for linear model. These and other function-specific objects have their own characteristic structure that allows them to be represented and analyzed in multiple ways.

2.6 Packages

R is a very powerful tool. But the functions built into R, or what we refer to as Base R, only go so far. Packages, however, use the components of Base R to enable additional tools and functions. As noted, R has a large community of users who have leveraged its opensource structure to create their own packages, each of which expands the capacities of R, creating new functions (and, in many cases, new classes of objects associated with those functions). This gives R possibly the greatest breadth of functionality of any statistical software.

R packages are published in multiple public repositories, the most prominent being CRAN. Just to give a sense of scope, there are 18,155 (!!) available packages on CRAN at the time of this writing. In fact, Joseph Rickert, an employee of RStudio, maintains a blog[1] that reports on the "Top 40" new R packages posted every month (in his opinion). Now, of course, as with any opensource or crowdsourced effort, these packages vary in their quality—some have limited utility, mistakes that will generate frequent errors, or are just poorly constructed, but others enable the most cutting-edge tools in statistics and visualization.

We will see a variety of packages in this book, including for working with graphics (**ggplot2** (Wickham et al., 2021); Chapter 3), character variables (**stringr** (Wickham, 2019)); Chapter 4), dates (**lubridate** (Spinu et al., 2021); Chapter 5), and spatial data and maps (**sf** (Pebesma, 2021); Chapter 8), and a whole bunch for advanced visualizations (Chapter 9). Here we want to learn about and install one called **tidyverse** (Wickham, 2021). Actually, **tidyverse** is not a single package but a suite of packages that have been designed to make coding in R more "tidy" by streamlining certain processes and introducing some new capacities. Don't worry, though, you won't have to install all these packages individually. **tidyverse** will install them all for you.

Enter the following code into your Syntax or Console:

```
install.packages('tidyverse')
```

[1] https://rviews.rstudio.com/tags/top-40/

You will see quite a bit of activity in red—that's R contacting a CRAN mirror site for each of the components of the `tidyverse`—and then some white text—that's R unpacking and confirming that all of the packages are ready to use.

We do not have much use for `tidyverse` now, but we will start using it in the next chapter. Because `tidyverse` intends to spruce up what Base R already does for us, we can replicate pretty much every task we conduct in Base R in `tidyverse`. Often, the latter is more straightforward, but there are times when Base R can be preferable, if just for personal style. There is also value to understanding the underpinnings of Base R. Thus, throughout the book, we will learn everything in both ways and you are welcome to use the approach that is most comfortable for you.

That said, we might as well activate `tidyverse` now, just to get in the practice. When we install packages, they are not automatically loaded into an R session. This is to avoid swamping your computer's memory with lots of extraneous tools. To activate a package we need to use either the `require()` or `library()` function (they do the same thing). Try it now:

```
require(tidyverse)
```

2.7 Learning R

As we move on to more serious material in the next chapter, keep in mind that part of the goal of this book is not so much to teach R but to teach you how to learn R. This is an important distinction. This book will teach you numerous packages, functions, and techniques in R. But no textbook can communicate all of what R can do, nor would that be a practical endeavor. Likewise, analysts have not memorized every function available in R. We all have to reference documentation from time to time, either to recall how to use old functions and packages or to find new ones. There are two tricks that come in handy.

Help. RStudio has a tab for Help that is in the bottom-right viewer by default. This searches and browses the detailed documentation available for all functions and packages. You can use `?` in the console to bring up the documentation for any function or package. (Try, for example, `?sqrt`.) You can also access these documentations from CRAN.

Google. It sounds silly to say, but an important tool for learning any coding language is Google. This is especially true when you are trying to solve a problem but do not yet know what functions or packages you should be using. In such cases the help documentation is not as helpful. But if you Google your problem, it is very possible someone else has encountered or solved the same issue. It is just a matter of entering a set of search terms that will be similar to the terms those who came before you might have used to describe the problem or the tool you need for it.

2.8 Summary

In this chapter, we have:

- *Installed R* from one of the mirror sites available at https://cran.r-project.org/mirrors.html;
- *Installed RStudio* from https://www.rstudio.com/products/rstudio/download/.
- *Become familiar with the RStudio interface*, including:
- *Creating a new project*,
- *The Console*, where data are submitted and results reported,
- *The Environment*, where objects in your project are visible,
- *Creating a new Script* for building code.
- *Used R as a calculator* that can execute arithmetic equations.
- *Used R as a data management software* that can store, manipulate, and examine:
- *Variables*, or single values;
- *Vectors*, or collections of values of a single type;
- and *Data Frames*, or sets of observations (rows) with values on each of two or more variables (columns).
- *Accessed packages*, specifically those composing `tidyverse`, which expand the capabilities of R.
- *Learned how to learn R* by using Help and Google.

2.9 Exercises

2.9.1 Problem Set

1. In the next chapter we will learn a variety of new functions. Learn a bit about them now by using ?. Describe in your own words what each does.
    a. `?nrow`
    b. `?length`
    c. `?nchar`
    d. `?read.csv`
2. Classify each of the following statements as describing a variable, vector, data frame, two of them, or all three.
    a. Can only contain one type of data (e.g., numeric, character).
    b. Consists of multiple values.
    c. Consists of multiple vectors.
    d. Can be given any name starting with a letter.
3. Define each of the following terms and their importance for analysis with R:
    a. Object-oriented programming
    b. Function
    c. CRAN

2.9.2 Exploratory Data Assignment

Just as we did in this chapter, create a new data frame from a series of vectors. Before you start, map out the elements.

- What are the observations?
- What will the variables be that describe them?
- Generate a `View` of the final product (take a screenshot) and describe the contents in no more than a paragraph or two.

3

Telling a Data Story: Examining Individual Records

In the early 2000's, New York City's 311 system received a call from a constituent about a "strange maple syrup smell." Needless to say, this did not fit a standard case type, like graffiti, streetlight outage, or pothole. And then the City received another such call. And another. And then dozens more. While complaining about a smell reminiscent of pancakes and waffles might seem a little silly, its unfamiliarity combined with a post-9/11 mentality convinced some that it was a harbinger of a chemical attack on the city. For 4 years, the smell and the resultant calls came and went, appearing in one neighborhood on one day, in another neighborhood a few weeks later, and so on.

In January 2009, when the smell appeared again in northern Manhattan, New York City's government took decisive action. It realized that each of these 311 reports was a data story communicating precise information about when and where the odor was occurring. As a composite, these many complaints might allow them to pinpoint what the smell was and where it was coming from. As illustrated in Figure 3.1, they mapped out the reports, combined them with detailed weather data, and voila! It seemed apparent that the smell was coming from a set of factories in northern New Jersey, which, as it happened, fabricated (among other things) a chemical extract from fenugreek seeds. And what, might you ask, did that tell them? Fenugreek seeds are the primary source of flavoring for synthetic maple syrup. Case closed.

This tale encapsulates so much about the practice of urban informatics . A team of analysts leveraged a naturally occurring data set—one that was intended to support administrative processes, not necessarily research ventures—in creative ways to answer a real-world problem that was unnerving (if not quite threatening) the neighborhoods of New York City. They became modern age detectives by looking closely at the data records generated by the 311 system and recognizing the data story they could tell. Each record is a discrete piece of information, rife with details and clues about an event or condition occurring at a place and time. Understanding the information locked within this content is the secret not only to interpreting individual records, but also knowing what can be learned from the thousands or even millions that populate a full data set.

In this chapter, we are going to learn the basic skills needed to "tell a data story" from individual records. It might seem a little simplistic to spend an entire chapter looking at individual records—the whole point of this book is to analyze big, complex data sets, right? But becoming acquainted with the content of individual records is a crucial first step to being able to make sense of the opportunities (and challenges) of the full data set. Rest assured that we will build toward that by the next chapter.

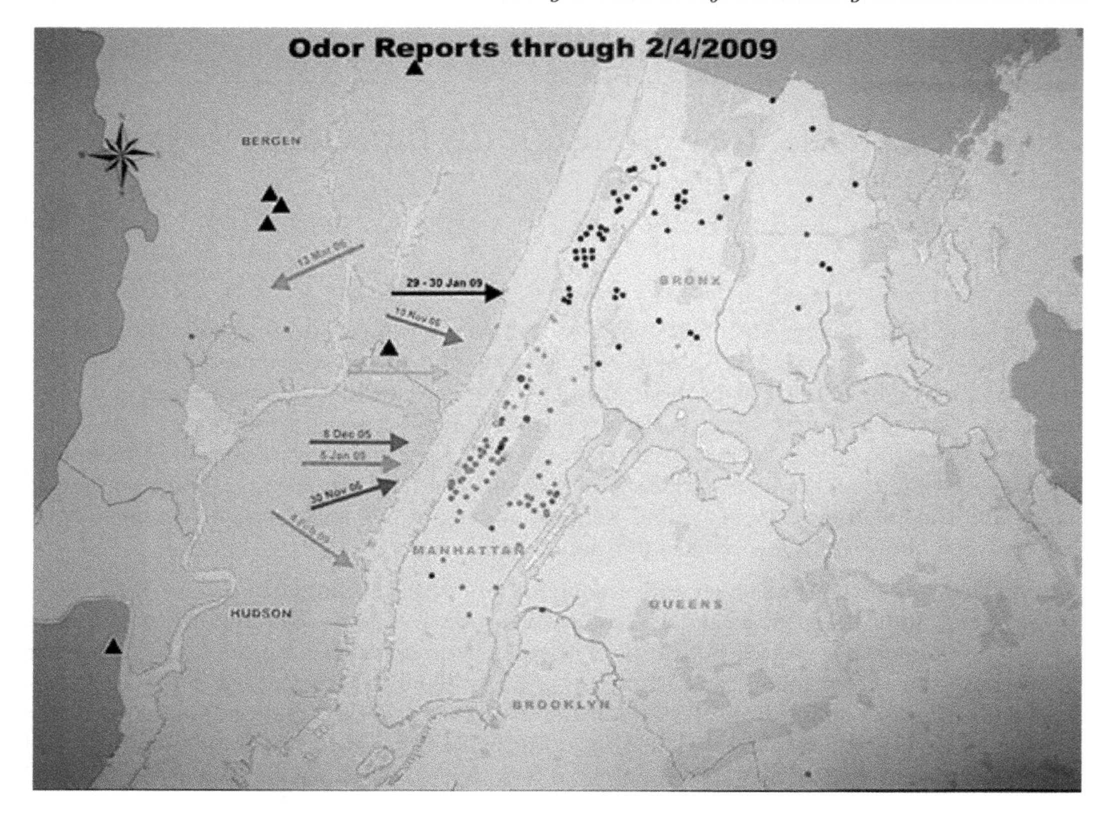

FIGURE 3.1
After years of investigation, in 2009 New York City officials combined 311 reports of a mysterious maple syrup smell (dots) with historical data on wind direction and speed (arrows) to determine that the odor was coming from factories that process artificial flavoring for maple syrup. (Credit: Gothamist)

3.1 Worked Example and Learning Objectives

Boston has a 311 system as well. Though it has never been used to solve a Maple Syrup Mystery, it has all sorts of other content that lends itself to investigation. We will use it to learn how to tell data stories, including the following skills:

- Use RMarkdown as a special scripting file for sharing both our code and results;
- Access and import datasets into R;
- Subset data to scrutinize specific records in both base R and `tidyverse`;
- Leverage other tools for systematically exposing records, including sorting;
- Think critically about the content of a data record and its interpretation.

3.1.1 Getting Started - A Reminder

This and all worked examples following it will assume that you have booted up RStudio and opened the desired project. You may want to have a separate project for each chapter or unit, or just have a single project running through the book—whatever works best for your style of organization. If you need to refresh on how to create a project, you are always welcome to flip back to Chapter 2. This will be the last time this reminder appears in the book.

3.2 Introducing R Markdown

In Chapter 2, we learned about scripts as a way to write and save code. Scripts allow us to write out multistep processes and send them to the console for execution at our desired pace. They also make it possible to easily share our manipulations and analyses with our colleagues (and the general public, if so desired). But imagine if you could share not only your code but also the results. A reader could then rerun the code and know if the results were the same or read the results and know precisely how you got there. There is a tool that does precisely this.

R Markdown is a special type of script in R that embeds executable chunks of code, each followed by the results they generate, in a single document. This process is referred to as "knitting," as it interweaves the code and its results together. As you might imagine, it is powered by additional packages, most prominently **knitr**. It can export the resulting document in HTML and pdf formats, among others. This book, in fact, is written in an extension of R Markdown called Bookdown. Getting started with Markdown is simple. Just as we did for a new script, click in the upper left-hand corner on the 'New' button

this time selecting "R Markdown...". (You will probably be prompted to install a bunch of packages. Just click Yes.) You will see an interface like this:

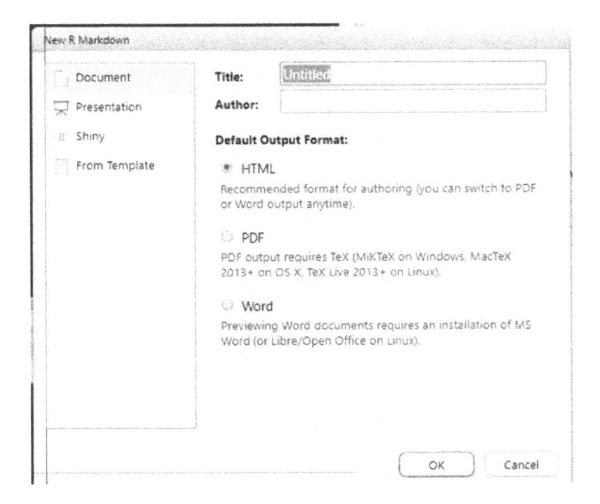

Name your document (give it an author if you like). I tend to keep HTML as the standard output; this is ideal especially if you want to post on the web at any point. The left-hand options give a sense of the other type of formats that Markdown can support, but we will stick with Document.

You should now have a new script, but one that looks a little different:

````
---
title: "Chap 3 Example"
output: html_document
---

```{r setup, include=FALSE}
knitr::opts_chunk$set(echo = TRUE)
```

## R Markdown

This is an R Markdown document. Markdown is a simple formatting syntax for authoring HTML, PDF, and MS Word documents. For more details on using R Markdown see <http://rmarkdown.rstudio.com>.

When you click the **Knit** button a document will be generated that includes both content as well as the output of any embedded R code chunks within the document. You can embed an R code chunk like this:

```{r cars}
summary(cars)
```

## Including Plots

You can also embed plots, for example:

```{r pressure, echo=FALSE}
plot(pressure)
```
````

This R Markdown, or .Rmd, script comes prepopulated with some text. A few things to note.

1. There is a title and output as you specified when you created the file. These will inform how the final document is knitted (e.g., the title will be placed at the top).
2. The document consists of plain text (black background in the figure) and code (on a lighter gray-blue background in the figure). Three backwards accents or tick marks (```) followed by brackets with the letter and then additional text (`{r}`) indicates the start of a chunk using R syntax and the title for the chunk, and the next three backwards accents (```) indicate the end of a chunk. R will execute each chunk as a single block of code and report all results in order underneath it. The plain text will appear as text in between the chunks in the final document.
3. Third, the default summary text references clicking the Knit button just above the script as the step for generating a document. This button will offer the standard format options for knitting.

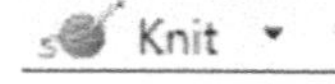

3.3 Access and Import Data

Let's get started for real! First, we need to visit the City of Boston's Open Data Portal and download 311 data. When you go to Analyze Boston[1], you will be presented with a search box:

Enter "311" into the box. Your first option should be "311 Service Requests." This will lead us to a collection of data files included under 311 Service Requests—you can think of the latter as a file folder holding all of these files. Note that the City of Boston has smartly divided the data into annual files. If they were combined they would be rather large. They also have published a documentation file in a .pdf, called the CRM Value Codex (CRM, or Constituent Relationship Management, system is another name for 311). We will not need that here, but if you want to dig into the data further, I suggest you read it.

For this exercise, we will download 311 Service Requests – 2020. The default title of the data set is a bit messy. If you choose to rename it (which I have done), you will want to do so by right-clicking on the file (or clicking with two fingers on a Mac) and selecting rename. **DO NOT ATTEMPT TO OPEN IN EXCEL AND RESAVE AS THIS CAN ALTER THE CONTENT OF THE DATA SET.** Save the data set in the directory of the project with which you are working in RStudio.

To import the data, return to your RStudio. There is a point-and-click way to import data, but we are going to do so using code as this is good practice. Remember that code allows us to specify exactly what we want R to do and allows us to communicate to others what we did. Importing data occurs through the `read.csv` command, because the data are in a .csv or comma separated values format, which is the most accessible format for transferring data across programs:

```
read.csv('Boston 311 2020.csv' , na.strings=c(''))
```

(I have had to add the additional argument `na.strings` owing to a quirk in the Boston data file. This ensures that blank values will be converted to NA, which is the standard in R.) If your file is in the same folder as your project, then simply entering its name will be sufficient. If it is in another folder, you will need to enter the folder directory as well:

```
read.csv('C:/Users/dtumm/Documents/Textbook/Unit 1 - Information/Chapter 3
- Telling a Data Story/Example/Boston 311 2020.csv', na.strings=c(''))
```

[1] data.boston.gov

You may be surprised here to see the use of / rather than \ in the file directory. This is because R syntax uses \ for other purposes, so if you copy-and-paste your directory you will need to flip the slashes.

We can enter this line of code from your Rmd using ctrl+Enter or by pressing the green arrow at the top-right of the chunk (the gear on the left is for options, and the down arrow is for running all chunks).

In either case, you should see the first 34 of 251,374 rows (or observations). This is rather large so I have not printed it here. But notice that nothing appeared in our Environment. This is because we did not store the imported data in an object. Let's try again:

```
bos_311<-read.csv('C:/Users/dtumm/Documents/Textbook/Unit 1 -
Information/Chapter 3 - Telling a Data Story/Example/Boston 311 2020.csv',
na.strings=c(''))
```

This time, nothing was outputted in the Markdown, but a data frame with 251,374 observations and 29 variables with the name `bos_311` appeared in the Environment. Success! (You are welcome to name the data frame anything you want when you import.) Importantly, if we are using `read.csv()`, R assumes we will store the imported content as a data frame. Also, note that if `read.csv()` is instructed to store the imported data in a new object, nothing is printed out, which is why we do not see any output.

`read.csv()` is the most commonly used import command in R and the one we will use almost exclusively in this book. There are other versions of the read command for other data formats (e.g., `read.table()`, `read.fortran()`, `read.delim()`). It may be disappointing to learn, however, that files from Excel (.xlsx) and other proprietary programs (including SAS, Stata, and SPSS) are not easily imported into R because of their highly specific encoding. As such, if you want to transfer data from these programs into R, you are best off saving them as .csv and importing them that way.

3.4 Getting Acquainted with a Data Frame's Structure

When we first import a data set, we may want to get a better sense of its structure to guide our explorations. Some tools for this include

```
dim(bos_311)
```

```
## [1] 251374     29
```

This tells us what we already know from the Environment, the data set contains 251,374 observations and 29 columns. We can also break this up with the `nrow()` and `ncol()` commands, which produce each of those pieces of information separately.

We may also want a preliminary sense of the types of information contained in the data set. The `names()` function gives us exactly that.

```
names(bos_311)
```

```
##  [1] "case_enquiry_id"
##  [2] "open_dt"
##  [3] "target_dt"
##  [4] "closed_dt"
##  [5] "ontime"
##  [6] "case_status"
##  [7] "closure_reason"
##  [8] "case_title"
##  [9] "subject"
## [10] "reason"
## [11] "type"
## [12] "queue"
## [13] "department"
## [14] "submittedphoto"
## [15] "closedphoto"
## [16] "location"
## [17] "fire_district"
## [18] "pwd_district"
## [19] "city_council_district"
## [20] "police_district"
## [21] "neighborhood"
## [22] "neighborhood_services_district"
## [23] "ward"
## [24] "precinct"
## [25] "location_street_name"
## [26] "location_zipcode"
## [27] "latitude"
## [28] "longitude"
## [29] "source"
```

From here we can see there are multiple variables about timing, including when a case was opened (`open_dt`) and closed (`closed_dt`), whether it was completed on time (`ontime`), the type of issue reported (`type`), where the issue occurred (`location`) and its neighborhood (`neighborhood`), among other things.

We can also find out more about each of these variables using the `class()` command. This requires isolating a single variable with what is called dollar-sign notation, taking the form `df$column` (we will do much more of this later in the chapter).

```
class(bos_311$neighborhood)
```

```
## [1] "character"
```

tells us that the `neighborhood` variable is, unsurprisingly a character variable and

```
class(bos_311$case_enquiry_id)
```

```
## [1] "numeric"
```

tells us that **case_enquiry_id** is a numeric variable, reflecting its role as a serial number to identify unique reports.

If we want to see all cases from the top of the data frame, we can use

```
head(bos_311)
```

```
##   case_enquiry_id             open_dt           target_dt
## 1    101003157986 2020-01-13 08:58:29 2020-04-12 08:58:29
## 2    101003158274 2020-01-13 11:53:00 2020-02-12 11:53:46
## 3    101003152474 2020-01-06 14:15:00 2020-01-07 14:15:39
## 4    101003154625 2020-01-08 15:39:00 2020-02-07 15:39:25
## 5    101003160351 2020-01-15 11:09:00 2020-01-16 11:09:05
## 6    101003160360 2020-01-15 11:10:00 2020-05-14 11:10:27
##   closed_dt  ontime case_status closure_reason
## 1      <NA> OVERDUE        Open
## 2      <NA> OVERDUE        Open
## 3      <NA> OVERDUE        Open
## 4      <NA> OVERDUE        Open
## 5      <NA> OVERDUE        Open
## 6      <NA> OVERDUE        Open
##                            case_title                 subject
## 1  Rental Unit Delivery Conditions   Inspectional Services
## 2   Heat - Excessive  Insufficient   Inspectional Services
## 3      Unsafe/Dangerous Conditions   Inspectional Services
## 4                  Rodent Activity   Inspectional Services
## 5              Pick up Dead Animal Public Works Department
## 6 Unsatisfactory Living Conditions   Inspectional Services
##                     reason                             type
## 1                  Housing  Rental Unit Delivery Conditions
## 2                  Housing   Heat - Excessive  Insufficient
## 3                 Building      Unsafe Dangerous Conditions
## 4 Environmental Services                  Rodent Activity
## 5          Street Cleaning              Pick up Dead Animal
## 6                  Housing Unsatisfactory Living Conditions
##                                   queue department
## 1                  ISD_Housing (INTERNAL)        ISD
## 2                  ISD_Housing (INTERNAL)        ISD
## 3                 ISD_Building (INTERNAL)        ISD
## 4 ISD_Environmental Services (INTERNAL)        ISD
## 5                          INFO_Mass DOT       INFO
## 6                  ISD_Housing (INTERNAL)        ISD
##   submittedphoto closedphoto
## 1           <NA>        <NA>
## 2           <NA>        <NA>
## 3           <NA>        <NA>
## 4           <NA>        <NA>
## 5           <NA>        <NA>
## 6           <NA>        <NA>
```

```
##                                                         location
## 1                       40 Stanwood St  Dorchester  MA  02121
## 2                           9 Wayne St  Dorchester  MA  02121
## 3                   21-23 Monument St  Charlestown  MA  02129
## 4                       98 Everett St  East Boston  MA  02128
## 5 INTERSECTION of Frontage Rd & Interstate 93 N  Roxbury  MA
## 6                        6 Rosedale St  Dorchester  MA  02124
##   fire_district pwd_district city_council_district
## 1             7           03                     4
## 2             9          10B                     7
## 3             3           1A                     1
## 4             1           09                     1
## 5             4           1C                     2
## 6             8           07                     4
##   police_district                          neighborhood
## 1              B2                               Roxbury
## 2              B2                               Roxbury
## 3             A15                           Charlestown
## 4              A7                           East Boston
## 5              C6 South Boston / South Boston Waterfront
## 6              B3                            Dorchester
##   neighborhood_services_district    ward precinct
## 1                             13 Ward 14     1401
## 2                             13 Ward 12     1207
## 3                              2  Ward 2     0204
## 4                              1  Ward 1     0102
## 5                              0       0     0801
## 6                              9 Ward 17     1703
##                          location_street_name location_zipcode
## 1                              40 Stanwood St             2121
## 2                                  9 Wayne St             2121
## 3                           21-23 Monument St             2129
## 4                               98 Everett St             2128
## 5 INTERSECTION Frontage Rd & Interstate 93 N               NA
## 6                               6 Rosedale St             2124
##   latitude longitude                source
## 1  42.3093  -71.0801     Constituent Call
## 2  42.3074  -71.0853     Constituent Call
## 3  42.3778  -71.0598     Constituent Call
## 4  42.3674  -71.0341     Constituent Call
## 5  42.3594  -71.0587 Citizens Connect App
## 6  42.2926  -71.0722     Constituent Call
```

which shows the first 6 cases (though that can be modified; try, for example, `head(bos_311,n=10)`). The `tail()` command does the same for the end of the data set.

The `str()` command combines nearly all of these tools, reporting the number of rows and columns and a list of all variable names, their class, and the first few values for each. (Try it on your own. I've omitted it because it gets a little lengthy.)

Last, you may want to browse through the observations a little as though it were an Excel spreadsheet. You can do this with `View()`, as we did in the previous chapter.

3.5 Subsetting Data Objects

3.5.1 Subsetting Data: How and Why?

Recall that R is an object-oriented programming language , meaning functions are applied to data objects that can take a variety of forms (see Chapter 2 for a refresher). In practice, this means that all of R's many functions are intended to manipulate and examine data objects to reveal or expose content in flexible ways. The most basic way to do this is to isolate and observe a subset of an object. This can be done for any data object. For now, we are going to learn how to do so in brief for vectors, and then in more depth for data frames. Importantly, R offers various tools that allow us to create subsets that reflect exactly the observations (and variables) we want to look at more closely.

Why is it useful to subset our data? The most basic answer is that we want to be able to scrutinize content more directly. We might want to make sense of the information contained in a single row or group of rows. We might want to check for anomalies that communicate the extremes of a data set or even errors that might be in it. We might want to use a closer look at multiple records to better understand the broader opportunities and challenges presented by the full data set. Those are our goals in this chapter, but they also lay the groundwork for ways that subsets will be useful in future chapters, often delimiting and targeting our analyses and visualizations to provide the specific insights that we are looking for.

3.5.2 Subsetting in Base R: Vectors

In base R, subsets are indicated by placing brackets after an object's name, with the brackets containing criteria denoting the desired cases. The simplest example is with vectors. To illustrate, let us return to a vector constructed in Chapter 2:

```
bos_event_types<-c('311','Permits','Major Crimes')
bos_event_types[1]
```

```
## [1] "311"
```

The bracket `[1]` instructed R to print the first element in the vector. Try again with `[2]` or `[3]`. What do you think will happen?

3.5.3 Subsetting in Base R: Data Frames

3.5.3.1 Subsetting by Values

The logic for subsetting data frames is similar to that for vectors, except that data frames have two dimensions on which we can subset: rows (observations) and columns (variables). Thus, the brackets need to specify what we want from each, with a comma in between. That is, we want `[row,column]`, for instance,

```
bos_311[5,1]
```

```
## [1] 101003160351
```

This command gives us the `case_enquiry_id` (a serial number identifying the report; the first column) for the fifth observation (`View()` or click on the data frame in the Environment to check, if you like). We might also see its `open_dt` (the date and time the report was received; column #2).

```
bos_311[5,2]
```

```
## [1] "2020-01-15 11:09:00"
```

which was January 15th, 2020 at 11:09 am.

What if we want to see the entirety of the case to tell its story? We need to enter a row without columns specified. The following code will generate the desired case and its values for all columns

```
bos_311[5,]
```

```
##   case_enquiry_id            open_dt           target_dt
## 5    101003160351 2020-01-15 11:09:00 2020-01-16 11:09:05
##   closed_dt  ontime case_status closure_reason
## 5      <NA> OVERDUE        Open
##             case_title                     subject          reason
## 5 Pick up Dead Animal Public Works Department Street Cleaning
##                   type         queue department submittedphoto
## 5 Pick up Dead Animal INFO_Mass DOT       INFO           <NA>
##   closedphoto
## 5        <NA>
##                                                       location
## 5 INTERSECTION of Frontage Rd & Interstate 93 N  Roxbury  MA
##   fire_district pwd_district city_council_district
## 5             4           1C                     2
##   police_district                              neighborhood
## 5              C6 South Boston / South Boston Waterfront
##   neighborhood_services_district ward precinct
## 5                              0    0     0801
##                              location_street_name location_zipcode
## 5 INTERSECTION Frontage Rd & Interstate 93 N               NA
##   latitude longitude                 source
## 5  42.3594  -71.0587 Citizens Connect App
```

Now we are getting somewhere, and possibly somewhere interesting. We see that the case was reported on January 15th, 2020 (`open_dt`), and the City indicated that it should be closed by the next day (`target_dt`). However, it was never closed (`closed_dt` is empty) and it is flagged as being OVERDUE (`ontime`) and Open (`case_status`).

Rmd limits to the first six variables. If we wanted to learn even more, we could do `View(bos_311[5,])` and find out that it was a request to Pick up Dead Animal (`type`) at the INTERSECTION of Frontage R & Interstate 93 N (`location`) in South Boston / South Boston Waterfront (`neighborhood`). Note how these pieces of information fit together to tell a story. What do you think happened here?

You will notice that R needed us to put a comma after the 5 to know that we wanted row 5. If you entered `bos_311[5]` it would have assumed that you wanted column 5. That is, as a default, R is designed to assume that analysts are more likely to want to see all cases on a subset of variables than all variables on a subset of cases. `bos_311[,5]` would accomplish the same outcome.

Selecting a single value or a single row or column is only so useful. There are multiple ways for us to expand this while still working exclusively with numbers.

1. Colons allow us to indicate a series of consecutive values. For instance, `bos_311[1:4,]` would generate all variables for the first four cases, and `bos_311[4:10,]` would generate the same for cases 4-10.
2. `c()`, or the combine command, can put together non-consecutive values. For example, `bos_311[c(1,2,4:6),]` would return all variables for cases 1, 2, 4, 5, and 6. Meanwhile, `bos_311[c(1,2,4:6)]` would generate variables 1, 2, 4, 5, and 6 for all cases.

If you like, this is a good time to play around with some options and to check them against the data set to confirm that you are comfortable with subsetting by row and column values.

3.5.3.2 Subsetting Columns

There are two other ways to subset columns that can feel more accessible. The first is by name.

```
head(bos_311['neighborhood'])
```

```
##                             neighborhood
## 1                                Roxbury
## 2                                Roxbury
## 3                            Charlestown
## 4                            East Boston
## 5 South Boston / South Boston Waterfront
## 6                             Dorchester
```

See how this lists the neighborhood where each event reported to 311 occurred. We can also combine multiple variables using this notation.

```
head(bos_311[c('neighborhood','type')])
```

```
##                             neighborhood
## 1                                Roxbury
## 2                                Roxbury
```

```
## 3                               Charlestown
## 4                               East Boston
## 5 South Boston / South Boston Waterfront
## 6                                Dorchester
##                                type
## 1  Rental Unit Delivery Conditions
## 2   Heat - Excessive  Insufficient
## 3      Unsafe Dangerous Conditions
## 4                  Rodent Activity
## 5              Pick up Dead Animal
## 6 Unsatisfactory Living Conditions
```

generates the list of neighborhoods and the case types for all observations. This notation is compatible with subsets of observations as well:

```
bos_311[5,c('neighborhood','type')]
```

```
##                              neighborhood                type
## 5 South Boston / South Boston Waterfront Pick up Dead Animal
```

generates these two pieces of information for observation #5.

The second way is with dollar-sign notation. This notation takes the form `df$column` and can only isolate a single variable:

```
head(bos_311$neighborhood)
```

```
## [1] "Roxbury"
## [2] "Roxbury"
## [3] "Charlestown"
## [4] "East Boston"
## [5] "South Boston / South Boston Waterfront"
## [6] "Dorchester"
```

generates the same information as `bos_311` (though the first stays within Rmd, the latter generates more extensive results in the Console).

Though it is not possible to select multiple variables this way, it is possible to simultaneously subset by rows and columns. Remember that each column in a data frame is actually a vector. So, when we subset "rows" with dollar-sign notation, we are actually subsetting a vector. This means we do not need the comma.

```
bos_311$neighborhood[5]
```

```
## [1] "South Boston / South Boston Waterfront"
```

This tells us again that this case occurred in the South Boston / South Boston Waterfront neighborhood.

3.5.3.3 Subsetting with Logical Statements

The last way to subset is by using logical statements based on the content of the data. This can be very powerful when we want to target our analyses based on specific criteria. For example, what if we want to see all cases that were still open when the data set was posted? (Limiting to the first six variables for brevity's sake.)

```
head(bos_311[bos_311$case_status=='Open',1:6])
```

```
##    case_enquiry_id             open_dt           target_dt
## 1     101003157986 2020-01-13 08:58:29 2020-04-12 08:58:29
## 2     101003158274 2020-01-13 11:53:00 2020-02-12 11:53:46
## 3     101003152474 2020-01-06 14:15:00 2020-01-07 14:15:39
## 4     101003154625 2020-01-08 15:39:00 2020-02-07 15:39:25
## 5     101003160351 2020-01-15 11:09:00 2020-01-16 11:09:05
## 6     101003160360 2020-01-15 11:10:00 2020-05-14 11:10:27
##    closed_dt  ontime case_status
## 1       <NA> OVERDUE        Open
## 2       <NA> OVERDUE        Open
## 3       <NA> OVERDUE        Open
## 4       <NA> OVERDUE        Open
## 5       <NA> OVERDUE        Open
## 6       <NA> OVERDUE        Open
```

You should see in your Rmd the first 6 of 34,836 rows and 6 of 29 columns, as well as a 7th column at the front. This is the row name, which is most often an index of the original order of cases. Also note that `case_status` indeed is equal to Open for all cases. How did this work? First, we needed to indicate the variable that would be the basis for the criterion, `bos_311$case_status`. We then needed to define the criterion as being equal to Open, which is in single quotations because it is text. In R, "equal to" in a criterion is not represented with a single equals sign but with two in a row (==). This is because a single equals sign sets one thing equal to another. We want to actually evaluate whether the two things are equal. In a sense, you might think of it as asking the computer, "Is it true (i.e., equal to reality) that these two things are equal?" Then, of course, we need a "," at the end to indicate that we want all rows equal to this criterion.

From a technical perspective, here is what is happening. R is evaluating whether `case_status` is equal to 'Open' for every case in the data set. It then assigns each a TRUE or FALSE value and creates a list of observation numbers for which TRUE is the case. It then uses this list to create the subset. Thus, R figured out that what we wanted was `bos_311[c(1:7, 19, 22, 41, ...),]`. You, of course, would not have known to ask for this unless you had evaluated every case to know whether it was still Open or not, but R did this for you.

An advantage of logical statements is the ability to create more than one criterion. One striking thing here is that case 22 appears to be ONTIME although still Open since January 1st, 2020. How common is it for these two things to be true? Let's find out by using the `&` operator to indicate multiple criteria.

```
head(bos_311[bos_311$case_status=='Open' &
               bos_311$ontime=='ONTIME',1:6])
```

```
##      case_enquiry_id             open_dt           target_dt
## 22      101003148513 2020-01-01 15:02:00                <NA>
## 62      101003148723 2020-01-02 07:03:00                <NA>
## 72      101003148792 2020-01-02 08:09:00                <NA>
## 83      101003148854 2020-01-02 08:55:00                <NA>
## 124     101003149184 2020-01-02 11:53:00 2021-12-22 11:53:26
## 152     101003149460 2020-01-02 14:45:00 2021-12-22 14:45:03
##      closed_dt ontime case_status
## 22        <NA> ONTIME        Open
## 62        <NA> ONTIME        Open
## 72        <NA> ONTIME        Open
## 83        <NA> ONTIME        Open
## 124       <NA> ONTIME        Open
## 152       <NA> ONTIME        Open
```

There are 13,512 such cases. Looking at the top of the data set, it appears that many of these had no **target_dt** to be benchmarked against. Hard to be OVERDUE if you had no due date. But there are some others for which the targets are very far away, in December 2021, meaning they had not been due yet at the time of the export of the data (though if you are doing this exercise after December 2021, the data might look a little different).

Is it possible this is true for certain case types? We could create a subset to examine this, too. Maybe we can learn more about this by limiting to those cases without target dates. This will use a new function called `is.na()`, which identifies all cases with the value of NA for a given variable:

```
head(bos_311[bos_311$case_status=='Open' &
               bos_311$ontime=='ONTIME' &
               is.na(bos_311$target_dt),1:6])
```

```
##      case_enquiry_id             open_dt target_dt closed_dt
## 22      101003148513 2020-01-01 15:02:00      <NA>      <NA>
## 62      101003148723 2020-01-02 07:03:00      <NA>      <NA>
## 72      101003148792 2020-01-02 08:09:00      <NA>      <NA>
## 83      101003148854 2020-01-02 08:55:00      <NA>      <NA>
## 177     101003149654 2020-01-02 19:04:00      <NA>      <NA>
## 214     101003150009 2020-01-03 09:29:00      <NA>      <NA>
##      ontime case_status
## 22   ONTIME        Open
## 62   ONTIME        Open
## 72   ONTIME        Open
## 83   ONTIME        Open
## 177  ONTIME        Open
## 214  ONTIME        Open
```

When we do this, we find that quite a few of those without target dates are about animals or are general informational requests, which might mean that it is hard to set a target when you cannot be sure that said animal will still be there or for something informational.

```
head(bos_311[is.na(bos_311$target_dt),c('target_dt','case_status',
                                        'ontime','type')])
```

```
##    target_dt case_status ontime                  type
## 22      <NA>        Open ONTIME Animal Generic Request
## 26      <NA>      Closed ONTIME          Needle Pickup
## 27      <NA>      Closed ONTIME          Needle Pickup
## 29      <NA>      Closed ONTIME          Needle Pickup
## 30      <NA>      Closed ONTIME          Needle Pickup
## 31      <NA>      Closed ONTIME          Needle Pickup
```

Note that only one of these at the top is animal related, and most are needle pickups. Happily, those were all Closed.

3.5.3.4 Extra Tools for Subsetting

This exercise has shown a good bit about how to subset, but there are two final skills we want to leave with. The first is the `!` operator, which means "not equal" in R. In the spirit of the double equals sign, though, it needs to be applied to something, in this case an equals sign. For instance,

```
head(bos_311[bos_311$case_status!='Open',1:6])
```

```
##    case_enquiry_id             open_dt           target_dt
## 8     101003172959 2020-01-19 06:44:00 2020-01-28 08:30:00
## 9     101003148263 2020-01-01 03:27:00 2020-01-02 03:27:09
## 10    101003148269 2020-01-01 06:19:00 2020-01-03 08:30:00
## 11    101003148271 2020-01-01 07:02:00 2020-01-03 08:30:00
## 12    101003148342 2020-01-01 10:50:00 2020-01-31 10:50:14
## 13    101003148276 2020-01-01 07:56:41 2020-01-03 08:30:00
##              closed_dt  ontime case_status
## 8  2020-02-03 14:04:40 OVERDUE      Closed
## 9  2020-01-06 08:03:45 OVERDUE      Closed
## 10 2020-01-02 06:10:56  ONTIME      Closed
## 11 2020-01-01 07:07:17  ONTIME      Closed
## 12 2020-01-03 10:57:38  ONTIME      Closed
## 13 2020-01-02 05:08:04  ONTIME      Closed
```

unsurprisingly generates a subset of all cases that are Closed, meaning not equal to Open.

Or

```
head(bos_311[!is.na(bos_311$target_dt),1:6])
```

```
##   case_enquiry_id             open_dt           target_dt
## 1    101003157986 2020-01-13 08:58:29 2020-04-12 08:58:29
## 2    101003158274 2020-01-13 11:53:00 2020-02-12 11:53:46
## 3    101003152474 2020-01-06 14:15:00 2020-01-07 14:15:39
## 4    101003154625 2020-01-08 15:39:00 2020-02-07 15:39:25
## 5    101003160351 2020-01-15 11:09:00 2020-01-16 11:09:05
## 6    101003160360 2020-01-15 11:10:00 2020-05-14 11:10:27
##   closed_dt  ontime case_status
## 1      <NA> OVERDUE        Open
## 2      <NA> OVERDUE        Open
## 3      <NA> OVERDUE        Open
## 4      <NA> OVERDUE        Open
## 5      <NA> OVERDUE        Open
## 6      <NA> OVERDUE        Open
```

generates all cases for which `target_dt` has a value, some of which have closed dates, some of which do not.

Second, sometimes we do not want the combination of criteria but their intersection. This requires an 'or' statement, which uses the | symbol (over on the far right under Backspace on most English keyboards). Suppose we want to know about all cases that are either OVERDUE or have never been closed, because the latter can slip by unnoticed if there was no target date.

```
head(bos_311[is.na(bos_311$closed_dt) |
              bos_311$ontime=='OVERDUE',1:6])
```

```
##   case_enquiry_id             open_dt           target_dt
## 1    101003157986 2020-01-13 08:58:29 2020-04-12 08:58:29
## 2    101003158274 2020-01-13 11:53:00 2020-02-12 11:53:46
## 3    101003152474 2020-01-06 14:15:00 2020-01-07 14:15:39
## 4    101003154625 2020-01-08 15:39:00 2020-02-07 15:39:25
## 5    101003160351 2020-01-15 11:09:00 2020-01-16 11:09:05
## 6    101003160360 2020-01-15 11:10:00 2020-05-14 11:10:27
##   closed_dt  ontime case_status
## 1      <NA> OVERDUE        Open
## 2      <NA> OVERDUE        Open
## 3      <NA> OVERDUE        Open
## 4      <NA> OVERDUE        Open
## 5      <NA> OVERDUE        Open
## 6      <NA> OVERDUE        Open
```

These are just some initial examples. We can of course get into more complicated criteria, with any number of `&`s and `|`s, though it takes some critical thinking to determine exactly what order you want them to be in and whether you need parentheses to set some off from others, or whether a `!` should be in there somewhere.

3.5.3.5 Summary

Because we have gained a variety of skills for subsetting in R through a worked example, it is useful to pause and reflect on what we learned.

1. We can subset vectors and data frames.
2. Subsetting data frames can occur using values in brackets with `[row,column]`.
3. The values in the brackets can be expanded using a colon for consecutive values, and `c()` for non-consecutive values.
4. We can isolate single variables with `df$column` and multiple variables with quotations within brackets.
5. We can use logical statements to create subsets according to criteria, including:
   a. `==` for equal to
   b. `!=` for not equal to
   c. `is.na()` for blank cases
   d. `&` to combine criteria and `|` to identify cases in which either of two criteria are true (an or statement).
   e. All of these can be combined flexibly to define the exact subset that you want.

3.5.4 Subsetting in tidyverse

As discussed in Chapter 2, many of the skills we will learn in Base R have parallels in `tidyverse`, a set of packages designed to make coding in R cleaner (or more "tidy," as the name implies). These parallels are often simpler to work with. We will see this for the first time by learning how to subset in `tidyverse`. If you have not already installed `tidyverse` and its underlying packages, you can do so with the command `install.packages('tidyverse')`. To get started with this part of the worked example, you will then need to

```
require(tidyverse)
```

First, it is worth noting that tidyverse commands convert dataframes into "tibbles." Tibbles are essentially data frames with a few small technical tweaks that we do not need to delve into here, but you can learn more about them in the tidyverse documentation at www.tidyverse.org.

Turning to subsetting in tidyverse, there are two main functions that we can use: `filter()` and `select()`. The first subsets rows, the latter subsets columns. We will also need to learn a third skill called "piping" to combine them. We can illustrate all three with code from our worked example. (Note: for brevity, I will not have the subsets in this section print out. You are welcome, though, to check that the Base R and `tidyverse` versions agree.)

3.5.4.1 filter()

When we wanted to see all Open cases, we entered

```
head(bos_311[bos_311$case_status=='Open',])
```

We can do the same with the `filter()` command, which takes the form `filter(df, criteria)` as

```
head(filter(bos_311, case_status=='Open'))
```

You will notice that we no longer have brackets after the name of the data frame. Instead, we are telling R the name of the data frame in the first argument, and then giving it the criteria in the second argument. Because filter specifically handles rows, we also no longer need the comma at the end. Last, because the first argument already indicates the data frame of interest, we can refer directly to variables therein without dollar-sign notation.

The same process would apply to one of our more complex criteria:

```
head(bos_311[is.na(bos_311$closed_dt) | bos_311$ontime=='OVERDUE',])
```

becomes

```
head(filter(bos_311, is.na(closed_dt) | ontime=='OVERDUE'))
```

But one other difference is that each `&` is replaced by a comma. You can think of it as entering a list of shared criteria.

```
bos_311[bos_311$case_status=='Open' & bos_311$ontime=='ONTIME',]
```

becomes

```
filter(bos_311, case_status=='Open', ontime=='ONTIME')
```

3.5.4.2 select()

The `select()` command subsets according to columns, and is structured in the same way as `filter()` with `select(df, variable names)`. Thus

```
bos_311['neighborhood']
```

and

```
bos_311$neighborhood
```

in Base R are equivalent to

```
select(bos_311,neighborhood)
```

in `tidyverse`.

And

```
bos_311[c('neighborhood','type')]
```

in Base R is equivalent to

```
select(bos_311,neighborhood, type)
```

in `tidyverse`. Again, the first argument, which is the name of the data frame, removes the need for dollar-sign notation or quotation marks. R already knows that it is looking for variables within that data frame.

There are also multiple valuable helper functions that can be useful, especially in a data set with many variables that have similar structures (e.g., `ends_with(dt)` could be useful for isolating all variables that end with the suffix `_dt`, thereby referring to a date and time; `starts_with()` and `contains()` offer similar opportunities).

An additional trick is if you want to select all columns but exclude one or more, you can use a minus sign. For instance

```
View(select(CRM, -neighborhood, -type))
```

selects all variables except for neighborhood and type.

3.5.4.3 Piping

Unfortunately, `select()` and `filter()` cannot be directly combined the way that brackets allow us to subset rows and columns at the same time. This can be solved, however, with a special capacity of `tidyverse` called "piping," which "passes" the product of one command directly to another command. This allows the analyst to combine multiple manipulations in a series. Often it is possible to do something similar in Base R in a single line by nesting functions within each other, but it can get complicated and difficult to ascertain whether everything is just as we want it. Piping allows this process and its outcomes to be more easily examined.

Piping depends on the pipe symbol, `%>%`. We simply place it at the end of a line and the product of that code is passed to the next line. R does not conclude the steps until it reaches a line that does not end with a pipe. In theory, you can connect any number of lines with pipes, but it is typically best practice to keep them under 10; otherwise the complexity may merit breaking things up into pieces.

Let us demonstrate piping with `select()` and `filter()` for a few of the subsets we created above.

```
bos_311[bos_311$case_status=='Open' &
          bos_311$ontime=='ONTIME',
        c('target_dt','ontime','type')]
```

becomes

```
bos_311 %>%
  filter(case_status=='Open', ontime=='ONTIME') %>%
  select(target_dt,ontime,type)
```

The first line of a series of piped commands is often just the data frame of interest. When we say that pipes "pass" the product of the previous command, we mean the product of one command becomes the first argument of the next command. Thus, we do not need to restate the data frame for `filter()`. We skip straight to the criteria for rows that we want. After `filter()` is completed, the subset of rows is passed to `select()`, which narrows down to our three desired variables.

This is also an illustration of the importance of order. If we changed the order of `filter()` and `select()`, we get an error. Why? Feel free to think about it for a moment before reading on.

If we give `select()` first, we have already removed all variables except for `target_dt`, `ontime`, and `type`. As such, we can no longer `filter()` based on the content of `case_status` because no such variable exists.

This simple example demonstrates the elegance of piping, which we will use regularly when coding with tidyverse.

3.5.5 Combining Subsets with Other Tools

Immediately after importing our data, we learned a series of tools for exposing the structure of a data frame. We can utilize these effectively for any subset, as well, which could be informative. The most obvious might be for viewing a subset more closely

```
View(bos_311[bos_311$case_status=='Open' &
               bos_311$ontime=='ONTIME',
             c('target_dt','ontime','type')])
```

Now we can browse through all 13,512 rows in our subset rather than being limited to the first few.

Even simpler

```
nrow(bos_311[bos_311$case_status=='Open' &
               bos_311$ontime=='ONTIME',
             c('target_dt','ontime','type')])
```

```
## [1] 13512
```

tells us that there are 13,512 rows in the subset without requiring us to generate the full subset.

To do the same thing in tidyverse, we add the new command to the end of the existing commands, though they will remain empty because we are passing the desired data frame directly to them:

```
bos_311 %>%
  filter(case_status=='Open', ontime=='ONTIME') %>%
  select(target_dt,ontime,type) %>%
  nrow()
```

```
## [1] 13512
```

Try to do the same with `View()`.

3.6 Sorting

A final skill we might want to leverage when exploring a new data set is sorting, which allows us to get a sense of the content of variables, especially their extreme values. Doing so in Base R is more complicated than you might expect, however. Let's say we want to see the top of our data frame in order of the open dates for all cases, we need to enter

```
head(bos_311[order(bos_311$open_dt),1:6])
```

```
##        case_enquiry_id             open_dt           target_dt
## 220012    101003148236 2020-01-01 00:13:06 2020-01-03 08:30:00
## 45692     101003148237 2020-01-01 00:20:09 2020-01-03 08:30:00
## 9196      101003148238 2020-01-01 00:24:00 2020-01-03 08:30:00
## 15497     101003148239 2020-01-01 00:28:07 2020-03-09 08:30:00
## 220013    101003148240 2020-01-01 00:29:06 2020-01-03 08:30:00
## 754       101003148242 2020-01-01 00:31:00 2020-01-09 08:30:00
##                  closed_dt  ontime case_status
## 220012 2020-01-02 03:12:46  ONTIME      Closed
## 45692  2020-01-02 03:10:50  ONTIME      Closed
## 9196   2020-01-01 01:20:40  ONTIME      Closed
## 15497                 <NA> OVERDUE        Open
## 220013 2020-01-02 03:11:25  ONTIME      Closed
## 754                   <NA> OVERDUE        Open
```

Substantively we have learned that the 2020 data begin early in the morning on January 1st, 2020, which would seem a bit obvious, but is at least a good sanity check on the data's veracity. But why did we just create a subset with brackets in order to sort? The answer is that `order()` \index{order()@*order()* did not simply sort `bos_311` according to the variable `open_dt`. Instead, it determined the sorted order of `open_dt`, and then passed this along as an ordered list through brackets, as if that list were a subset.

`order()` can also accommodate reverse sorting, as in

```
head(bos_311[order(bos_311$open_dt,decreasing = TRUE),1:6])
```

```
##        case_enquiry_id                  open_dt           target_dt
## 216255     101003578870 2020-12-31 23:56:03 2021-01-05 08:30:00
## 216254     101003578868 2020-12-31 23:40:28 2021-01-05 08:30:00
## 176508     101003578867 2020-12-31 23:38:09 2021-01-05 08:30:00
## 216253     101003578856 2020-12-31 23:02:00 2021-01-05 08:30:00
## 209957     101003578854 2020-12-31 23:01:00 2021-01-05 08:30:00
## 216237     101003578853 2020-12-31 22:56:00 2021-01-05 08:30:00
##                  closed_dt ontime case_status
## 216255 2021-01-01 00:28:55 ONTIME      Closed
## 216254 2021-01-02 12:47:46 ONTIME      Closed
## 176508 2021-01-02 12:48:25 ONTIME      Closed
## 216253 2020-12-31 23:40:43 ONTIME      Closed
## 209957 2020-12-31 23:55:28 ONTIME      Closed
## 216237 2021-01-01 02:08:25 ONTIME      Closed
```

and even multiple variables with sorting occurring by the first one first, and then ties decided by the second, and so on.

```
head(bos_311[order(bos_311$ontime, bos_311$open_dt),1:6])
```

```
##        case_enquiry_id                  open_dt           target_dt
## 220012     101003148236 2020-01-01 00:13:06 2020-01-03 08:30:00
## 45692      101003148237 2020-01-01 00:20:09 2020-01-03 08:30:00
## 9196       101003148238 2020-01-01 00:24:00 2020-01-03 08:30:00
## 220013     101003148240 2020-01-01 00:29:06 2020-01-03 08:30:00
## 756        101003148244 2020-01-01 00:38:53 2020-01-03 08:30:00
## 220014     101003148245 2020-01-01 00:55:00 2020-01-03 08:30:00
##                  closed_dt ontime case_status
## 220012 2020-01-02 03:12:46 ONTIME      Closed
## 45692  2020-01-02 03:10:50 ONTIME      Closed
## 9196   2020-01-01 01:20:40 ONTIME      Closed
## 220013 2020-01-02 03:11:25 ONTIME      Closed
## 756    2020-01-02 03:07:18 ONTIME      Closed
## 220014 2020-01-01 01:35:42 ONTIME      Closed
```

The complexity of `order` \index{order()@*order()* is precisely the kind of thing that the `tidyverse` sought to eliminate. In this case, the parallel command is `arrange()` \index{arrange()@*arrange()*.

```
head(arrange(bos_311,open_dt))
```

is identical to our first command with `order`,

```
arrange(bos_311,desc(open_dt))
```

creates a descending order, and

```
arrange(bos_311,ontime, open_dt)
```

parallels our last command with `order`, as the multiple variables are read as the order to be applied for sorting. \index{arrange()@*arrange()*}

Sorting, especially when getting started with a new data set, is often effectively combined with `View()` and subsets. We have not elaborated much on our worked example in this section, but now would be a good time to practice sorting with some of the other variables in the data set to see what else you can learn about its contents.

3.7 Summary

In this chapter we have started to get to know the database generated by the City of Boston's 311 system. As we moved along our subsets, we began to better understand the types of content it holds and how we might use it to "tell a data story"—from when a report was received, to the type of issue it referenced, to where that issue was, to when it was resolved (and if that was done on time or not). We clearly have only scratched the surface of these data, but it should be taken as one example of how we might utilize these tools to get a better sense for the stories that a data set can tell, both through its individual records and through the analysis of them as a composite.

Along the way, we have:

- *Created a script in R Markdown* to document our code and the results;
- *Downloaded and imported* a data set from an open data portal;
- *Assessed the structure and content* of a dataset;
- *Subset data in base R* with
- Observation and variable numbers,
- Variables based on their names, in dollar-sign notation and with quotations,
- And observations, through logical statements of varying complexity;
- *Subset the data in* `tidyverse` with `filter()` and `select()` commands;
- *"Piped" (%>%) commands in* `tidyverse` to link them together, in particular to combine `select()` and `filter()` commands.
- *Combined subsets with commands for exploring structure*, in this case of subsets of the data frame.
- *Sorted the data* according to the values of select variables.

3.8 Exercises

3.8.1 Problem Set

1. For each of the following, identify the error and suggest a fix. Assume `bos_311` as used throughout the chapter when relevant.

a. `nbhd<-c("Dorchester","Jamaica Plain","Charlestown","East Boston","Roxbury") ncol(nbhd) + nrow(nbhd) = ??`
b. `bos_crime<-read.csv(Boston 911 2020.csv)`
c. `bos_311[bos_311$case_status=='Open' | bos_311$ontime=='ONTIME']`
d. `bos_311 %>% filter(case_status=='Open') select(target_dt,ontime, type) %>% View()`
e. `order(bos_311$caes_enquiry_id)`
f. `bos_311 %>% filter(is.na(target_dt)) %>% bos_311[c('neighborhood', 'type') %>% View()`

2. You've imported a new 311 data frame called CRM that has the content below. What will R return for each of the following commands?

TYPE	propId	CASE_ID	OPEN_DT	location	neighborhood	Source
Traffic Signal Repair	I1739	101000729429	1/1/2013	Bowdoin St & Bowdoin St	Dorchester	Self Service
Request for Snow Plowing	I1921	101000729431	1/1/2013	Brent St & Talbot Ave	Dorchester	Constituent Call
Snow Removal	A43627	101000729432	1/1/2013	268 E Cottage St	Dorchester	Citizens Connect App
Snow Removal	A219808	101000729433	1/1/2013	276 E Cottage St	Dorchester	Citizens Connect App
Snow Removal	I1281	101000729436	1/1/2013	Belmore Ter & Boylston St	Jamaica Plain	Citizens Connect App
Snow Removal	A131824	101000729437	1/1/2013	45 Sullivan St	Charlestown	Citizens Connect App
Notification	A112520	101000729439	1/1/2013	48 Pratt St	Allston / Brighton	Constituent Call
Notification	A112520	101000729439	1/1/2013	48 Pratt St	Allston / Brighton	Constituent Call
Requests for Street Cleaning	A14574	101000729441	1/1/2013	1124 Bennington St	East Boston	Constituent Call
Requests for Street Cleaning	A158377	101000729445	1/1/2013	895 Massachusetts Ave	Roxbury	Constituent Call

a. `CRM[ncol(CRM),nrow(CRM)/2]`
b. `nrow(CRM[CRM$neighborhood == 'Dorchester' & CRM$TYPE == 'Snow Removal',])`
c. `nrow(CRM[CRM$neighborhood == 'Dorchester' | CRM$TYPE == "Snow Removal",])`

3. Translate each of the following commands into `tidyverse`.
a. `CRM[ncol(CRM),nrow(CRM)/2]`
b. `nrow(CRM[CRM$neighborhood == 'Dorchester' & CRM$TYPE == 'Snow Removal',])`
c. `nrow(CRM[CRM$neighborhood == 'Dorchester' | CRM$TYPE == "Snow Removal",])`
d. `CRM[order(CRM$neighborhood, CRM$TYPE),]`

3.8.2 Exploratory Data Assignment

Import a data set of your choice into R. This is the first exploratory data assignment in this book that works with a full data set drawn from an outside source. It is suggested that to get the most from these assignments you use the same data set in all of them, though that is not required. Once you have imported the data set, complete the following:

1. Briefly describe the structure of your whole data set or of some meaningful subset based one or more criteria (number of rows, columns, names of relevant variables).
2. Describe at least three observations (rows) from this set/subset using the information contained in the variables you see as relevant.
3. What have you learned from these observations individually, and from the differences between them?
4. How do they illustrate what we can learn from these data more generally?
5. Can you infer things about the observations that are not explicitly in the data?

4

The Pulse of the City: Observing Variable Patterns

You have probably heard much about how cell phone records contain detailed information about people's movements, their social ties, and habits. MIT's Senseable City Lab was one of the first research centers to reveal the rich potential for these novel data. In one paper, they used cell phone records from three different metropolises—London, New York City, and Hong Kong—to illustrate the daily rhythms of communication in different parts of the city (Grauwin et al., 2014). They found that calls, texts, and downloads were more common in downtown and business districts during weekdays and that they spiked during the evening in residential areas. They also found that certain specialized areas, like airports, had rather distinct patterns. In this way, the cell phone data had revealed the *pulse of the city* —that is, the predictable patterns that characterize the activities of an urban area and its communities.

I would guess that you are of one of two minds about this work. You might be deeply impressed by the predictability of human behavior, saying something like, "Wow, cities have pulses just like humans!" Or you could be more skeptical, grumbling that "We didn't need cell phone data to know that people play on their phones at workplaces during the day and at home at night." Both perspectives are correct. These are obvious findings. But the methodology itself, the ability to observe the pulse of the city, is a critical advance. Let's take the metaphor of the "pulse" seriously. Because we can easily measure our own body's pulse (and other bodily functions), we can quantify and explain variation between individuals, identify aberrations and their relationship to health, and invent tools for responding to emergencies.

Observing the pulse of the city through cell phone data was just the first step. As illustratd in Figure 4.1, analysts have since used the data for all nature of insight. They have redrawn the districts of nations based on communication and movement connections (Ratti et al., 2010). They have identified the "signature of friendship" based on the frequency with which people are near each other and contact each other by call or text (Eagle et al., 2009). They have shown that communities whose residents have more connections to other communities tend to also have higher incomes; and that people of different racial backgrounds not only tend to live in separate communities, they also tend to move through completely different spaces (Wang et al., 2018). And during the COVID-19 pandemic, there were regular reports showing how stay-at-home orders effectively "shut down" the pulse of the city, at least in terms of people leaving home (Abouk and Heydari, 2021). All of these discoveries (and more!) were possible thanks to the initial work demonstrating the reliability of cell phone data to capture the daily rhythms of communities.

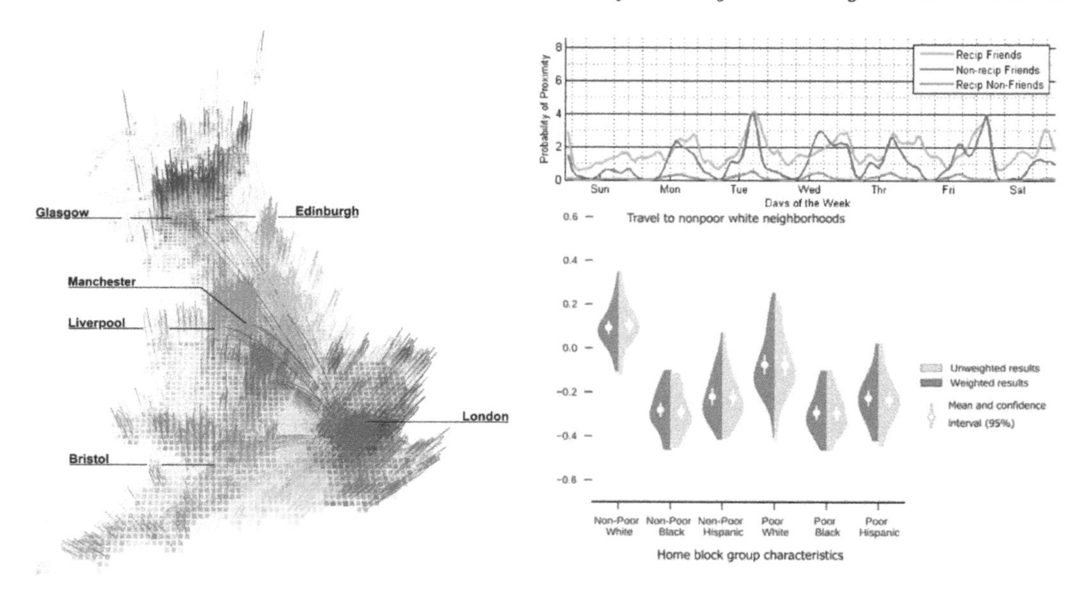

FIGURE 4.1
Cell phone records and related data have been used to reveal the patterns of society in many ways, including to map the "true" districts of the United Kingdom based on calling patterns (left), the "signature" of friendship based on frequency of proximity (upper right) and how racial segregation extends to the kinds of places people go (lower right). (Credit: The MIT Senseable City Lab, The MIT Media Lab, Wang et al. (2018))

4.1 Worked Example and Learning Objectives

Cell phone data are fascinating but quite difficult to work with (see Chapter 13 for an inkling). But, as noted, many other data sets capture their own characteristic "pulse." In this chapter, we will explore the pulse of the housing market as captured by Craigslist postings from Massachusetts in 2020-2021. A fun aspect of this time period is that we will not only be able to observe the pulse of the housing market but also how the pandemic altered it. These data were gathered and curated by the Boston Area Research Initiative as part of our COVID in Boston Database, an effort to track conditions and events across the city's communities before, during, and after the onset of the pandemic.

Through this worked example, we will learn the following skills for revealing the pulse of any data set:

- Identify and describe the multiple classes of variable we might encounter;
- Summarize the contents of a variable;
- Graph the distribution of a single variable (using the package `ggplot2`);
- Replicate summaries and graphs for a variable under multiple conditions (say, comparing years);
- Inferring aspects of the data-generation process and interpreting the data appropriately.

Because we have learned how to download and import data in Chapter 3, starting with this worked example I will provide the link for the data set of interest here at the beginning of the chapter and indicate the name I gave it when I imported it into RStudio (using `read.csv()`). The link also contains variable-by-variable documentation.

Link: https://dataverse.harvard.edu/dataset.xhtml?persistentId=doi:10.7910/DVN/52WSPT

Data frame name: `clist`

```
clist<-read.csv("Unit 1 - Information/Chapter 4 - Pulse of the
City/Example/Craigslist_listing.final.csv")
```

4.2 Summarizing Variables: How and Why?

When working with a new data set, one of the first things you want to do is summarize your variables. This includes generating summary statistics, tables of counts of values, and simple graphs. These tools all illustrate the content and distribution of the variable. This might seem a bit pedestrian or perfunctory, but it is a crucial step in conducting an analysis. Just as the skills in Chapter 3 allowed us to become familiar with the content and interpretation of individual records, summarizing variables allows us to become familiar with the content and interpretation of the data set as a whole.

There are three specific things we might gain from summarizing variables:

1. *We can reveal basic patterns*, or what we have called the pulse of the city . This tells us what a "typical" value is and how much variability there is around it, or how the data are distributed across multiple categories. It also gives us direct access to the variation that we will eventually want to study further with more sophisticated techniques.
2. *We can identify and understand outliers* that stand apart from the other values in the data. Are these special cases that are to be expected, like a very tall person or a property that generates an inordinate amount of crime? Or are they errors in the data arising from typos, glitches, or other processes?
3. From these observations *we can learn about the data-generation process* and how we should interpret the data. If some of these values are potentially errors, how do we treat them? What does that tell us about the kinds of biases that might exist in the data? Is there information that we do not expect to be there or that is incorrect in some way? Or information that should be that might be missing? These conclusions will be important for couching how we communicate our results.

4.3 Classes of Variable

As we have learned in previous chapters, R is an object-oriented programming language, and every object is of a specific "class"—meaning it has certain defining characteristics that govern the information therein, how that information is structured, and the way that functions operate upon it. Data frames are a specific class of object composed of observations (rows) that each have values for a series of variables (columns). Each variable contains a specific type of data that can be categorized as one of a variety of classes. (Yes, this can be a little confusing. Whole objects have classes at the object level. The subcomponents of an object—like the variables of a data frame—have their own set of classes by which they are categorized.)

4.3.1 The Five Main Classes of Variable

There are five main classes of variable that we will work with in this book:

- *Numerical* variables contain numbers. A subtype of numerical variables is *integer* variables. These are constrained to values that are whole numbers, whereas other numerical variables can accommodate any number of digits after the decimal point.
- *Logical* variables can only take the values of `TRUE` or `FALSE`. They also can operate as pseudo-numerical variables as R will interpret `TRUE == 1` and `FALSE == 0` if a logical variable is placed in an equation. Logical variables are crucial intermediaries when conducting analyses. In fact, every time you create a subset, you are actually asking R to evaluate a logical statement (e.g., in Chapter 3, we asked "was this 311 case closed?"). These are analogous to a special type of integer variable called *dichotomous* variables that only take the values of 0 or 1.
- *Character* variables contain text in an unconstrained format, meaning they can contain any variety of values of any length.
- *Factor* is a special type of character variable in which each value is treated as a meaningful category or "level." This organization can be helpful as functions can treat the categories as such, whereas other character variables are treated as having freeform content.
- *Date/time* variables comprise many different classes of variable. This is because of the many different ways that one might organize this sort of information—date, date followed by time, time followed by date, month and year only, and time only are just a handful of examples. And then there are questions like whether the time has seconds or not, whether it includes time zone, whether the date is month-day-year or day-month-year, etc. For these reasons date and time can be tricky to work with and we will return to them more in Chapter 5.

4.3.2 Example Variables from Craigslist

Let's get started with our data frame of Craigslist postings by looking at the classes of variable it contains. We should probably begin by learning what variables are there:

```
names(clist)
```

```
##  [1] "X"             "LISTING_ID"    "LISTING_YEAR"
##  [4] "LISTING_MONTH" "LISTING_DAY"   "LISTING_TIME"
##  [7] "RETRIEVED_ON"  "BODY"          "PRICE"
## [10] "AREA_SQFT"     "ALLOWS_CATS"   "ALLOWS_DOGS"
## [13] "ADDRESS"       "LOCATION"      "CT_ID_10"
```

You can read more about each of these variables in the documentation (see link above), though many of the names are self-explanatory. Let's see what classes some of them are.

```
class(clist$LISTING_DAY)
```

```
## [1] "integer"
```

```
class(clist$LISTING_MONTH)
```

```
## [1] "character"
```

```
class(clist$BODY)
```

```
## [1] "character"
```

```
class(clist$PRICE)
```

```
## [1] "integer"
```

```
class(clist$ALLOWS_DOGS)
```

```
## [1] "integer"
```

Unsurprisingly, `LISTING_DAY` is an integer, as days of the month are always whole numbers. `LISTING_MONTH`, however, is a character, suggesting that it might be the name of the month rather than a number associated with it.

Meanwhile, the variables `BODY` and `PRICE` will tell us more about the posting itself. The former is a character, which makes sense if it is the body of the post, including all the text that the landlord or property manager might have included. PRICE is an integer, which also makes sense as a price should be a whole number.

Last, I have thrown in `ALLOWS_DOGS` just for fun and we find that it is an integer. Why do you think that is the case (we will find out in Section 4.4)?

4.3.3 Converting Variable Classes: `as.` Functions

Sometimes we want to coerce a variable into a format that it wasn't originally in. This can happen when we intend a variable to have a certain structure, but this is lost when importing a .csv. For instance, if `LISTING_MONTH` is indeed a list of months by name, it may be more useful as a factor—that is, treated as a set of categories. We can do this with the `as.factor()` function:

```
clist$LISTING_MONTH2<-as.factor(clist$LISTING_MONTH)
```

Note that we have created a new variable, `LISTING_MONTH2`, which is a factor version of `LISTING_MONTH`. We will learn more about creating new variables in the next chapter, but for current purposes just know that this variable is now a part of the `clist` data frame (use `names()` to check if you like).

There are `as.` functions for all classes of variable (e.g., `as.numeric`, `as.character`, etc.), but they are to be used with caution. Data stored as a certain class are often stored that way because it reflects their content. Trying to squeeze it into another format can generate unexpected results. For instance, if we try to coerce the month variable, which is a character variable, into a numeric, we get an error:

```
clist$LISTING_MONTH3<-as.numeric(clist$LISTING_MONTH)
```

```
## Warning: NAs introduced by coercion
```

Any value that was not numerical was replaced by an NA! We will see soon the implications of that. One particularly tricky situation is if your numeric or integer variable was read by R as a factor. See the next subsection for more on why and how to handle it.

4.3.3.1 Extra: From Factor to Numeric

Suppose you have a numeric variable stored as a factor.

The Intuitive Process: Use `as.numeric()` to convert it.

But R does not think those values are numbers! They are categories that have been organized into alphanumeric ordering that you cannot see. As a result, if the smallest value in the variable is 50 and the second smallest is 100, these are actually considered by R to be categories 1 and 2. If you run `as.numeric()`, it will convert them to these values, rather than keeping the values you actually want.

The Solution: Make your factor variable a character variable first, and then convert to a numerical variable: `as.numeric(as.character(df$factor))`.

4.4 Summarizing Variables

There are a lot of classes of variable, and the tools needed to summarize each can differ. Numerical variables call for summary statistics. Factor variables call for counts of categories. The rest of this chapter walks through these skills and their nuances.

4.4.1 Making Tables

One simple step is to create a table of all values in a variable. For example

```
table(clist$LISTING_MONTH)
```

```
##
##      April    August  December  February   January      July
##      22604     17830     14025      9209      9141     21715
##       June     March       May  November   October September
##      24910     14262     24271     13078     14028     20372
```

tells us how many postings occurred in each month. You might notice some simple variations here already. For instance, January and February are a bit lower, which can be explained by the fact that BARI started scraping postings in March 2020. But why might December be lower? Does that possibly tell us something about the cycle of postings—or the pulse of the housing market? Also, March has fewer postings than April or May. Could that be an effect of the pandemic?

One way to answer that last question is to create a cross-tab, which is a table that looks at the intersections of two variables. This is done by adding an argument to the table command:

```
table(clist$LISTING_MONTH,clist$LISTING_YEAR)
```

```
##
##              2020  2021
##    April     13981  8623
##    August    15973  1857
##    December   7886  6139
##    February    392  8817
##    January       0  9141
##    July      19478  2237
##    June      16467  8443
##    March      4298  9964
##    May       16122  8149
##    November   9348  3730
##    October   12608  1420
##    September 13401  6971
```

Look at the difference between March 2020 and March 2021! The latter had two times as many postings as the former. Interestingly, April 2020 had many more postings than April 2021, suggesting that many postings that were held back in March were posted then. Still, the monthly numbers for December seem low relative to other parts of the year.

We can also see indications of the data-generation process. There are no cases in January 2020. Why? Because BARI was not scraping data then. We do have a small number of cases from February 2020, though. It appears that when BARI started scraping data, some postings from February were still up on the site. There are also lower numbers in August, October, and November 2021, when the BARI scraper stopped collecting data for a time.

We can use `table()` to look at numerical variables as well, though this is typically only useful for integer variables that take on a small number of values; otherwise the table would be hard to interpret. Thus, we would not want to use it for the `PRICE` variable, but maybe for `LISTING_DAY`.

```
table(clist$LISTING_DAY)
```

```
##
##    1    2    3    4    5    6    7    8    9   10   11   12   13
## 6075 6941 7313 7196 6830 6369 7622 7700 7364 6722 6742 6275 6401
##   14   15   16   17   18   19   20   21   22   23   24   25   26
## 6694 7287 7528 7049 6745 6580 6695 6753 6006 6750 6556 5671 5788
##   27   28   29   30   31
## 6979 7146 6734 5665 3269
```

Here we see how cases are distributed across the days of the month. Not too much of a story, however, except for the low count for the 31st, which only occurs in seven months.

Last, let's learn a little more about the `ALLOWS_DOGS` variable.

```
table(clist$ALLOWS_DOGS)
```

```
##
##      0      1
## 131022  71845
```

Here we see that it is indeed a 0/1, or dichotomous variable (the numerical version of a logical), and that way fewer postings are for housing that allows dogs.

This also gives us an excuse to learn another command that can be applied to tables, called `prop.table()`, which calculates the proportion of cases in each cell:

```
prop.table(table(clist$ALLOWS_DOGS))
```

```
##
##         0         1
## 0.6458517 0.3541483
```

R has precisely confirmed our observation: only 35% of postings allow dogs. Note that `prop.table()` can do a lot more with tables that examine the intersections of multiple variables, like the one above between month and year. Use `?prop.table` to learn more.

4.4.2 The summary() Function

Conveniently, one of the most versatile functions in R for summarizing data is called `summary()`. This function can be applied to any object. All classes of object and variable have embedded in them instructions for how summary will expose its content to the analyst. For example, when we run regressions in Chapter 12, it will provide clean, organized results from a regression model. For this reason, it will pop up throughout the book. Here we will see what `summary()` returns for variables.

Let's start with the numerical variable PRICE

```
summary(clist$PRICE)
```

```
##     Min. 1st Qu.  Median    Mean 3rd Qu.    Max.    NA's
##       10    1425    1895    2026    2500    7000       5
```

It appears to have given us the minimum, maximum, mean (arithmetic average), median (i.e., middle value), and the 1st quartile and 3rd quartile (i.e., the points at which 25% of the values are lower or higher, respectively). Some of this information makes a lot of sense: the mean is $2,026, which would seem reasonable for the Boston area, which can be quite expensive. The maximum is even more daunting at $7,000.

There is something a little strange, though. How is the minimum $10? Who is renting out an apartment for that little money? It is possible this is an error or typo when entering the price. It could also be a deliberate manipulation of the system. Think like a clever landlord for a moment. Some people will sort by price to get a sense of what kind of deal they can get. If they do that and you put a falsely low number in the price box, your listing comes up first. Either of these is possible, and each may be true for different cases. It does not seem like either is a predominant strategy, though, as the 25th percentile is $1,425, which is reasonable for a low-rent listing. We will spend more time later in the chapter trying to determine how frequent these erroneous cases are and what we should do with them.

Let's look at another numerical variable, AREA_SQFT, which is the size of the apartment or home for rent.

```
summary(clist$AREA_SQFT)
```

```
##     Min. 1st Qu.  Median    Mean 3rd Qu.    Max.    NA's
##        0     718     925    1304    1184 7001500  120361
```

Again, we see a few strange things. The maximum is 7,001,500 sq. ft., which seems, well, impossible. This is another possible error or manipulation.

Also, this summary generated a new piece of information: a count of NAs. Note that 120,361 cases (or 59%) had no information for sq. ft. This is because posters do not need to indicate the square footage of the apartment, though they do have to enter a price. This raises questions as to whether landlords of certain types of apartments are more likely to post the size—maybe listings for larger apartments include their size more often, or landlords of higher-rent apartments assume their clientele wants to know this information, or something else.

We can also return to the ALLOWS_DOGS variable:

```
summary(clist$ALLOWS_DOGS)
```

```
##     Min. 1st Qu.  Median    Mean 3rd Qu.    Max.    NA's
## 0.0000  0.0000  0.0000  0.3541  1.0000  1.0000    2583
```

Note that the mean is 0.35, which perfectly reflects the 35% of postings permitting a dog. This is because the mean of a dichotomous variable is the proportion of 1's. Also note that the NAs here are far fewer than the square footage.

The summary function is less useful for character variables, however.

```
summary(clist$LISTING_MONTH)
```

```
##     Length     Class      Mode
##     205450 character character
```

This is because character variables are assumed to be freeform and thus are difficult to summarize in any systematic way. But, if we return to our factor version of the LISTING_MONTH variable, we get more information.

```
summary(clist$LISTING_MONTH2)
```

```
##      April    August  December  February   January      July
##      22604     17830     14025      9209      9141     21715
##       June     March       May  November   October September
##      24910     14262     24271     13078     14028     20372
##       NA's
##          5
```

In fact, we have generated the same table from above! This is just further validation that table() is the most fundamental way to look at a factor variable.

4.4.3 Summary Statistics

The summary() function is very powerful and often all that we need to get to know a numerical variable better. But what if we want to know information that summary() does not offer? Or what if, for whatever reason, we want to specifically access (and maybe store as an object) one of the statistics that summary() generates, like the mean or the maximum? There are a host of summary statistics that R can calculate, only a few of which are included in the default summary() output. Some of these are included in Table 4.1. Here we want to illustrate how they work.

Let us return to two of our numerical variables, PRICE and AREA_SQFT.

```
mean(clist$PRICE)
```

```
## [1] NA
```

Wait! Why did that happen? It turns out that summary functions do not know how to handle NAs. In a sense, they say "If there are any missing cases, it is impossible to accurately calculate a summary statistic." It can also act as a warning that your data contain NAs, if you had not already noticed that.

There is a workaround, however.

TABLE 4.1
Some summary statistics built into R.

Function	What It Calculates
max(x)	Maximum value
median(x)	Median value
min(x)	Minimum value
range(x)	Range of values (max – min)
sd(x)	Standard deviation across values
sum(x)	Sum of all values

```
mean(clist$PRICE, na.rm=TRUE)
```

```
## [1] 2026.146
```

By adding the argument `na.rm=TRUE` we have told R to remove (`rm`) NAs from the analysis. And with that, we have our mean properly calculated. And, unsurprisingly (and comfortingly), this is the same value that the `summary()` function generated above—an average price of $2,026—though with a bit more precision after the decimal point.

4.5 Doing More with Summaries

Now that we have a set of tools for summarizing variables, there are a variety of ways we can extend them to do more sophisticated, thorough, and precise analyses.

4.5.1 Summarizing Multiple Variables with `apply()`

What if we wanted to calculate statistics for two variables with a single line of code? Let's see what happens if we try to do this on another statistic, the maximum:

```
max(clist[c('PRICE','AREA_SQFT')], na.rm=TRUE)
```

```
## [1] 7001500
```

Why did we only get one value? Summary functions are based on all of the data entered, so what it reported was the maximum value across the two variables. As we already know, this is the maximum size of an apartment, which is a much larger value than any of the prices.

To solve this problem, the `apply()` function executes (or "applies") another function across multiple elements.

```
apply(clist[c('PRICE','AREA_SQFT')],2,max, na.rm=TRUE)
```

```
##      PRICE AREA_SQFT
##       7000   7001500
```

It worked! The structure of the function's arguments is: `apply(df, dimension, FUN)`, where `FUN` stands for function. Importantly, dimension refers to whether apply should execute the function for variables or cases. If dimension is set equal to 2, it executes the function separately for each variable. If set equal to 1, it executes it separately for each case. (Remember that the dimensions of a data set are represented `[rows, columns]`.) Try entering 1 and see what happens.

Note that `apply()` is part of a large family of commands that can do the repetition of analyses on the subcomponents of various data structures. We will not touch on them much in this book but more complex situations sometimes call for them.

4.5.2 Summarizing across Categories in One Variable: `by()`

We might be interested in how a variable's characteristics vary based on another condition. For example, are prices for apartments stable across the year? This can be accomplished with the `by()` command, which has three arguments: the numerical variable whose statistics you want to calculate, the categorical variable by which you want to split the analysis, and the function you want.

```
by(clist$PRICE,clist$LISTING_MONTH,mean)
```

```
## clist$LISTING_MONTH: April
## [1] 2039.955
## ------------------------------------------------------
## clist$LISTING_MONTH: August
## [1] 2042.148
## ------------------------------------------------------
## clist$LISTING_MONTH: December
## [1] 1871.895
## ------------------------------------------------------
## clist$LISTING_MONTH: February
## [1] 1872.817
## ------------------------------------------------------
## clist$LISTING_MONTH: January
## [1] 1805.201
## ------------------------------------------------------
## clist$LISTING_MONTH: July
## [1] 2165.895
## ------------------------------------------------------
## clist$LISTING_MONTH: June
## [1] 2168.869
## ------------------------------------------------------
## clist$LISTING_MONTH: March
```

```
## [1] 2054.44
## ------------------------------------------------
## clist$LISTING_MONTH: May
## [1] 2147.718
## ------------------------------------------------
## clist$LISTING_MONTH: November
## [1] 1871.635
## ------------------------------------------------
## clist$LISTING_MONTH: October
## [1] 1858.777
## ------------------------------------------------
## clist$LISTING_MONTH: September
## [1] 1997.779
```

As you can see, the mean shows some shifts across the year, with average rents for postings from April to August topping $2,000 and even nearing $2,200 in the middle of summer, and average rents closer to $1,800 through the fall and winter. There are multiple possible interpretations here. It could be that the competition for housing is greater over the summer as many leases being in August or September, especially in a city like Boston which has many colleges and universities. It might also be that the summer postings tend to cater to the specific population of students and their contemporaries, which command higher prices. Also, how much are the waves of the pandemic affecting this dynamic? From the current analysis we do not yet know.

4.6 Summarizing Subsets: The Return of Piping

Thus far our analyses have found some potentially interesting things, but they are also surfacing concerns about whether some of the information is erroneous or complicating, in which case we might want to set that content aside. What do we do with apartments that claim a price of $10? How do we isolate our analysis to the pandemic year to be confident in the interpretation? This requires us to subset before conducting our analysis.

There are three ways to subset before analysis. One is to create a new data frame.

```
clist_2020<-clist[clist$LISTING_YEAR==2020,]
```

This can fill the Environment with intermediate datasets, however, making things messy and taking up lots of memory.

We can also subset inside a function.

```
mean(clist$PRICE[clist$LISTING_YEAR==2020], na.rm=TRUE)
```

```
## [1] 2024.37
```

Note that the subset is of the variable we are analyzing, removing the need for a comma. This can be an effective approach but can make for a complicated line of code if there are lots of criteria. The third technique is to use piping, which is what we will concentrate on here (reminder, you will need to enter `require(tidyverse)` if you are following along with the code). We will need to introduce the `group_by()` and `summarise()` commands in tidyverse for these purposes.

4.6.1 Summary statistics with summarise()

First, let's suppose we want to remove postings with non-sensical prices from the data set. This sounds easy enough, but where should we draw the line? This requires some further scrutiny and a bit of a judgment call. For instance:

```
mean(clist$PRICE<100, na.rm=TRUE)
```

```
## [1] 0.0001557594
```

What I have done here is take the mean of the logical statement that the price of a posting is less than $100. R determines whether that statement is true or false for every case, with `TRUE==1`, and then takes the mean (which is also the percentage for a logical or dichotomous variable). Thus, 0.02% of cases have a price under $100. That is very few cases. We are probably safe calling these outliers.

But $100 is also a ridiculously low price for an apartment. What if we raise the threshold?

```
mean(clist$PRICE<500, na.rm=TRUE)
```

```
## [1] 0.005509991
```

```
mean(clist$PRICE<1000, na.rm=TRUE)
```

```
## [1] 0.100387
```

So, 0.6% are under $500, which is still a good indicator of outliers. But 10% are under $1,000. It would seem then that there are plenty of legitimate postings in between $500 and $1,000, possibly for sublets or private rooms within an apartment or house.

Moving forward, then, we will exclude cases with price less than $500 (and the NAs for good measure). How does this change things? Remember that to answer this question we will need the `filter()` function from `tidyverse`. Also, we will need to use `summarise()`, another `tidyverse` function that sidesteps limitations of certain traditional functions.

```
clist %>%
  filter(PRICE>500, !is.na(PRICE)) %>%
  summarise(mean_price = mean(PRICE))
```

```
##   mean_price
## 1   2036.989
```

The average has gone up slightly, from $2,024 to $2,037. Even though those outliers were so small, it seems that 0.6% of the data do not have that large of an effect on totals. Nonetheless, we can be more confident now in our results.

4.6.2 Summary Statistics by Categories with group_by()

Let us replicate the analysis by month with our subset. This will require the `group_by()` function, which instructs R to organize the data set by the categories in a particular variable for all proceeding steps of the pipe.

```
clist %>%
  filter(PRICE>500) %>%
  group_by(LISTING_MONTH) %>%
  summarise(mean_price = mean(PRICE))
```

```
## # A tibble: 12 x 2
##    LISTING_MONTH mean_price
##    <chr>              <dbl>
##  1 April              2055.
##  2 August             2053.
##  3 December           1883.
##  4 February           1881.
##  5 January            1814.
##  6 July               2175.
##  7 June               2175.
##  8 March              2066.
##  9 May                2163.
## 10 November           1883.
## 11 October            1869.
## 12 September          2008.
```

The results largely look the same as before (though they are cleaner when reported by `tidyverse`). The theme of higher prices in the summer and lower prices in the fall and winter remains, though some of the numbers have inched upwards with the removal of the outliers.

If you want to go further, you might limit the data to either 2020 or 2021 by adding an additional criterion to the `filter()` command.

4.7 Tables as Objects

You might look at a table and say, "I see counts across categories, I have what I want." Or you might say, "I'd love to analyze *those* numbers." As I have said numerous times, R is an object-oriented software, meaning that it applies functions to objects, each being of a class with defining characteristics. I have also alluded that many functions generate new objects that we can then analyze further. This is our first opportunity to do this.

The `table()` command does not itself generate an object, but it does generate something that looks a lot like a data frame if we examine its product with `View()`.

```
View(table(clist$LISTING_MONTH))
```

	Var1	Freq
1	April	23045
2	August	15608
3	December	7873
4	February	8544
5	January	9205
6	July	21375
7	June	25686
8	March	15223
9	May	24727
10	November	8790
11	October	12688
12	September	12980

Could we turn it into one?

```
Months<-data.frame(table(clist$LISTING_MONTH))
```

We now have a data frame of 12 observations and 2 variables in the Environment. You can use `View()` and `names()` to check that it is the same as before.

Before we analyze this table as a data frame, though, remember that representation across the years is a little funny because the scraping started in February of 2020. We could use piping to create a desired subset, though, and then analyze the final product.

```
table(clist$LISTING_MONTH[clist$LISTING_YEAR==2020 &
                            clist$LISTING_MONTH!='February']) %>%
  data.frame() %>%
  summary()
```

```
##         Var1         Freq
##  April   :1   Min.   : 4298
##  August  :1   1st Qu.:10163
##  December:1   Median :13691
##  July    :1   Mean   :12956
##  June    :1   3rd Qu.:16085
##  March   :1   Max.   :19478
##  (Other) :4
```

Now we have a much stronger insight as to what listings per month looked like in 2020. The lowest month had 4,298 and the maximum had 19,478, whereas the mean month had 12,956.

4.8 Intro to Visualization: ggplot2

We have accomplished quite a bit using summary statistics, but some visuals would probably be helpful when communicating the patterns we are seeing. To do this, we are going to use ggplot2 (Wickham et al., 2021), which is a package designed to enable analysts to generate a vast array of visualizations. Base R does include some tools for basic visualizations, but the breadth of ggplot2's capabilities has made it the go-to package for making graphics. Further, ggplot2 makes it straightforward to customize all aspects of a visualization, from naming the axes to labeling values to designing the legend to setting the color scheme and so on. It is premised on *The Grammar of Graphics*, which is a framework for layering additional details atop each other as you design a graphic (Wilkinson, 2005). What this means practically is that ggplot2 commands consist of multiple pieces that you "add" to each other. This will make more sense as we go along.

There is a lot to ggplot2. We will learn additional skills in it in nearly every chapter of the book. I will not be able to share all of the different graphics that ggplot2 supports nor all the tools for customization. As you move along, you may want to consult either the ggplot2 cheat sheet from https://www.rstudio.com/resources/cheatsheets/ or the package documentation so that you can create the precise visuals you want. Also, ggplot2 is part of the tidyverse, so you may not need to install it to get started here.

4.8.1 Histograms: The Most Popular Univariate Visualizations

We are going to get started with ggplot2 in the simplest way we can: generating a series of graphics that describe the distribution of a single variable. These are also known as univariate graphics. This adds a visual dimension to everything we have done thus far.

Two features of ggplot2 will shape how we write code for it. First, it generates plots as objects, which means we can save them. Second, it builds those objects through sequential commands. A graphic always starts with the ggplot() function that specifies the data frame we want to visualize and an aes() argument that specifies the variables for each axis. If we enter this on its own, however, it just generates a blank graph:

```
ggplot(clist, aes(x=PRICE))
```

Essentially, we have told ggplot2 what we want to visualize but not how we want to visualize it. If we store this in an object, though, we can layer any number of visualizations on top of it.

For example, if we want a basic histogram, which organizes cases into "bins," or ranges, and then counts the number of cases in each bin, we can do the following.

```
base<-ggplot(clist, aes(x=PRICE))
base+geom_histogram()
```

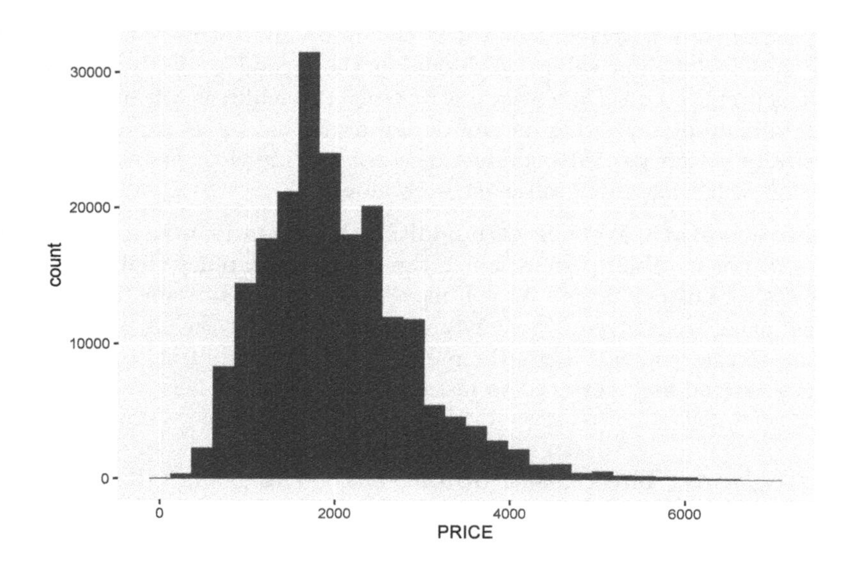

This is a visual version of the summaries we have already conducted, but it gives us a richer set of insights than a simple mean, or even minimum and maximum. The left tail does indeed get very close to $0, reflecting our minimum of $10, but it ramps up quickly—which verifies the interpretation that anything with `PRICE>$500` might very well be legitimate. On the other end of the spectrum, though, it is noteworthy that the distribution has a quick drop-off after $3,000, but then a long tail stretching out to $7,000.

There are two things to note about the code. First, `geom_histogram()` has no additional arguments. There are ways to customize the command, but if it is empty, it simply borrows information from the `ggplot()` command regarding the data frame and variable of interest. Second, we added `geom_histogram()` to the object `base`, which contains the original `ggplot()` command. That means we have not altered this original object and can easily layer other visual elements on top of it.

4.8.2 Additional Univariate Visualizations

Let us try layering other visuals on top of our base `ggplot()` command. For example, a density graph gives us a smoother representation that tries to avoid the blockiness of bins. It also reports proportions rather than counts.

```
base+geom_density()
```

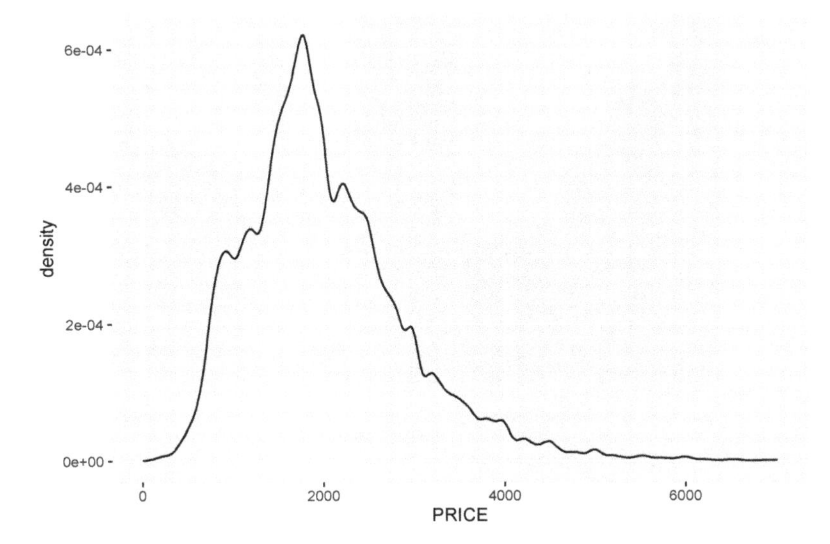

A frequency polygon is a merger of these two approaches, representing counts with a curve rather than boxes.

```
base+geom_freqpoly()
```

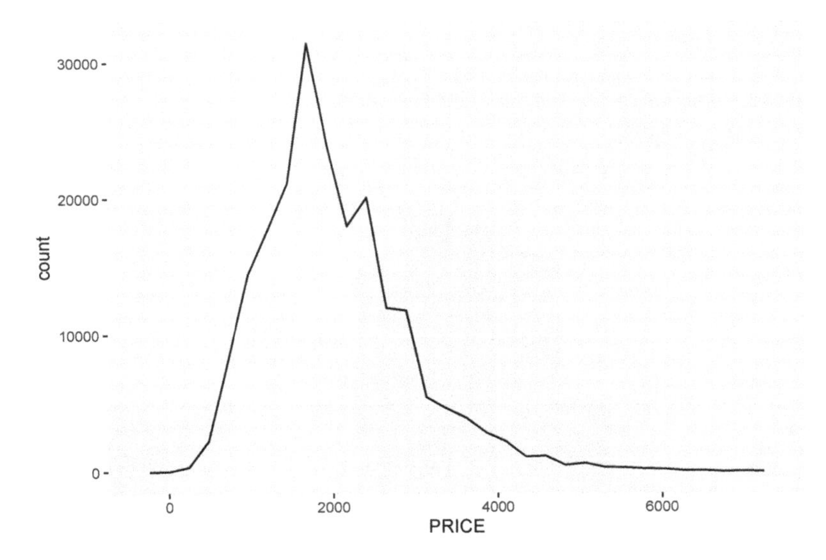

We might also visualize categorical variables in a similar way; though this requires a distinct function for a bar chart, `geom_bar()`, because a histogram assumes a numeric variable.

```
base_month<-ggplot(clist, aes(x=LISTING_MONTH))
base_month + geom_bar()
```

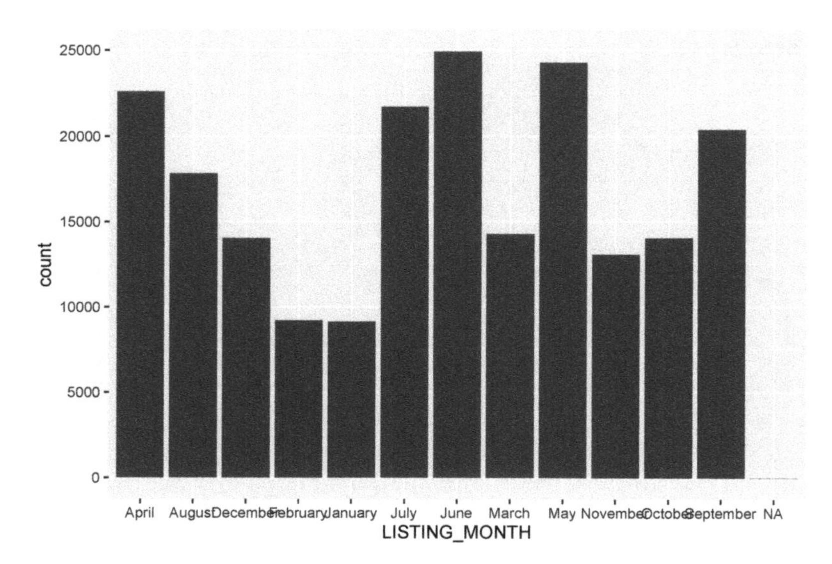

This is a visual version of the table we previously created, but it is possible that we would get more out of reordering these by the calendar rather than their alphabetical names. This requires an extra trick that we will use to re-create our `LISTING_MONTH2` variable.

```
clist$LISTING_MONTH2<-factor(clist$LISTING_MONTH,
                             levels = month.name)
```

This told R that the factor levels have an order that corresponds to the months of the year. Now we can try again.

```
base_month<-ggplot(clist, aes(x=LISTING_MONTH2))
base_month + geom_bar()
```

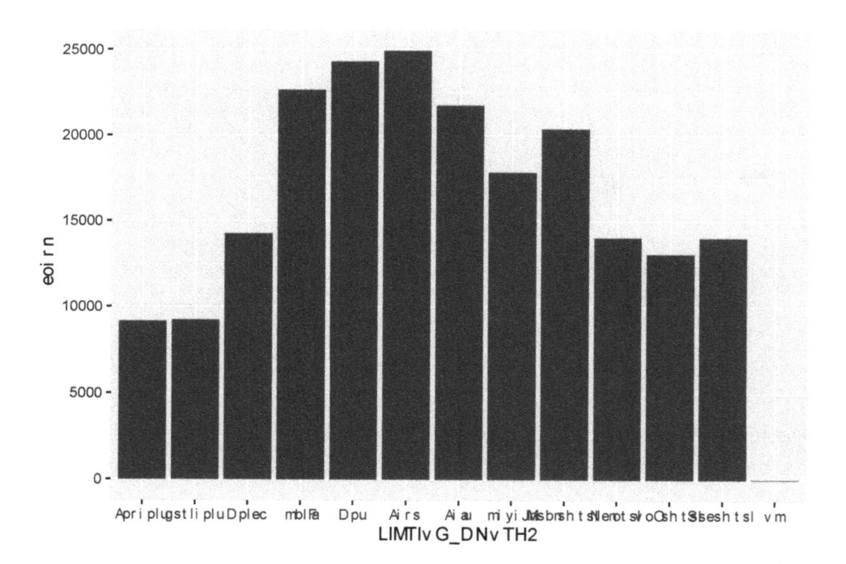

There we have it! Now I have not added additional code to clean up the graph, including replacing the variable name with a more interpretable axis name and moving the axis labels so that they do not overlap. We will learn more about customizing in the next chapter, but for now we will content ourselves with some nice graphics whose details are a little rough around the edges.

4.8.3 Incorporating Pipes into ggplot2

To close out this initial visual analysis, let us return to our subsetting criteria above and use pipes to generate a more easily interpreted graphic.

```
base_month_pipe<-clist %>%
  filter(clist$LISTING_YEAR==2020,
         clist$LISTING_MONTH!='February') %>%
  ggplot(aes(x=LISTING_MONTH2))
base_month_pipe + geom_bar()
```

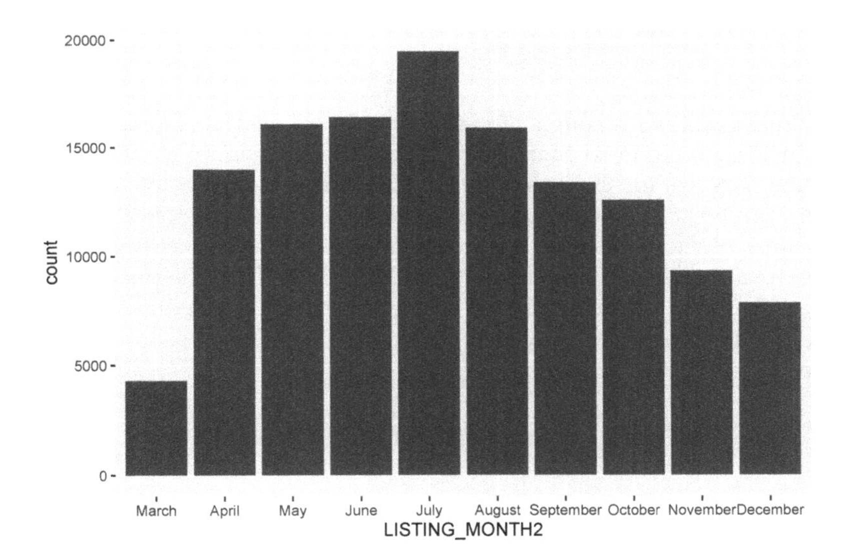

Here we see more clearly the dramatic impact of the pandemic on listings in March 2020 and the seemingly quick rebound in April. Also, see how we stored the full pipe through the `ggplot()` command in the object `base_month_pipe` and were able to layer the bar graph on top of it.

4.8.4 Stacked Graphs: One Variable across Categories

As a last step, let us compare the trends across the two years, which will enable us to evaluate more closely the impact of the pandemic. This can be done with a stacked bar graph, wherein the counts are split across a categorical variable, as indicated by the `fill=` argument. We want to return to the full 2020-2021 data set (note that we have to tell `ggplot` to treat `LISTING_YEAR` as a factor).

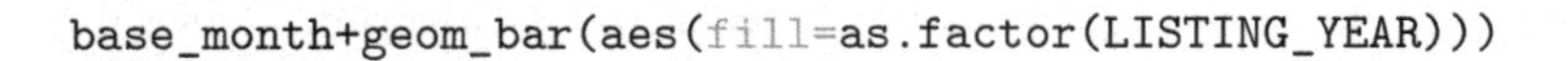

```
base_month+geom_bar(aes(fill=as.factor(LISTING_YEAR)))
```

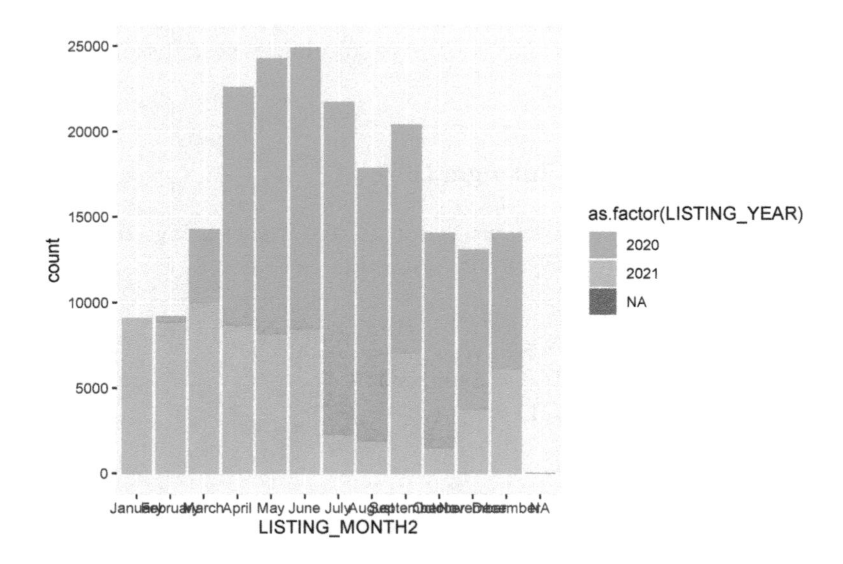

This is a striking graph for a couple of reasons. First, we see that March 2020 really was an aberration, at least relative to March 2021. But it is interesting to note how much taller the April-June bars are for 2020 than 2021. This suggests that a lot of postings that did not occur in March then flooded onto the market. It is also possible that there was more turnover because of the pandemic, leading to additional listings as well. Last, as we have discussed, BARI did not start scraping data until March 2020, so the small number of listings in February of that year are an artifact of the data-generation process. And scraping paused in July-August and October-November in 2021, creating some gaps.

We can do the same with histograms for numeric variables.

```
base+geom_histogram(aes(fill=as.factor(LISTING_YEAR)))
```

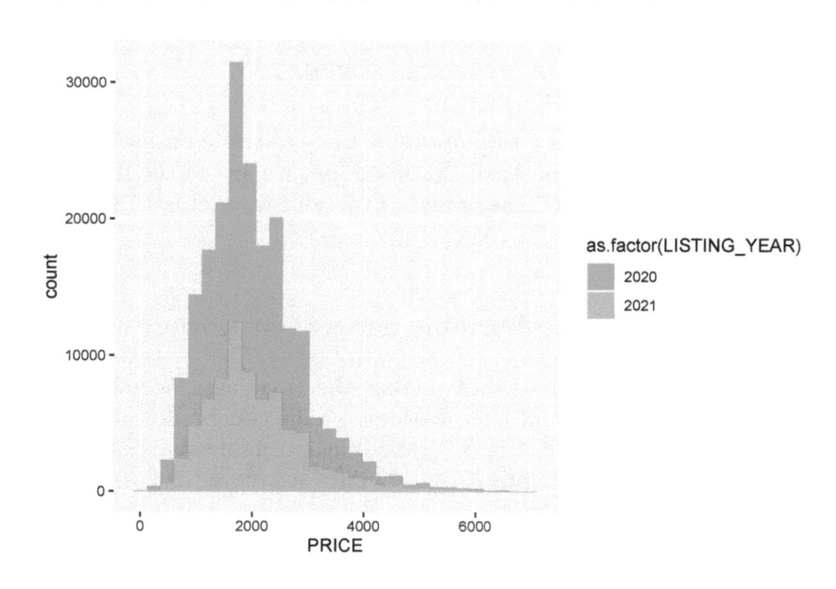

Here we see that, unlike listings by month, the distribution of prices was largely the same between the two years.

4.9 Summary

In this chapter we have learned how to reveal the pulse captured by a given data set, including both the features of the pulse itself and the ways in which the data-generation process needs to be taken into consideration when interpreting a naturally occurring data set. We saw how, in general, the housing market in Massachusetts appears to be more active in the summer, with more listings and somewhat higher prices than the rest of the year. We also saw how the pandemic dramatically diminished listings in March, but that this rebounded rather quickly with listings possibly even above average in the following months.

In terms of interpretation, however, we observed a few details that required us to take pause, or at least exclude cases whose meaning was unclear (or downright incorrect). We saw that any system that requires user input can have mistakes, like fake prices and missing or false square footage for apartments or houses. We came up with a strategy for setting a threshold for the former, though. We also saw that the timing of the data-generation process—the scraping of Craigslist, in this case—influenced how listings looked in certain months. In sum, we learned a lot about the pulse of the housing market in Massachusetts during the pandemic as well as the strengths and limitations of scraped Craigslist data to tell us about it.

We also practiced a variety of skills that can easily be applied to other data sets. We:

- *Identified classes of variables* so that we would know how to work with them;
- *Made tables* of counts of values across categories;
- *Generated summary statistics* for numeric variables using `summary()` and other commands;
- *Summarized multiple variables* using `apply()`;
- *Organized summaries by categories* provided by another variable using `by()`;
- *Summarized subsets of data* using piping, including the new commands `group_by()` and `summarise()`;
- *Analyzed tables as data frames*;
- *Visualized data with the package* `ggplot2`, including
- *Making histograms, bar graphs, and related visuals* of the distribution of a single variable and
- *Making stacked histograms* that split the distribution according to categories from another variable;
- *Critiqued the interpretation* of a data set based on potential errors and biases arising from its origins.

4.10 Exercises

4.10.1 Problem Set

1. Each of the following commands will generate an error or give you something other than what you want. Identify the error, explain why it did not work as expected, and suggest a fix.
   a. `by(clist$LISTING_YEAR,clist$PRICE,mean)`
   b. `mean(clist$ALLOWS_DOGS)`
   c. `summary(clist$BODY)`
2. Classify each of the following statements as true or false. Explain your reasoning
   a. All factors are character variables, but not all character variables are factors.
   b. All numeric variables are integer variables, but not all integer variables are numeric.
   c. An integer variable with only the values 0 and 1 is effectively equivalent to a logical variable.
   d. There is only one class of date-time variable.
3. Describe in about a sentence what information R will return for the following commands.
   a. `median(clist$PRICE)`
   b. `clist %>% filter(LISTING_MONTH=='July') %>% summarise(mean = mean(ALLOWS_CATS)`
   c. `by(clist$AREA_SQFT,clist$LISTING_MONTH,mean,na.rm=TRUE)`
   d. `ggplot(clist, aes(x=LISTING_YEAR)) + geom_freqpoly`
4. Which function would you use for each of the following tasks?
   a. Calculate the mean for multiple variables.
   b. See a set of basic statistics for one or more variables.
   c. Calculate the means on a single variable across multiple groups.
5. Which function in `ggplot2` would you use for each of the following?
   a. Create a smoothed curve of a variable's distribution.
   b. Visualize a variable's distribution organized into rectangular bins.
   c. Visualize a variable's distribution with a curve fit to bins.
   d. Visualize the distribution of observations across categories.

4.10.2 Exploratory Data Assignment

Working with a data set of your choice:

1. Generate at least three interesting pieces of information or patterns from your data set and describe the insights they provide regarding the dynamics of the city.
2. Find at least two things in your data set that are strange or do not make sense. What might they tell you about the data-generation process and any errors or biases it might generate?
3. Redo your first analysis excluding cases that do not make sense, if necessary.

4. Your findings can be communicated as numbers, graphs, or both and should convey what you see as the overarching information contained in the data set. You must include at least one graph.

5

Uncovering Information: Making and Creating Variables

"Life and health can never be exchanged for other benefits within the society." So reads the mission statement for Vision Zero, a philosophy that treats pedestrian and cyclist deaths from automobile collisions not as inevitable events that should be reduced, but as tragedies that could be entirely eliminated by better design and management of roads. The idea was first introduced in Sweden in 1997, where traffic fatalities have since plummeted by over two-thirds. As of 2021, dozens of countries in Europe and cities in the United States and Canada have also adopted Vision Zero as a guiding principle in their transportation policy.

Vision Zero and the expanded use of data in public policy and practice—including in transportation—have evolved in tandem over the last two decades. But they have merged in the use of indicators for tracking a city's success in eliminating traffic fatalities. San Francisco, for example, has a Transportation Safety Scorecard that includes traffic citations for the five violations responsible for the most collisions, including speeding, running stop signs, and disregarding the pedestrian right of way (e.g., ignoring a crosswalk; see top of Figure 5.1). New York City has a similar collection of metrics, also including their successes in revamping infrastructure, such as building new bike lanes, and outreach they have done around safety at schools and senior centers. Through these sorts of measures, cities can assess their progress, identify weak spots, and create awareness in the broader public.

Indicators have become a popular tool in urban informatics and are certainly not confined to questions of transportation. Indicators are central to performance management efforts that evaluate the effectiveness of operations across departments. These are often the basis of "dashboards" that can be part of a city's open data portal or an internal interface for managers—or even the mayor—to quickly take stock of current conditions. There are also efforts to use indicators to guide social service provision, as embodied in the National Neighborhood Indicators Partnership , a learning network coordinated by the Urban Institute on the use of data to advance equity and well-being across neighborhoods (see bottom of Figure 5.1).

The creation of indicators itself is a skill. An analyst—often in conversation with colleagues and community members—has to first determine, "What specifically do we want to know? What information needs to be tracked?" The answers to these questions are often lurking in the available data, but they need to be uncovered. This calls for numerous types of data manipulation: maybe tabulating relevant cases through the creation of categories, such as running stop signs; calculating one or more new variables or combining them, as in the case of an "opportunity index" that quantifies assets and deficits across communities; or even extracting content from lengthy descriptions of events through text analysis. Each of these processes (and others) help us to coax the desired information out of the data set. These are the skills we will learn in this chapter.

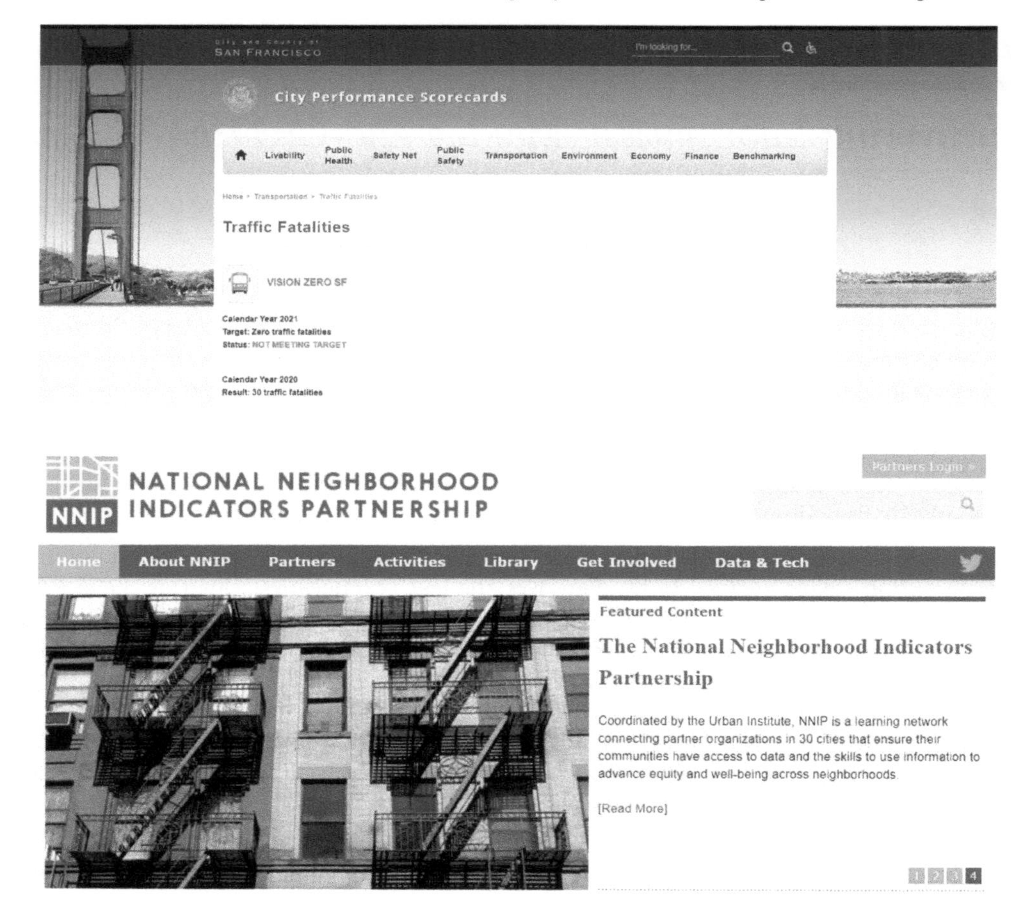

FIGURE 5.1
The City of San Francisco maintains a Transportation Safety Dashboard that tracks numerous metrics, including its goal of achieving zero traffic fatalities (i.e., Vision Zero; top). Meanwhile, National Neighborhood Indicators Partnership is an effort of the Urban Institute to coordinate partners in 30 cities all dedicated to the use of metrics to inform social service design and provision (bottom). (Credit: City of San Francisco, National Neighborhood Indicators Partnership)

5.1 Worked Example and Learning Objectives

In this chapter, we will work with a database of bicycle collisions occurring in Boston from 2009 to 2012. The data were developed by Dahianna Lopez, who was a graduate student at the Harvard T.S. Chan School of Public Health at the time, through a series of internships with the Boston Police Department. She worked with raw police transcripts to extract numerous pieces of information about each collision, resulting in a detailed report on bicycle safety from the mayor's office in 2014. The data were then published by the Boston Area Research Initiative .

These data give us an opportunity to observe the types of details that might be included in a data set to enable crucial indicators, like whether collisions resulted in medical treatment or transportation to a hospital. They also present the opportunity to manipulate data to create new variables (or simply edit old ones), thereby learning and practicing the following skills:

- Calculate numeric variables;
- Manipulate strings (often using the package `stringr`);
- Create categories based on criteria;
- Summarize and visualize the content of lengthy blocks of text;
- Work with date and time variables (using the package `lubridate`);
- Extend the ability to customize graphs in `ggplot2`.

Link: https://dataverse.harvard.edu/dataset.xhtml?persistentId=doi:10.7910/DVN/24713. You may also want to familiarize yourself with the data documentation posted there.

Data frame name: `bikes`

```
bikes<-read.csv("Unit 1 - Information/Chapter 5 - Revealing
Information/Example/Final Bike Collision Database.csv")
```

5.2 Editing and Creating Variables: How and Why?

There are three reasons why we might want to edit and create variables. Two speak to the tasks we can accomplish when working with variables. The third is the inspiration behind why we would want to manipulate variables in the first place.

1. *Adjusting issues that make analysis difficult.* This could include errors in data entry that need to be fixed before we proceed. We saw in the previous chapter that some Craigslist postings have inexplicably low prices. We excluded them there by subsetting, but what if we want to replace them with NA's instead? There might also be outliers that interfere with broader interpretations and we need to establish some "ceiling" or "floor" to the distribution of a variable. For instance, in some cases analysts will set incomes over some threshold to a single highest value (e.g., all incomes >$500,000 would be set to $500,000).
2. *Making certain information more accessible for analysis.* This could entail reorganizing the content in a single variable—maybe recoding it to be more interpretable or translating a numerical variable into a series of categories. It might also entail combining information contained in multiple variables. Many indexes, in fact, involve summing or averaging a collection of sub-indicators. In such a situation, you might need to calculate the sub-indicators first and then combine them.
3. *Exposing the information you actually want.* Analyses are driven by the decisions of the analyst. As such, you want to enter every analysis asking the question, "What do I want to know?" The answer to that question will determine what variables you will create, analyze, and visualize.

5.3 Variables in the Bike Collisions Dataset

Before we get going, let's take a quick look at the data set so that we can decide what direction our analysis will take.

```
names(bikes)
```

```
##  [1] "ID"          "YEAR"        "DATE"        "DAY_WEEK"
##  [5] "TIME"        "TYPE"        "SOURCE"      "XFINAL"
##  [9] "Xkm"         "YFINAL"      "Ykm"         "Address"
## [13] "Main"        "RoadType"    "ISINTERSEC"  "TRACT"
## [17] "CouncilDIS"  "Councillor"  "PlanningDi"  "OIF1"
## [21] "OIF2"        "OIF3"        "OIF4"        "BLFinal"
## [25] "CS"          "LIGHTING"    "Indoor"      "Light"
## [29] "LightEng"    "WEATHER"     "PrecipCond"  "AtmosCondi"
## [33] "DayNight"    "Tmax"        "Tmin"        "Tavg"
## [37] "Temprange"   "SunriseTim"  "SunsetTime"  "SnowFall"
## [41] "PrecipTota"  "Fault"       "Doored"      "HelmetDocu"
## [45] "TaxiFinal"   "hitrunfina"  "AlcoholFin"  "INJURED"
## [49] "TRANSPORTE"  "TREATED"     "GENDER"      "ETHNICITY"
## [53] "AGE"         "Narrative"
```

A quick glance at variable names (which could be confirmed by reading the data documentation published alongside the data) reveals that there are a lot of variables describing the weather, including temperature (`Tmax`, `Tmin`, `Tavg`, `Temprange`) and precipitation (`PrecipCond`, `SnowFall`, `PrecipTota`), and the circumstances surrounding the collision (`Doored`, `TaxiFinal`, `hitrunfina`), its outcomes (`INJURED`, `TRANSPORTE`, `TREATED`), and the characteristics of the biker (`GENDER`, `ETHNICITY`, `AGE`). There is also an extended police `Narrative` that we might want to leverage. As we move forward, we will look for ways to better specify the way the weather might have shaped the event and also to capitalize on the rich content of the police narrative.

5.4 Calculating (and Recalculating) Numeric Variables

We will start with functions that calculate numerical variables. Notably, any tool we learn here or throughout the chapter can create a "new" variable or recreate an old one by writing over it. The only distinction is the variable name that we indicate on the left side of the arrow. Overwriting can be risky, though. If we make a mistake, it can be difficult to recover the original content.

5.4.1 One Variable, One Equation

The simplest way to calculate a new variable in base R is by storing or assigning an equation to a new variable, using our friends `<-` and `$`. For instance, let's say we want to know the average temperature in Celsius rather than Fahrenheit.

```
bikes$Cels_Tavg<-(bikes$Tavg-32)*5/9
```

Note that this does not generate any output, but that the number of variables in `bikes` just increased from 54 to 55 in the Environment.

The new variable is `Cels_Tavg` (use `names()` if you would like to check). Let's take a quick look at the variable using `summary()`.

```
summary(bikes$Cels_Tavg)
```

```
##     Min. 1st Qu.  Median    Mean 3rd Qu.    Max.
##   -10.00   11.67   17.78   16.42   22.78   33.33
```

Those values would seem to check out as plausible in a city like Boston. Now, if you were dedicated to analyzing the data in Celsius only and overwriting the existing `Tavg` variable, you could have done that by (note that I am not actually executing this command, just showing it):

```
bikes$Tavg<-(bikes$Tavg-32)*5/9
```

These commands will store the output of any equation on the right side in the variable named on the left side. If that variable already exists, it will be overwritten.

5.4.2 Multiple Variables, One Equation

`Tavg` is not our only temperature variable. We also have a minimum and maximum. What if we want to convert them all to Celsius at once? First, let's dispose of our newly created variable so that we can recalculate it as part of this new process:

```
bikes<-bikes[-55] #(i.e., remove column 55)
```

Now we are ready to do this:

```
bikes[55:57]<-(bikes[c(34:36)]-32)*5/9
```

Why does this work? Let's unpack it, starting with the equation. We have subset bikes to columns 34-36. If you check with `names()`, these are the original temperature variables. We have then calculated the Fahrenheit-to-Celsius conversion on each of them and then stored them in three new columns, 55-57.

This type of calculation can be efficient from a coding perspective but is error prone. For instance, you will want to be confident you know the number of columns in your data frame (use `ncol()`). It also makes for code that can be difficult to replicate if there are slight differences between versions of data that change the order of variables. Nonetheless, it is important to know this is possible.

5.4.2.1 Renaming Variables

Take a look at names(). The last three variables are a little odd looking (e.g., Tmax.1, Tmin.1). We need to rename them if they are going to be meaningful. But how do we do that? It turns out that the names of a data frame are an editable object as well! Thus, names() not only tells us what the variable names are but also allows us to change them.

```
names(bikes)[55:57]<-c('Cels_Tmax','Cels_Tmin','Cels_Tavg')
```

Thus, we have given each of these columns a name matching the original variable but with 'Cels_' in front of it.

5.4.3 Multiple Variables, Different Equations

What if we want to calculate multiple variables but each one requires its own equation? The transform() function is built for this very purpose. We still need to recreate our range variable in Celsius. But, for the sake of illustration, maybe we want to check that the average was calculated correctly while we are at it.

```
bikes<-transform(bikes,Cels_Temprange=Cels_Tmax-Cels_Tmin,
                 Cels_Avg=(Cels_Tmin+Cels_Tmax)/2)
```

transform() requires two arguments: the name of the data frame of interest and one or more variables to be calculated. Here we have calculated the range (*max-min*) and the average (*(min+max)/2*). Note that I have given the average variable a distinct name so that we do not overwrite the existing one. Also, though the output is the entire bikes data frame, transform() keeps all existing variables and adds only the new ones.

You can take a look at the range variable using summary(). And if you use View() you can see that our two average variables are identical (minus some occasional rounding). You can remove the new version if you like with the following code.

```
bikes<-bikes[-59]
```

5.4.4 Calculating Variables in tidyverse

Next we want to learn how to do the same tasks in tidyverse (remember to require(tidyverse) for this segment). The single variable-single equation calculations do not require tidyverse, but the multiple calculations might benefit from it. It turns out that the mutate() command in tidyverse is almost identical to the transform() command.

```
bikes<-mutate(bikes,Cels_Temprange=Cels_Tmax-Cels_Tmin,
              Cels_Avg=(Cels_Tmin+Cels_Tmax)/2)
```

Because the first argument in the mutate() function is the name of the data frame of interest, it can be incorporated into piping, as in this trivial example:

```
bikes<-bikes %>%
  mutate(Cels_Temprange=Cels_Tmax-Cels_Tmin,
      Cels_Avg=(Cels_Tmin+Cels_Tmax)/2)
```

Note that, as we have seen in previous chapters, the argument bikes is not included in the `mutate()` command in the second line of the pipe because it is being drawn from the previous line. Also, note that we are replacing the bikes data frame with a new version that has all previous variables plus the new ones.

The one difference between `transform()` and `mutate()` is the ability of the latter to reference variables within it that do not yet exist. For example, we could rewrite the previous command:

```
bikes<-mutate(bikes,Cels_Temprange=Cels_Tmax-Cels_Tmin,
              Cels_Avg=Cels_Tmin+Cels_Temprange/2)
```

That is, the average is equal to the minimum plus half of the range. This might not be the most intuitive way to write this equation, but it illustrates an important capability of `mutate()` that can come in handy: one variable in the command can be calculated using another variable calculated in the same command. `transform()` would generate an error in this case because `Cels_Temprange` does not already exist in the bikes data frame.

5.5 Manipulating Character Variables: `stringr`

The most powerful tool for working with character variables, including factors, is the `stringr` package (Wickham, 2019) . You likely will not need to install and `require()` it at this time because it is included as part of `tidyverse`.

`stringr` comprises a plethora of functions, including for mutating character values, also known as “strings” of text (i.e., systematically changing their values), joining and splitting strings, detecting matches, isolating and altering subsets of strings, and others. We will work through examples of the first three of these, but I also recommend you spend some time with the `stringr` cheat sheet that the folks at `tidyverse` have been kind enough to provide (https://github.com/rstudio/cheatsheets/blob/master/strings.pdf).

You will also want to familiarize yourself with “regular expressions,” also known as regex. These are on pg. 2 of the stringr cheat sheet. Regular expressions allow you to specify more general patterns, like something being at the beginning or end of a word, whether it is alphanumeric or only contains numbers, and others. We will avoid these in our examples here, but they can all be expanded with regular expressions.

5.5.1 Mutating Strings

Sometimes strings are messy. The simplest issue can be R’s sensitivity to capitalization. It is good practice to make your character variables all uppercase or lowercase before moving forward using the `str_to_lower()` or `str_to_upper()` functions.

```
bikes$WEATHER<-str_to_lower(bikes$WEATHER)
bikes$WEATHER<-str_to_upper(bikes$WEATHER)
```

Use the `table()` function after each command to see how the values have been mutated. We have overwritten the `WEATHER` variable with these lines of code, but we could also create two new variables using `mutate()`.

```
bikes<-mutate(bikes,
          WEATHER_lower=str_to_lower(bikes$WEATHER),
          WEATHER_upper=str_to_upper(bikes$WEATHER))
```

5.5.2 Joining Strings

The `str_c()` is useful when we want to combine sets of strings. Importantly, it will align the strings of two vectors (i.e., lists of strings) of the same length, creating a vector of the same length as the first two.

For instance, we earlier created four new variables for temperature in Celsius. We named these variables according to a convention with `'Cels_'` as a prefix to the previous variable name. `str_c()` can make this simpler by:

```
names(bikes)[55:58]<-str_c('Cels',names(bikes)[c(34:37)],sep='_')
```

Let's unpack this. The first argument was `'Cels'`. Because it is a single value, rather than a vector, it becomes a prefix for all values in the second argument, which is the names of variables 34-37 in `bikes`. Last, the `sep=` argument allows us to put a consistent separator between our arguments. The results are exactly what we would expect (you can check with `names()`) : `Cels_Tmax`, `Cels_Tmin`, `Cels_Tavg`, and `Cels_Temprange`. `str_c()` will also work if the first argument were itself a vector, say, if we wanted a variable that combined the minimum and the maximum in one place

```
bikes$Cels_MinMax<-str_c(bikes$Cels_Tmin,bikes$Cels_Tmax,sep='-')
```

Use `summary()` to see the results.

5.6 Creating Categories

Creating new categorical variables can be an effective way to simplify data and make it more immediately accessible to analysis. Here we will walk through three examples: categorizing by textual content, categorizing more general by any combination of criteria, and categorizing numerical variables by their levels.

5.6.1 Categorizing by Content: `str_detect()`

Let's return to our `WEATHER` variable, which we will work with as being all uppercase. Take a look at the different values therein:

```
table(bikes$WEATHER)
```

```
##
##                                          CLEAR              CLEAR-COLD
##                   156                       11                       1
##             CLEAR-DAY          CLEAR-DAYLIGHT            CLEAR-DRY-DAY
##                    12                        1                       1
##         CLEAR-MORNING                 CLEAR -            CLEAR - DAY
##                     1                        2                       7
##         CLEAR - NIGHT     CLEAR   /   DUSK         CLEAR AFTERNOON
##                   344                        1                       1
##            CLEAR COOL               CLEAR DAY           CLEAR EVENING
##                     1                        7                       1
##           CLEAR NIGHT     CLEAR SUNNY WARM CLEAR SUNNY WARM DAY
##                     6                        1                       1
##             CLEAR/DAY               CLEAR/DRY                  CLOUDY
##                     1                        1                     191
##            CLOUDY/WET          COOL AND CLEAR                    DARK
##                     1                        1                       1
##                   DAY            DAY - CLOUDY               DAY/SUNNY
##                     4                        1                       1
##               DRIZZLE               DRIZZLING                    DUSK
##                     1                        3                       1
##                   FOG              HEAVY RAIN              HEAVY SNOW
##                     1                        2                       1
##            LIGHT RAIN                     N/A                   OTHER
##                     1                        4                       8
##              OVERCAST           PARTLY CLOUDY                    RAIN
##                     1                        2                     104
##          RAIN - NIGHT                 RAINING             RAINY, WINDY
##                     1                        1                       1
##                 SLEET                    SNOW         SNOW/SLEET/RAIN
##                     2                       10                       1
##             SPRINKLNG                    SUNN                   SUNNY
##                     1                        1                       2
##           SUNNY - DAY           SUNNY - WARM        SUNNY 75 DEGREES
##                   689                        1                       1
##             SUNNY DAY                 TORNADO                 UNKNOWN
##                     3                        1                       4
##                  WARM          WARM AND CLEAR                     WET
##                     1                        1                       1
```

There are a lot of different values. It might help us to simplify this list. For instance, there are numerous different versions of "CLEAR." The `str_detect()` function can be useful here.

```
bikes$Clear<-as.numeric(str_detect(bikes$WEATHER, "CLEAR"))
```

What did we do? str_detect has two arguments: the variable (or list of text) within which we are looking for text, and the text (or pattern) we are looking for. We have sought to find every value in `bikes$WEATHER` with "CLEAR" in it. `str_detect()` generates a logical (`TRUE`/`FALSE`) variable, so we convert it to a dichotomous numeric (0/1) variable.

```
table(bikes$Clear)
```

```
##
##    0    1
## 1205  403
```

There are 403 cases for which the description of the weather included the word "clear." We could do the same for "rain," which appears in a number of the descriptions.

```
bikes$Rain<-as.numeric(str_detect(bikes$WEATHER, "RAIN"))
table(bikes$Rain)
```

```
##
##    0    1
## 1497  111
```

There are 111 cases for which the description of the weather included the word "rain."

5.6.2 Categorizing by Criteria: `ifelse()`

Categorizing by content is one specific case of the broader ability of R to create categories based on specific criteria. This is crucial when creating custom variables and indicators as the analyst must decide what the criteria are that define the desired information.

The `ifelse()` function is able to translate criteria into new variables. It has three arguments: (1) the criterion or criteria of interest; (2) the result you want if the criteria are `TRUE`; (3) the result you want if the criteria are `FALSE`. First, let us replicate our `str_detect()` command for `bikes$Clear` from above with `ifelse()`.

```
bikes$Clear<-ifelse(str_detect(bikes$WEATHER, "CLEAR"),1,0)
```

Instead of converting the result of `str_detect()` with `as.numeric()`, we stated that `TRUE` should be coded as `1` and `FALSE` as `0`. This is a trivial example, but illustrates how `ifelse()` works.

We might expand our criteria. We know that the weather was described as clear, but does that guarantee that there was no recent precipitation and the street was dry? We might instead want the following

```
bikes$ClearNoPrecip<-ifelse(str_detect(bikes$WEATHER, "CLEAR")
                            & bikes$PrecipTota == 0, 1, 0)
table(bikes$ClearNoPrecip)
```

```
##
##    0    1
## 1322  286
```

We now have only 286 cases for which this was true, down from 403.

We might go further, though, by nesting `ifelse()` commands to create a more nuanced but simplified weather variable.

```
bikes$WeatherCateg<-ifelse(str_detect(bikes$WEATHER, "CLEAR") &
                               bikes$PrecipTota == 0, "CLEAR",
                             ifelse(bikes$PrecipTota>0,
                                    "PRECIPITATION","CLOUDY"))
```

What did we do? The first `ifelse()` command includes our previous criteria, but the new value for `TRUE` is `"CLEAR"`. Then, instead of assigning a value to FALSE, we conduct another `ifelse()` on those remaining cases. If they have any precipitation, then the value is `"PRECIPITATION"`. If not, the final value is `"CLOUDY"`. We have now created a three-category variable:

```
table(bikes$WeatherCateg)
```

```
##
##          CLEAR        CLOUDY PRECIPITATION
##            286           714           608
```

5.6.3 Categorizing by Levels

We can create categories by levels of numerical variable using `ifelse()` as we saw in the examples using `bikes$PrecipTota`. We might want to create more categories based on systematic splits in the data. A classic example of this is splitting a data set into its quartiles—that is, evenly splitting the data into four groups from highest to lowest. This is done with the `ntile()` function (which is from the `dplyr` package in `tidyverse`).

```
bikes$Tavg_quant<-ntile(bikes$Tavg,n=4)
```

This divides the average temperature variable into four evenly sized groups by their order.

```
table(bikes$Tavg_quant)
```

```
##
##   1   2   3   4
## 402 402 402 402
```

You could do the same thing with any other n—if you want 5, 10, 100 groups, you simply need to alter the `n=` argument.

5.7 Text Analysis

If we want to take our efforts on categorizing data to the next level, we might pursue text analysis, also known as *text mining*. This is when an analyst explores a large corpus of text for patterns, especially quantifying which words and combinations of words are the most common. This is a bit more advanced of a topic than most of the other content in this chapter, but for those who are interested you should be perfectly capable of following along and replicating this example based on what you have learned so far. This section will use the `tm` (Feinerer and Hornik, 2020) package, which was designed for text mining, to work through five steps:

1. Preparing the data
2. Creating a corpus, or analyzable body of text
3. Cleaning the text
4. Creating a document term matrix, or the frequency of each word in each record in the corpus
5. Tabulating word frequencies

We will use the products of Step 5 to then inform new categorical variables that might otherwise have been just guessing at.

```
require(tm)
```

5.7.1 Step 1: Preparing the Data

The first step is to make sure that your data is ready to be analyzed. We want to analyze the variable `Narrative` as it has such rich content. It is a character variable (check using `class()`), which is what we need. If you take a quick look at the content, though (using `head()` or `View()`) you will notice a lot of strings of x's. This is because the data were redacted to remove any potentially identifiable information about the individuals in the collision. We want to eliminate all of these using `str_replace_all()`, which searches a particular set of text for a certain pattern of text and then replaces it with another pattern of text.

```
bikes$Narrative <- str_replace_all(bikes$Narrative, "xx", "")
bikes$Narrative <- str_replace_all(bikes$Narrative, "xxx", "")
bikes$Narrative <- str_replace_all(bikes$Narrative, "xxxx", "")
bikes$Narrative <- str_replace_all(bikes$Narrative, "xxxxx", "")
bikes$Narrative <- str_replace_all(bikes$Narrative, "xxxxxx", "")
bikes$Narrative <- str_replace_all(bikes$Narrative, "xxxxxxx",
                                   "")
bikes$Narrative <- str_replace_all(bikes$Narrative, "xxxxxxxx",
                                   "")
bikes$Narrative <- str_replace_all(bikes$Narrative, "x ", "")
bikes$Narrative <- str_replace_all(bikes$Narrative, "the", "")
bikes$Narrative <- str_replace_all(bikes$Narrative, "THE", "")
```

To illustrate, the first line searched `bikes$Narrative` for every instance of `"xx"` and replaced it with `""`, meaning a blank space. We also want to remove the word "the." While this might not be what we want if we were going to read something, it is okay here because we simply want R to be able to see all of the words.

5.7.2 Step 2: Creating a Corpus

Next, we need to create a corpus. What is a corpus? It is a collection of individual pieces of text in a workable format. These pieces of text are more or less organized as the elements of a vector, but in a unique structure that enables the commands of the `tm` package to be most efficient. Creating the corpus is where we will first use the functionality of `tm` with the `VectorSource` and `VCorpus` functions.

```
my_corpus <- VCorpus(VectorSource(bikes$Narrative))
```

Note that the class of `my_corpus` is indeed specific to `tm`:

```
class(my_corpus)
```

```
## [1] "VCorpus" "Corpus"
```

If you try to print `my_corpus`, you will not get very much information back. But you can look at individual cases with `writeLines`.

```
writeLines(as.character(my_corpus[[100]]))
```

```
On  at app Officer , x, responded to a r/c for a pedestrian struck by a
m/v at  Rd and Dudley St. Dispatch states that  x, x, and   are already on
scene.  Upon arrival Officer  spoke to Officer , x, who states that  victim
was riding his bike down St and entered  intersection of  Rd and St and was
struck by Mass reg . Officer  furr states that  victim had a strong odor
of alcohol coming from his breath. Officer  states that  victim suffered an
injured left leg and was transported by H+H Amb  to  BMC. Officer  states
that  victim told her that he had  green light on St and as he entered
intersection,  car came down  Rd and hit him.   Officer  spoke to  operator
of Mass reg , Butler, who stated he was driving down  Rd, he had  green
traffic light at St, he entered  intersection of  Rd and St and he was
almost through  intersection when a man on a bicycle came out of St, and
drove right in front of his m/v and that he struck  man on  bike. Officer
observed  passenger side of  windshield smashed.
```

The text here is that of the narrative from the 100th row.

5.7.3 Step 3: Cleaning the Text

Now that we have a corpus, we can start to work with it. Our next step will be to clean the text. You may be thinking, "Didn't we already get rid of all of the redacted text?" Yes, but there's a lot more to do to make the corpus intelligible for analysis. This will all be done with the `tm_map()` function, which takes the corpus in question and applies one of a variety of transformations to it.

We need to:

1. Remove punctuation: `my_corpus <- tm_map(my_corpus, (removePunctuation))`
2. Remove numbers: `my_corpus <- tm_map(my_corpus, (removeNumbers))`
3. Remove stop words, like "a", "the", "is", and "are", which are very common but contain little information: `my_corpus <- tm_map(my_corpus, content_transformer(removeWords), stopwords("english"))`
4. Replace hyphens with spaces, so that the pieces of a hyphenated word can be observed separately: `my_corpus <- tm_map(my_corpus, content_transformer(str_replace_all), "-", "")`
5. Transform to lower case for consistency: `my_corpus <- tm_map(my_corpus, content_transformer(tolower))`
6. And remove unnecessary spaces: `my_corpus <- tm_map(my_corpus, (stripWhitespace))`.

```
my_corpus <- tm_map(my_corpus, (removePunctuation))
my_corpus <- tm_map(my_corpus, (removeNumbers))
my_corpus <- tm_map(my_corpus, content_transformer(removeWords),
                    stopwords("english"))
my_corpus <- tm_map(my_corpus,
                  content_transformer(str_replace_all), "-", " ")
my_corpus <- tm_map(my_corpus, content_transformer(tolower))
my_corpus <- tm_map(my_corpus, (stripWhitespace))
```

Let's see what row 100 reads like now.

```
writeLines(as.character(my_corpus[[100]]))
```

```
on app officer x responded rc pedestrian struck mv rd dudley st dispatch
states x x already scene upon arrival officer spoke officer x states victim
riding bike st entered intersection rd st struck mass reg officer furr
states victim strong odor alcohol coming breath officer states victim
suffered injured left leg transported hh amb bmc officer states victim
told green light st entered intersection car came rd hit officer spoke
operator mass reg butler stated driving rd green traffic light st entered
intersection rd st almost intersection man bicycle came st drove right
front mv struck man bike officer observed passenger side windshield smashed
```

It is essentially a list of meaningful words without the intervening words and punctuation that make text readable, like punctuation, stop words, etc. But this is exactly what we want for text mining. These words are the basis for the analysis to follow.

5.7.4 Step 4: Creating a Document Term Matrix

We now need to create a new object called a document term matrix (of class `DocumentTermMatrix`). The document term matrix converts the corpus into a table that counts the incidences of each word in each item in the corpus. Let's create one from our corpus and then use the `inspect()` command to see what this means in practice for five rows.

```
dtm_bike <- DocumentTermMatrix(my_corpus)
inspect(dtm_bike[1:5,])
```

```
## <<DocumentTermMatrix (documents: 5, terms: 3088)>>
## Non-/sparse entries: 366/15074
## Sparsity           : 98%
## Maximal term length: 19
## Weighting          : term frequency (tf)
## Sample             :
##     Terms
## Docs and bicycle front left motor observed officer officers
##    1   6        1     2    0     3        1       4        0
##    2   0        1     0    2     0        0       2        0
##    3   0        2     1    0     0        2       5        0
##    4   0        1     1    2     0        0       0        1
##    5   0        4     3    4    12        6       7        8
##     Terms
## Docs stated vehicle
##    1      4       5
##    2      3       0
##    3      3       2
##    4      0       0
##    5      6      12
```

Reading this table, we can see that the document we have been looking at, row 100, contains the word "officer" 8 times, as does row 103. The word also appears in all three other cases. One might infer that row 103 involved an illegal action as the word "suspect" occurs 10 times and "victim" occurs 9 times. Overall, this is a simple way to interact with the content of the corpus.

The top of the output can also be useful. We have limited to five documents here that across them have 3,088 different terms (or words). The sparsity analysis tells us how consistent these terms are across documents. Of the 15,440 possible document-term combinations (3,088*5), only 329 are true. This makes sense-—3,088 terms is a lot of terms, and only some of them will repeat often, and many will be specific to the given case. If, however, we had a corpus with fewer total terms that were expected to be consistent across documents, we would expect lower sparsity.

5.7.5 Step 5: Examining Word Frequency

Oddly, to really analyze our word frequencies, we are better off converting the document term matrix into a data frame, summing across all columns to calculate the frequency of every term present in the corpus.

```
words_frequency<-data.frame(colSums(as.matrix(dtm_bike)))
```

This multi-function command does more or less what we want, but using the `View()` command, you might note that it is a little strange looking. There is no variable for the terms, but instead the terms are the names of the rows. The column has a funny name.

We want to pretty this up by:

1. Creating a new variable for the term: `words_frequency$words <- row.names(words_frequency)`
2. Renaming the frequency column: `names(words_frequency)[1] <- "frequency"`
3. Renaming the rows as numbers, which is more common: `row.names(words_frequency) <- 1:nrow(words_frequency)`
4. and flipping the order of out variables to the more customary setup of terms followed by frequencies: `words_frequency <- words_frequency[c(2,1)]`.

Now that we have this, we can look more closely at the frequency of certain words. Are words about weather common?

```
words_frequency$words<-row.names(words_frequency)
names(words_frequency)[1] <- "frequency"
row.names(words_frequency)<-1:nrow(words_frequency)
words_frequency<-words_frequency[c(2,1)]

words_frequency[words_frequency$words == "rain",]
```

```
##      words frequency
## 2077  rain         9
```

Only 9 cases of the word rain! What if we use `str_detect()` to look for any word containing the text “rain”, which could then include “rainy”, “rainstorm”, or otherwise.

```
words_frequency[str_detect(words_frequency$words, "rain"),]
```

```
##        words frequency
## 1647 moraine         6
## 2077    rain         9
## 2078 raining         3
## 2722   train        10
```

Here we see the potential weakness of this approach without paying close attention—“moraine” (which is a street name) and “train” are included as well.

If we look at the word “clear” we get 14 instances, but that could also include statements like, “The road was clear.”

Meanwhile, the most frequent words appear to be more about the actors in the collision and its resolution.

```
head(arrange(words_frequency,-frequency), n=10)
```

```
##      words frequency
## 1   victim      5386
## 2  officer      4929
## 3   stated      4709
## 4  vehicle      3627
## 5  bicycle      2917
## 6      and      2752
## 7      was      2184
## 8   struck      2134
## 9      the      2093
## 10  street      1841
```

We see here that some of the most common words are "victim" (5,386), "officer" (4,929), "vehicle" (3,627), and "bicycle" (2,918).

5.7.6 Summary and Usage

We have now tabulated the frequency of every word in the full corpus of police narratives on bicycle collisions in Boston for 2009-2012. This makes that information much more accessible, and anything we do with it better informed. We might use the list of frequent terms or a search for terms with similar content to create new, better-informed categorical variables. You are welcome to take a break from the book and go play with this now. We can also visualize these frequencies to better communicate them, which we will do in Section 5.9.3.

5.8 Dealing with Dates

Date and time variables come in many classes and can be quite tricky to work with. This is because of the many ways that one might organize this sort of information—date, date followed by time, time followed by date, month and year only, and time only are just a handful of examples. And then there are questions like whether the time has seconds or not, whether it includes time zone, whether the date is month-day-year or day-month-year, etc. Luckily, the package `lubridate` makes this much more straightforward (Spinu et al., 2021). Here we will focus specifically on analyzing dates.

First, we need to convince R that we have date data. Currently, it thinks that the `DATE` variable is a character (check with `class()`). But if we print the content, it looks like a date of the form year-month-day. `lubridate` has a series of functions for converting characters that are intended to be dates to be recognizable in the appropriate format. Here we need `ymd()` for year-month-day.

```
require(lubridate)
ymd(bikes$DATE)[1]
```

```
## [1] "2009-01-25"
```

We can then use a series of commands to extract the details of the date:

```
day(ymd(bikes$DATE))[1]
```

```
## [1] 25
```

```
month(ymd(bikes$DATE))[1]
```

```
## [1] 1
```

```
year(ymd(bikes$DATE))[1]
```

```
## [1] 2009
```

```
wday(ymd(bikes$DATE))[1]
```

```
## [1] 1
```

Most of these generate information we already knew from the date itself, but it isolates that information for further analysis. `wday()` does add something in that it identifies the day of the week, in this case it was a Sunday. We could also use this with `ifelse()` to create a new variable for collisions on weekends.

```
bikes$Weekend<-ifelse(wday(ymd(bikes$DATE))==1 |
                      wday(ymd(bikes$DATE))==7,
                      'Weekend', 'Weekday')
table(bikes$Weekend)
```

```
##
## Weekday Weekend
##    1262     346
```

We can also calculate differences between dates. For example, if we want to determine the full range of the corpus, we might try the following

```
max(ymd(bikes$DATE))-min(ymd(bikes$DATE))
```

```
## Time difference of 1427 days
```

That is, we subtract the earliest date (the minimum) from the latest date (the maximum) and find that it is about 4 years long (which we knew based on the 2009-2012 time range).

It is also possible to calculate new date variables using lubridate with equations, though we have no immediate use for this here. A simple example could be adding three days onto any event.

```
head(ymd(bikes$DATE)+days(3))
```

```
## [1] "2009-01-28" "2009-02-08" "2009-02-11" "2009-02-12"
## [5] "2009-02-16" "2009-02-16"
```

5.9 Returning to ggplot2

The spirit of this chapter has been to manipulate variables to expose the precise information that we want to learn from. Apart from the text mining, we have not engaged with many new data structures. Thus, our learning of `ggplot2` will move forward primarily to customize how we create our graphs as we get to know our content a bit better. The one exception to this will be in Section 5.9.3 when we learn how to visualize our text analysis in word clouds using the package `wordcloud` (Fellows, 2018).

5.9.1 Visualizing Weather and Injuries – Customizing Graphs

The goal of Vision Zero is to eliminate serious injuries and fatalities from vehicle collisions. Let's visually explore the relationship between injuries and the weather in this dataset. First, we have to set aside cases for which `bikes$Injured == 99`, which is often used in data to represent missingness.

```
base.no.na<-ggplot(bikes[bikes$INJURED!=99,], aes(x=Tavg))
```

Remember that the first step of working in `ggplot2` is a command that establishes the basis for the graph but does not actually create the graph. We will now build upon `base.no.na`.

```
base.no.na+geom_histogram(aes(fill=as.factor(INJURED)))
```

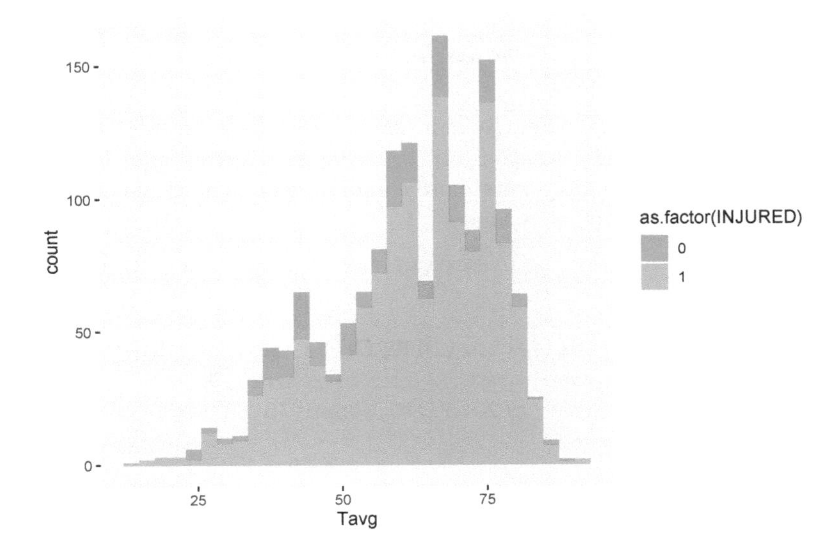

This is a nice graph, but it's not perfect. We might improve it by adding a series of operations to our `ggplot2` command:

1. Make things a bit smoother with smaller bins of 5 degrees each with the `binwidth=` argument: `geom_histogram(aes(fill=as.factor(INJURED)), binwidth=5)`
2. Rename the x-axis with `scale_x_continuous()`, while also setting the range to 20°-90° with `limits=`, and making the labels every ten degrees with breaks= : `+ scale_x_continuous("Average Temp.", limits = c(20,90), breaks = c(20,30,40,50,60,70,80,90))`
3. and rename the legend with `labs()`: `+ labs(fill = "INJURED")`

The whole thing looks like this:

```
base.no.na+
geom_histogram(aes(fill=as.factor(INJURED)), binwidth=5) +
scale_x_continuous("Average Temp.", limits = c(20,90),
                   breaks = c(20,30,40,50,60,70,80,90)) +
  labs(fill = "INJURED")
```

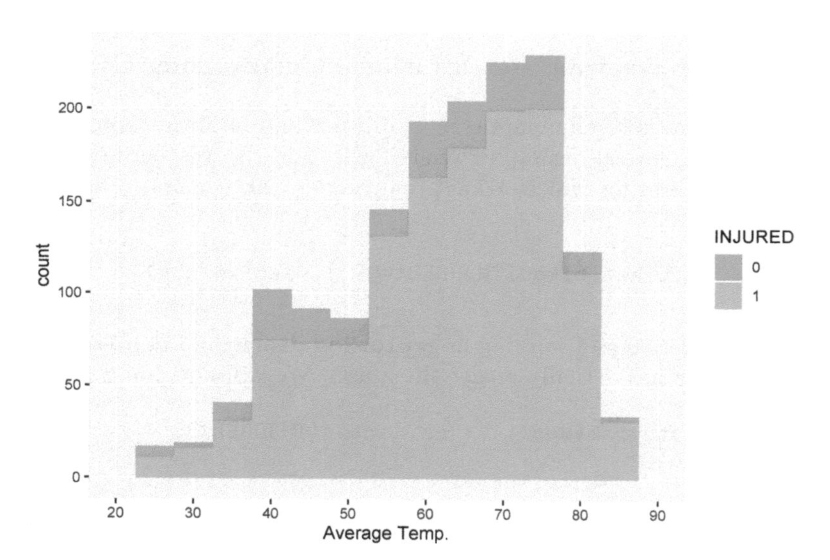

Now that we have the graph we want, we can take a closer look. In most cases there was an injury. It is not obvious, though, whether injuries were more common at some temperatures than others. It is possible that they were more common when it is very hot or very cold, but this graph does not definitively show that.

Before we move on, we are now going to save our current graph as a new object, `full_graph`.

```
full_graph<-base.no.na+
  geom_histogram(aes(fill=as.factor(INJURED)), binwidth=5) +
  scale_x_continuous("Average Temp.", limits = c(20,90),
                     breaks = c(20,30,40,50,60,70,80,90)) +
  labs(fill = "INJURED")
```

5.9.2 Visualizing a Third Variable: Facets

We have learned how to bring categorical variables in with the `fill=` argument. Another option is *faceting*, or creating the same graph repeatedly for all values of a categorical variable. Let's suppose we want to split our temperature-injury analysis by precipitation conditions:

```
full_graph+facet_wrap(~WeatherCateg)
```

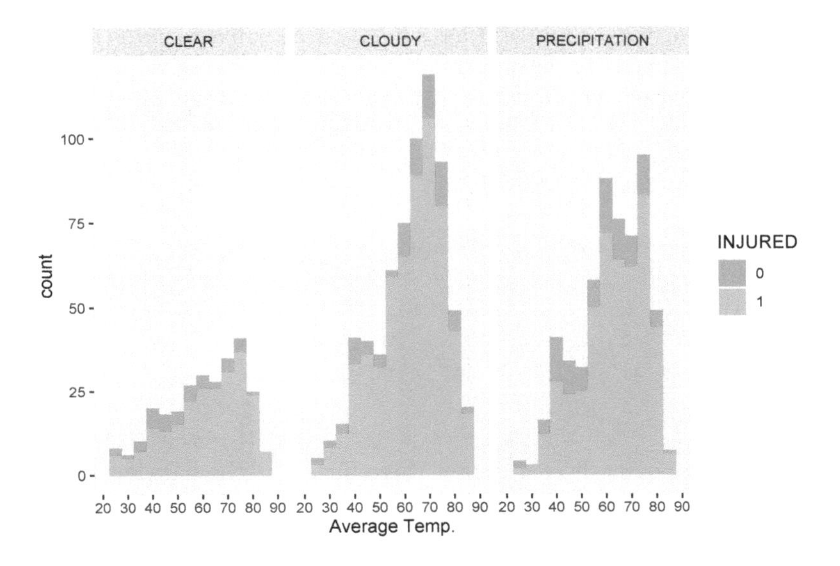

We could even go crazy and want to split by precipitation conditions and whether it was a weekend. This is possible with the `facet_grid()` command.

```
full_graph+facet_grid(WeatherCateg~Weekend)
```

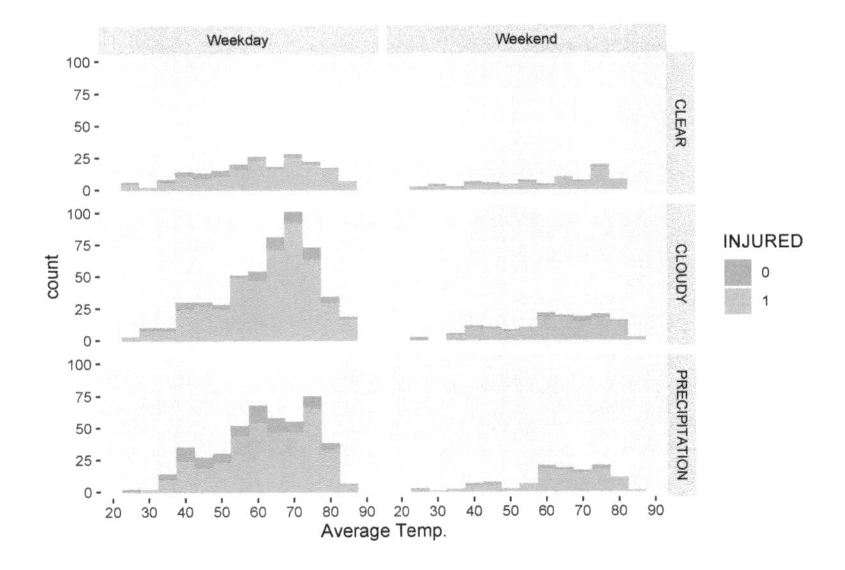

Interestingly, neither of these graphs communicates anything obvious about the relationship between injuries from collisions and weather. This reflects the concern that, regardless of the conditions, a collision between a car and a bicycle will almost always result in an injury for the bicyclist.

5.9.3 Visualizing Word Frequencies: Word Clouds

Remember the table of word frequencies we created from the police narratives? Provided you are in the same project, you should still have it in your Environment as `words_frequency`. We have all the tools we need to make a simple bar graph. Because we will want to limit the content—-2,971 words are way too many to visualize—we will use pipes.

```
pp <- words_frequency %>%
  arrange(desc(frequency)) %>%
  head(n=20) %>%
  ggplot(aes(x = reorder(words, -frequency), y= frequency)) +
  geom_bar(stat = "identity", fill = "red3") +
labs(title = "The top 20 most used words in bike
    collision narratives",
              x = "Words", y = "count") +
  coord_flip()
```

To explain what is happening here, the first three lines sort the data set by frequency with `arrange()` and then select the first 20 rows with `head()`, meaning the graph that follows will only visualize those 20 words. The `ggplot()` command then pre-orders our words by frequency, creates a bar graph (with `stat='identity'` instructing the graph to represent the y-value associated with each x and `fill=` setting a custom color), customizes the labels with `labs()`, and flips the coordinates to make the graph more readable with `coord_flip()`. The product is:

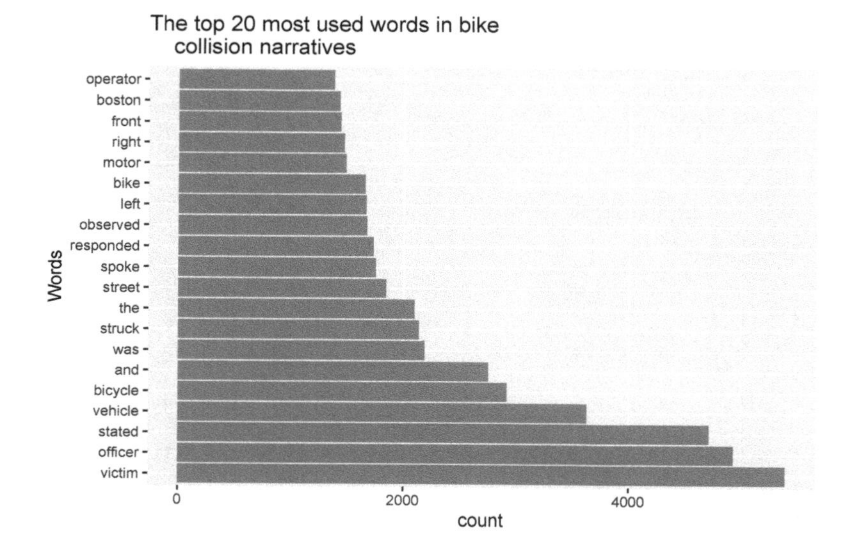

This is nice, but could we present the same information in a more engaging way? You may be familiar with word clouds, a graphic that represents the most common words in a corpus and sizes them proportional to their frequency. It is a fun and easily interpretable way of representing the prominent words in a corpus. This will require an additional package, `wordcloud`. You will need to install it.

```
require(wordcloud)
wordcloud(words = my_corpus, min.freq = 1, max.words=200,
      random.order=FALSE, rot.per=0.35,
      colors=brewer.pal(8, "Dark2"))
```

Unpacking this command, we have specified my_corpus as the source for words=; we have limited to words that appear at least once (`min.freq`) and to no more than the 200 most common words (`max.words`); we have turned off the `random.order=` option, which would grab words at random rather than by the order of their frequency; the `rot.per=` controls how many words are rotated to fit together; and the `colors=` command jazzed it up a bit (otherwise it would have been black and white). The product is:

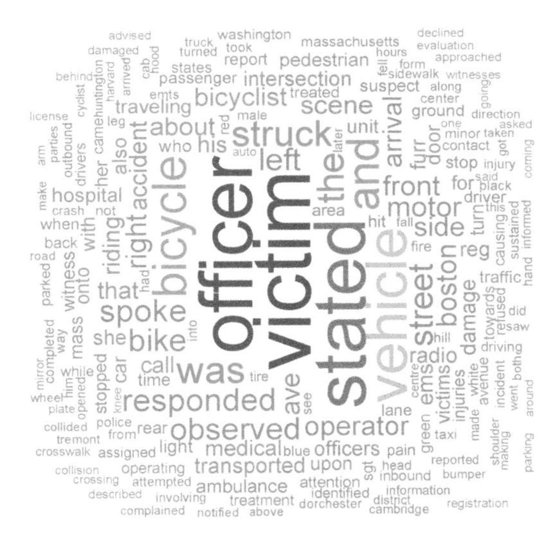

This image tells the story we largely already knew about the content of the corpus, but for someone who has not worked through it as thoroughly as we just did, it quickly highlights that the most common words center on the actors involved in the collision and its reporting—the victim, officer, vehicle, bicycle, and operator.

5.10 Summary

This chapter has worked through the creation of variables of various types using a database of bike collisions reported by the Boston Police Department from 2009 to 2012. We have focused on weather and, in the end, its relationship with injuries from collisions. In doing so, we have practiced the conceptual skill of thinking through the questions, "What do we want to know? How do we manipulate the data to expose that information?"

We practiced multiple technical skills for calculating variables, including:

- Calculated numerical variables in multiple ways:
  - One variable, one equation with `<-` and `$` notation,
  - Multiple variables, one equation with `[]` notation,
  - Multiple variables, multiple equations with `transform()` and `mutate()`, the latter being part of the `tidyverse`;
- Manipulated character variables with the package `stringr`, including:
  - Mutating strings to be easier to analyze,
  - Joining strings into a single varible;
- Created new categorical variables based on:
  - Text elements with `str_detect()`,
  - More general criteria with `ifelse()`,
  - Quantiles of numerical variables with `ntiles()`;
- Conducted text mining with the `tm` package,
- Dealt with dates with the `lubridate` package,
- Customized graphs in `ggplot2`, including the incorporation of facets,
- Created a wordcloud based on word frequencies using the `wordcloud` package.

5.11 Exercises

5.11.1 Problem Set

1. Assume that we are working with the following data frame titled `bikes`. What would be generated for each of the following commands?

ID	DATE	Address	WEATHER	Tmax	Tmin	INJURED	TRANSPORTE	TREATED	AGE
B1001	1/9/2009	1001 Commonwealth Ave	SUNNY - DAY	28	21	1	0	0	
B1002	1/16/2009	280 Brighton Ave	CLOUDY	39	25	1	0	0	47
B1003	1/25/2009	50 Kenilworth St	SUNNY - DAY	24	8	1	1	1	72
B1004	2/5/2009	584 Saratoga St	CLEAR/DRY	20	8	1	1	1	61
B1005	2/8/2009	222 Harrison Ave	RAIN	51	35	0	0	0	48
B1006	2/9/2009	59 Lindall St	CLEAR - NIGHT	39	25	1	1	1	26
B1008	2/13/2009	136 Tremont St	SUNNY - DAY	42	29	1	1	1	25
B1009	2/13/2009	420 Marlborough St	CLEAR - NIGHT	42	29	1	0	0	23
B1010	2/19/2009	87 Huntington Ave	RAIN	48	28	1	0	0	21
B1011	2/21/2009	259 Dartmouth St	SUNNY - DAY	39	27	1	1	1	23

a. `bikes$TRange<-bikes$Tmax-bikes$Tmin bikes$TRange[6]`
   b. `sum(str_detect(bikes$WEATHER,'SUNNY'))`
   c. `bikes <-mutate(bikes,Tavg=(Tmax+Tmin)/2) bikes$Tavg[3]`
   d. `day(ymd(bikes$DATE))[1] + month(ymd(bikes$DATE))[2] + year(ymd(bikes$DATE))[3]`

2. Using the data frame `bikes` from the worked example, write code to create each of the following categorizations. Check your work by executing the code and running tables on the resultant variables in R.
   a. Average temperature was below freezing.
   b. Temperature was below freezing at any time.
   c. A categorical variable for no precipitation, rain, and snow.
   d. Whether the narrative referenced the driver as a "suspect."
   e. Split precipitation by quartiles. Bonus: Do you think this is a useful variable? Why or why not? (Hint: Look at the values in each of the four quartiles.)
3. Write code for creating each of the following variables. These are general and could be applied to any data frame. If it is easier, describe the steps that you would do in place of or alongside the code.
   a. You have a date variable. You want to create a variable for season.
   b. You have a 'comments' field. You want to flag for each case whether Boston is mentioned.
   c. You have a variable for day vs. night (0/1) and weekend vs. weekday (0/1). You want a single variable for weekend night.
      i. You want a text variable for all four categories—weekend day and night, weekday day and night.
   d. You have a date variable and want to isolate the year. i.The computer won't recognize it as a date format.

5.11.2 Exploratory Data Assignment

Working with a data set of your choice:

1. Create at least three new variables, each of which must either,
   a. Make some angle of your data's content more interpretable, or
   b. Fix some issue in the data. Make sure to explain why these variables are useful.
2. Describe the contents of these variables using analysis and visualization tools learned in this chapter and Chapter 4.

Information: *Unit I Summary and Major Assignments*

Summary and Learning Objectives

Unit I has focused on the skills necessary to represent and understand the basic information contained in data. This has included both technical and interpretive skills.

Technical Skills

- Installing R and associated packages to enable specific functions.
- Working with data objects using functions.
- Isolating desired rows and columns using subsets and related tools.
- Summarizing the content of individual variables with tables, statistics, and univariate graphics.
- Creating new variables based on numerical calculations, transforming strings, and categorical criteria.
- Customizing graphics of various types.

The packages we have learned include:

- `tidyverse` for "tidier" coding when working with data.
- `knitr` for creating R Markdown scripts that then can be output as interwoven code and results.
- `ggplot2` for graphics.
- `stringr` for transforming, combining, and otherwise working with string data.
- `tm` for extended mining of rich textual data.
- `lubridate` for working with date and time variables.

Interpretive Skills

- Scrutinizing individual records to better understand the broader content of the data set.
- Inferring aspects of the data-generation process to frame interpretation of the data set.
- Planning variable manipulations and calculations to reveal the desired information.

Unit-Level Assignments

Community Experience Assignment

The community exploration assignments in this book are designed to align skills you have been learning with real-world contexts. They are most useful in conjunction with the Exploratory Data Assignments at the end of each chapter, especially when you have been working through them with a single data set. They provide an opportunity to "ground truth," or really evaluate the assumptions and objectives that have guided your analysis thus far. There will be one in each unit. These can also be combined with a service-learning or capstone oriented course.

For this first community city exploration assignment, please:

1. Select a neighborhood (or more localized place) based on something notable in the data you are working with this semester, e.g., the highest or lowest value on some variable of interest, a density of cases in one region, etc.
2. Visit and explore this neighborhood either in person or virtually, with an eye to the observations you made in the data and how these characteristics manifest themselves in the real world.
    - Though there is no strict guideline on how long you need to spend exploring a neighborhood, either in person or virtually, a visit that lasts less than a half-hour would be unlikely to generate enough observations to support a high-quality memo and presentation.
    - A virtual visit might start with Google Maps and Street View and similar searches, but might branch out anywhere the internet can take you.
3. Write a 3-5 page memo describing the logic for why you visited this place, what you discovered, and what this tells you about the interpretation of your data. This written document should include images from your walk (or from StreetView if visiting virtually) and maps with data describing the region.

Post-Unit Assignment: Read-Me Document

The first unit of this book has focused on turning data into information about the city, its people, and its neighborhoods. We have also learned how to clean and modify certain variables to make the data set easier to analyze. These skills learned here are the same ones necessary to create a high-quality, engaging Read-Me Document that acts as a first view into the contents of the data set and how it might be used.

Think about a person browsing an open data portal, looking for interesting data and what they might do with it. The Read-Me Document should fulfill this role, offering a quick look into the contents of the data and the analytic possibilities they might offer. It should be both informative *and* enticing, and should include:

- *A brief overview of what the data set contains.* This should be a short paragraph composed of short sentences describing the source of the data and the type of information it holds. This will look a lot like the opening paragraph of the Data Dictionary that you have, though possibly even shorter and a little less dry.

- *"Fun facts."* Provide about 5-10 pieces of information that are notable and illustrate the nature of the data set. These should be both engaging and potentially useful to someone looking to analyze the data. For example: "The 311 system categorizes requests for service into 218 types, the most common being General Request." "The neighborhood with the most 911 calls in 2013 was Dorchester."
- *Visualization.* Include at least two visualizations that illustrate some aspect of the data set that you find particularly noteworthy, complementing the "fun facts."
- *No more than two pages.* Remember that this is something that a visitor will want to read quickly to decide whether it is of interest.

Suggested Rubric (Total 10 pts.)

Communication of Data Content: 3 pts.

Facts: 2 pts.

Figures: 2 pts.

Structure: 2 pts.

Details (Grammar, etc.): 1 pt.

Unit II

Measurement

6

Measuring with Big Data

What can you measure with a data set that was not originally intended to measure anything? This was the subject of a meeting in April 2011 at Boston City Hall between a handful of researchers from the Boston Area Research Initiative (BARI; including myself) and members of the Mayor's Office of New Urban Mechanics and the City's performance management team. The data set of interest was the extensive corpus of records (more than 300,000 at that time) that had been generated by the City's relatively young 311 system.

We had no lack of ideas. The problem was that our two main ideas were mutually incompatible. First, there was the possibility that the data acted as the "eyes and ears of the city" and could enable measures of deterioration and dilapidation (also known as physical disorder or "broken windows") across neighborhoods. The second idea was that they gave us an insight into how constituents engaged with government services. But if the data were a pure reflection of either of these two things, it could not be the other. That is, if 311 perfectly communicated current conditions, then constituent engagement was robust and consistent across the city, and thus not worth measuring; and if it exactly reflected constituent engagement, then those tendencies would obscure the ability to "see" objective conditions. Obviously, the data reflected a combination of objective conditions and subjective patterns of engagement, meaning the two interpretations of the data seemed hopelessly entangled. And yet, in the months that followed, we solved the problem. BARI now releases annual metrics drawn from 311 data for both physical disorder *and* custodianship, or the tendency of a community to take care of public spaces (Figure 6.1).

The problem of measurement is not unique to 311 reports but is a general weakness of administrative data, social media, and the other "naturally occurring data" that drive urban informatics . Because these data were not created for the purposes of research they lack many of the features that we often take for granted in traditional data. What is it that they measure? Do the data capture what they appear to, or do they suffer from biases? What are the units of measurement (i.e., people?, neighborhoods?) that we can describe and how often (i.e., monthly?, annually?)? Answering such questions is an early step in unlocking the potential of a given data set. This chapter will focus on the conceptual tools that enable such efforts to develop original measures from novel data sets. As the first chapter in the unit on *Measurement*, it will also set the stage for the subsequent chapters, in which we will create and further work with new measures.

6.1 Worked Example and Learning Objectives

This chapter is unique in this book, as it will focus entirely on conceptual skills—you will not need to open RStudio once! We will walk through the process that my colleagues and I undertook to make sense of the 311 data to generate measures that describe neighborhoods,

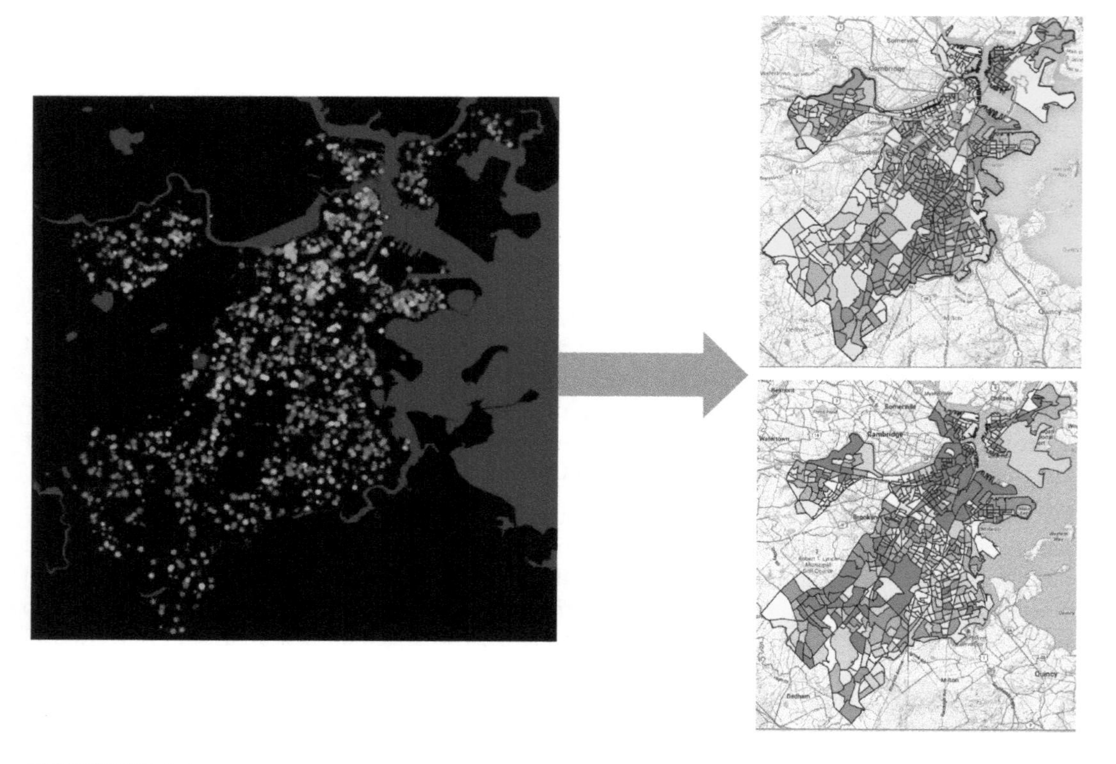

FIGURE 6.1
311 reports are discrete records that capture various events and conditions across the city (left), but they can be brought together to create measures of physical disorder (top right) and custodianship (bottom right) across neighborhoods.(Credit: Author)

but it is applicable to any naturally occurring data set and unit of analysis you might encounter. Along the way, we will learn to:

- Describe the "missing ingredients" of naturally occurring data;
- Specify the desired unit of analysis for measurement;
- Leverage the *schema* or organization of a data set to access the desired unit of analysis;
- Identify concepts that might be measured through a data set;
- Isolate the desired content to pursue a given measurement;
- Identify potential biases in the data and propose ways of addressing them.

Link: Though this chapter does not work through an example in R, you might be interested in visiting BARI's Boston Data Portal, where the 311 metrics described here are published annually: https://dataverse.harvard.edu/dataset.xhtml?persistentId=doi:10.7910/DVN/CVKM87 (or go to the crime and disorder page on the interactive map: https://experience.arcgis.com/experience/ce3975f9368841f791d1fc8891e6171b).

Note that there will be regular references to technical steps that must be taken to create a measure, especially aggregation of records and the merger of data sets. We have not learned how to do these skills yet. Not to worry, however, as we will learn all about them in Chapter 7. For now, we are just going to discuss them conceptually.

6.2 Data and Theory: The Responsibility of the Analyst

Before describing the missing ingredients of naturally occurring data, it is useful to reflect on the responsibility the analyst has when deciding how to properly use a data set. This speaks to a rather contentious debate that has arisen following the emergence of big data. In 2008, Chris Anderson, editor of *WIRED* magazine, wrote an essay presaging "The End of Theory" (Anderson, 2008). He argued that "with enough data, the numbers speak for themselves." In his view there is no longer any need for theory. The data will guide us through the questions that need to be asked and answered, and analysts just need to be capable of following along.

Needless to say, Anderson's essay has attracted numerous critics. One of my favorite retorts was from Massimo Pigliucci, a philosopher of biology, who argued that Anderson misses that the very purpose of science is to explain why things work the way they do (Pigliucci, 2009). The same knowledge of "why" is necessary for practitioners and policymakers to do their jobs and for companies to develop effective products. If we do not understand the meaning of the questions elicited by data, the answers will be useless to us. Pigliucci also encourages Anderson to recall the maxim, "there are no data without theory." Without some theory or model of the world, one has no basis for determining which data to collect or how to interpret them. Every decision that an analyst makes when working with data, from data cleaning to variable construction to model specification, is based on the knowledge, assumptions, and goals they bring to the data. To claim otherwise is to ignore the agency that the analyst brings to the data and the responsibility they hold for framing, interpreting, and communicating the results. Whether we call it "theory" or not, awareness of these concerns is critical to a well-informed analysis. This perspective has been a major thrust throughout this book and will be especially important as we start to design new measures.

6.3 Missing Ingredients of Naturally Occurring Data

Naturally occurring data were not collected according to a systematic research protocol but as the byproduct of some administrative or technological process. As such, the analyst must grapple with three "missing ingredients" that are traditionally baked into data collection processes.

1. *What units of analysis can be described?* Most naturally occurring data sets consist of records. They might reference units of analysis, such as neighborhoods or individuals, but describing them requires the composite analysis of all of the records together. First, the analyst must determine what the options are and which are of interest.
2. *What can the data measure?* This is a deceptively simple question. Often the answer is "lots of stuff," which can create additional complications. It is up to the analyst to decide which aspect or aspects of the desired unit of analysis are of interest and how to isolate the relevant information from the rest of the data set's content.
3. *Are the data biased (and what to do about it)?* An administrative system or online platform might capture all cases associated with it, but this is not a guarantee

that that information is a comprehensive reflection of all related information. Does it capture all of the events it purports to? Do some demographic groups avail themselves of a service more than others, creating an unbalanced view? What other aspects of the data-generation process might introduce biases into the data?

How well these three questions are answered will determine the usefulness of any measure based on naturally occurring data. The remainder of this chapter will walk through them in turn.

6.4 Unit of Analysis

In Unit I, we learned how to create new variables, or measures (or indicators, metrics, etc.). But those variables were either additional descriptors of individual records or summations of records. What if we want to describe a person, place, or thing referenced in the data set, or event to analyze *all* of the people, places, or things referenced in the data set? We would need to aggregate the records to a particular unit of analysis. Before we can do that, we need to know what units of analysis are available to us and to select from them the one that would best further our goals. An important tool for doing this is the *schema*, or organizational blueprint of how a data set's structure relates to other data sets at various levels of analysis.

6.4.1 Schema

A schema is an organizational blueprint not only for a single data set but for a series of related data sets. It is most often used in database management because databases often contain multiple datasets that link to each other on specific key variables, also known as *unique identifiers.* Sometimes the data sets within a database contain separate sets of variables all referring to the same unit of analysis to limit the number of variables in any single data set. For example, Figure 6.2 contains two different sets of variables describing Boston's 311 cases: one set describes the features of the event (date of report, case type, etc.); the other describes the geographic location of each case. These can be kept apart or merged on the `CASE_ENQUIRY` variable, which is the unique identifier for records.

A data set might also include variables that are the primary units of analysis in another data set, enabling linkage between them. This is apparent in Figure 6.3, which represents a more complex set of relationships between Boston's 311 records and other data sets. A fundamental `ID` data set is simply the list of cases, with the unique identifier of `CASE_ENQUIRY`. This unique identifier also links to the Report Description data set, which contains additional information on the case itself. The relationship between these two data sets is similar to that illustrated in Figure 6.2.

The `ID` data set contains two additional unique identifiers, though, that link it to other data sets with distinct units of analysis. First, the variable `LOCATIONID` is a unique identifier for all addresses and intersections in Boston that links to a separate `Geographic Information` data set. This data set is a list of all possible values for `LOCATIONID` with additional information

CASE_ENQUIRY	OPEN_DT	CLOSED_DT	REASON	TYPE	SOURCE
1.01001E+11	2/27/2013 0:00	2/28/2013	Highway M	REQUEST I	Citizens Cc
1.01001E+11	2/27/2013 0:00	2/28/2013	Highway M	REQUEST I	Citizens Cc
1.01E+11	6/29/2010 0:00	6/29/2010	Sanitation	SCHEDULE	Constituer
1.01E+11	9/10/2012 0:00	9/14/2012	Sanitation	SCHEDULE	NA
1.01001E+11	12/18/2014 0:00	12/18/2014	Enforceme	PARKING E	Constituer
1.01E+11	8/28/2011 0:00	11/14/2011	Trees	TREE MAII	Self Servic
1.01E+11	3/6/2011 0:00	3/7/2011	Sanitation	SCHEDULE	Self Servic
1.01001E+11	3/25/2013 0:00	4/25/2013	Recycling	REQUEST I	Constituer
1.01001E+11	7/2/2013 0:00	7/6/2013	Sanitation	SCHEDULE	Constituer

CASE_ENQUIRY	LOCATION	X	Y	TLID	BLK_ID_10	BG_ID_10
1.01001E+11	1 Soldiers Fiel	-71.13	42.37	85736470	2.5E+14	2.5E+11
1.01001E+11	1 Soldiers Fiel	-71.13	42.37	85736470	2.5E+14	2.5E+11
1.01E+11	42 Hichborn S	-71.14	42.36	85695024	2.5E+14	2.5E+11
1.01E+11	40 Waverly St	-71.14	42.36	85716345	2.5E+14	2.5E+11
1.01001E+11	INTERSECTION	-71.14	42.36	85718528	NA	2.5E+11
1.01E+11	37 Portsmoutl	-71.14	42.36	85718304	2.5E+14	2.5E+11
1.01E+11	37 Portsmoutl	-71.14	42.36	85718304	2.5E+14	2.5E+11
1.01001E+11	38 Hichborn S	-71.14	42.36	85695024	2.5E+14	2.5E+11
1.01001E+11	65 Waverly St	-71.14	42.36	85695299	2.5E+14	2.5E+11

CASE_ENQUIRY	OPEN_DT	CLOSED_DT	REASON	TYPE	SOURCE	LOCATION	X	Y	TLID	BLK_ID_10	BG_ID_10
1.01001E+11	2/27/2013 0:00	2/28/2013	Highway M	REQUEST I	Citizens Cc	1 Soldiers Field	-71.13	42.365	85736470	2.5E+14	2.5E+11
1.01001E+11	2/27/2013 0:00	2/28/2013	Highway M	REQUEST I	Citizens Cc	1 Soldiers Field	-71.13	42.365	85736470	2.5E+14	2.5E+11
1.01E+11	6/29/2010 0:00	6/29/2010	Sanitation	SCHEDULE	Constituer	42 Hichborn St	-71.14	42.356	85695024	2.5E+14	2.5E+11
1.01E+11	9/10/2012 0:00	9/14/2012	Sanitation	SCHEDULE	NA	40 Waverly St I	-71.14	42.361	85716345	2.5E+14	2.5E+11
1.01001E+11	12/18/2014 0:00	12/18/2014	Enforceme	PARKING E	Constituer	INTERSECTION	-71.14	42.358	85718528	NA	2.5E+11
1.01E+11	8/28/2011 0:00	11/14/2011	Trees	TREE MAII	Self Servic	37 Portsmouth	-71.14	42.359	85718304	2.5E+14	2.5E+11
1.01E+11	3/6/2011 0:00	3/7/2011	Sanitation	SCHEDULE	Self Servic	37 Portsmouth	-71.14	42.359	85718304	2.5E+14	2.5E+11
1.01001E+11	3/25/2013 0:00	4/25/2013	Recycling	REQUEST I	Constituer	38 Hichborn St	-71.14	42.356	85695024	2.5E+14	2.5E+11
1.01001E+11	7/2/2013 0:00	7/6/2013	Sanitation	SCHEDULE	Constituer	65 Waverly St I	-71.14	42.361	85695299	2.5E+14	2.5E+11

FIGURE 6.2
An illustration of a simple schema in which variables describing 311 case events (top left) and their geographic locations (top right) have been separated into distinct data sets but can be merged using CASE ENQUIRY (bottom).

on the location of each, including its street segment (`TLID`), census geographies (`BLK_ID_10`, `BG_ID_10`, `CT_ID_10`), its `CITY`, `STATE`, `ZIP` code, and `LAT` and `LONG`. Also, the variable `PARTYID` is the unique identifier for `User Account`. Users of the 311 system can create an account for tracking their cases and enabling the city to contact them by `EMAIL` or mailing `ADDRESS`. Again, this is all represented graphically in 6.3.

A schema like this one enables us to do two things. One is to merge information across data sets, for example bringing in additional geographic information associated with the address or intersection referenced in each record, as we have already done. The other is the ability to aggregate records to describe the units of analysis that they reference. For instance, we might count the number of cases referencing every address and intersection in the city, or the number of cases made by every user of the system. Thinking bigger, we might use the information contained in those cases to develop any number of measures describing addresses or users.

6.5 What We Did: Defining "Neighborhood"

The goal of me and my colleagues when approaching the 311 data set was to describe the conditions of neighborhoods. Such measures have a long intellectual history and are referred to as *ecometrics*, as they quantify the physical and social characteristics (*-metrics*) of a space (*eco-*). Though they have been traditionally measured with resident surveys and neighborhood audits known as systematic social observation (think the Google StreetView car replaced by people walking with clipboards), we believed there was an opportunity to achieve the same goal using 311. To do this we had to capitalize on 311's schema (as illustrated in 6.3) and the linkage between records and geographic information.

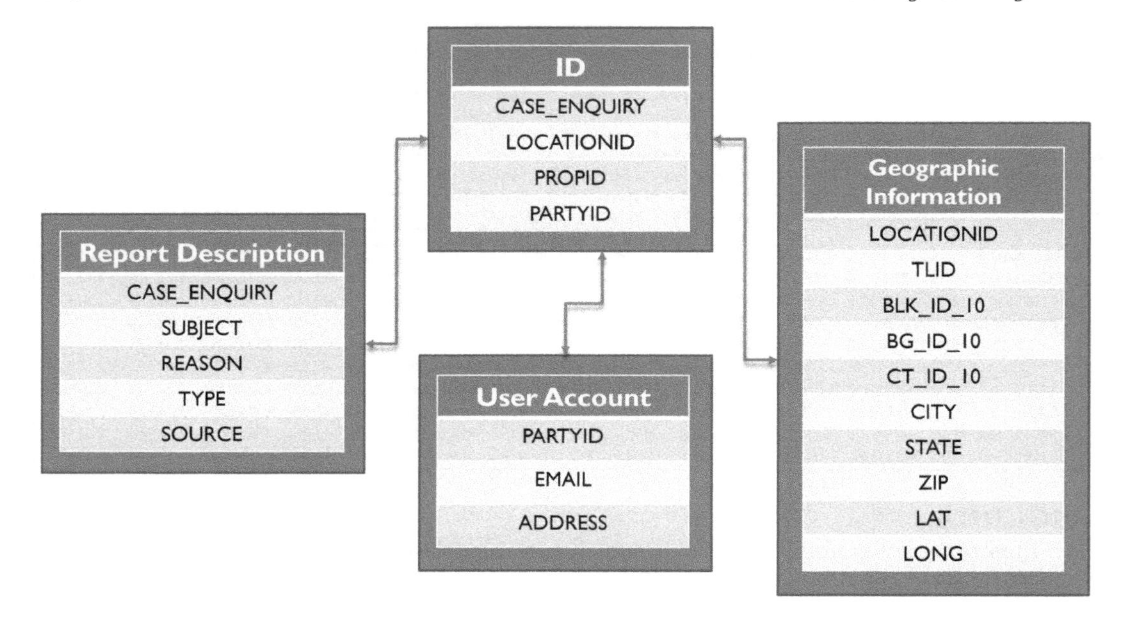

FIGURE 6.3
Schema for the data sets describing 311 records and related data sets for geographic locations and user accounts to which records can be linked.

Geography is a great example of an extended schema because it has so many scales of organization. To facilitate the analysis of the 311 data and related data sources describing Boston, BARI developed a database we call the *Geographical Infrastructure for the City of Boston* (GI)[1], which we curate and update annually. The GI compiles and connects the places and regions of the city at multiple levels: discrete locations, including properties, which are contained in land parcels, and intersections; street segments; census geographies (i.e., blocks, block groups, and tracts); and broad administrative regions (e.g., planning districts, neighborhood statistical areas, election wards, districts for public health, police, fire, and public works). Importantly, these different scales are nested within each other. Properties are contained in land parcels, which sit on streets and inside census geographies. Census geographies are designed to have blocks sit within block groups and block groups sit within tracts. The smaller census geographies are generally contained within the City's administrative regions. The database reflects this, with every property having a variable for its land parcel, every land parcel having variables for its street and census block, and so on.

311 records come with a `LOCATIONID`, which links to land parcels. We wanted to measure the characteristics of neighborhoods, which are often approximated with census geographies. Thus, we needed to link with the data set of land parcels, which includes variables for all three census geographies. We were then able to aggregate to any of these geographic scales we wanted. Depending on the goals of an analysis, our preference might be for census block groups or tracts. Thus, we did both (and have since published metrics for both annually).

[1]https://dataverse.harvard.edu/dataset.xhtml?persistentId=doi:10.7910/DVN/ZHTMIW

6.6 Isolating Relevant Content

You have probably taken many surveys in your life. Maybe as part of a psychology course or a doctor's visit or a public poll or a BuzzFeed quiz. Typically, a survey consists of dozens of questions. These questions are not randomly selected, however, nor does the person who designed that survey see them as discrete pieces of information. They are organized into groupings intended to capture specific *constructs*, like personality traits, health behaviors, political leanings, etc.

Like a traditional survey, naturally occurring data also contain lots of content, often far outstripping those dozens of questions. For instance, the 311 system in Boston has over 250 case types ranging from pick-ups for large trash items to downed trees to needle cleanup to cracked sidewalks. Though these diverse case types are unified in their need for government services, they describe many, many different aspects of the urban landscape. But unlike the survey they do not come equipped with any pre-established constructs for what they can measure. In other words, we do not really know what the "eyes and ears of the city" are seeing, hearing, and communicating about the urban landscape. Consequently, an analyst working with this or any other naturally occurring data sets needs to determine what they want to measure and how to isolate that information from all of the other content contained in the data.

6.6.1 Latent Constructs: A Guide for Measurement

To guide our approach to measurement, we will borrow a concept from psychology called the latent construct model. It is based in the philosophical assumption that the things we truly want to know about a person, place, or thing are underlying tendencies and propensities that are impossible to directly access. Instead, we have to infer them from the behaviors, events, and patterns that we can observe. In short, these underlying *latent constructs* are what we want to measure, but to do so we must quantify the *manifest variables* that are symptoms or reflections of these latent constructs in action.

Stepping away from urban informatics for a moment, let us take a classic example from personality psychology that will likely be familiar. Extraversion is a personality trait that has been measured for nearly a century, but you would not necessarily ask someone, "How extraverted are you?" Instead, as illustrated in Figure 6.4, a survey on extraversion might include items about being "talkative," "energetic," "bold," and "outgoing," or "enjoying parties". An observational measure might look for similar features. The combination of these discrete manifest variables then gives rise to a measure that reflects the latent construct we wanted to observe in the first place.

The latent construct model is a helpful tool for thinking about how we might extract information from naturally occurring data, though the process is a bit reversed from traditional survey development. Researchers developing surveys define a latent construct and then write items that can act as the manifest variables that capture that latent construct. Analysts working with naturally occurring data are presented with a multitude of manifest variables already present in the data. They must determine which of these represent their latent construct of interest (and which should be ignored). In either case, however, the process begins by defining the latent construct and the kinds of ways it might be manifest.

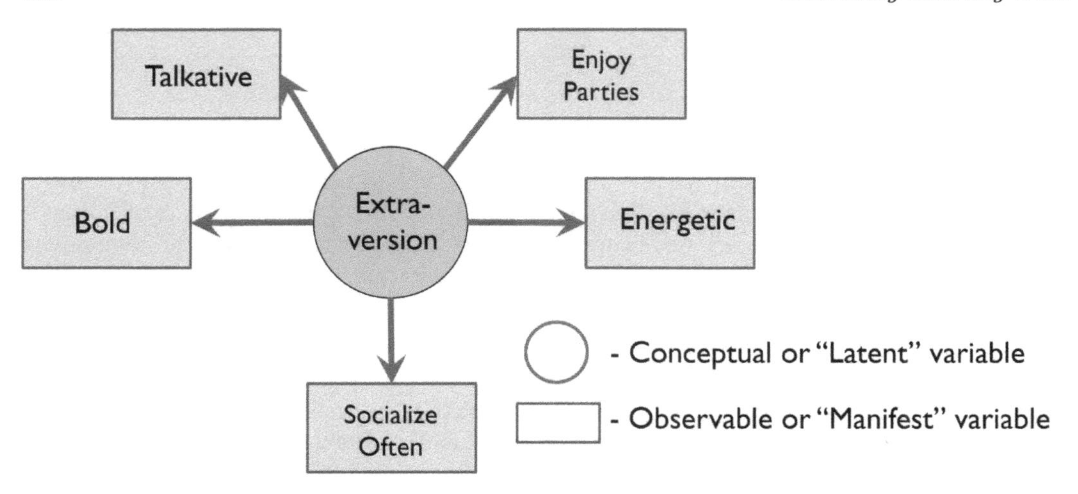

FIGURE 6.4
The latent construct of extraversion might be measured using a variety of observable or manifest variables that reflect being energetic, sociable, and outgoing.

6.6.2 What We Did: Isolating Physical Disorder

A large proportion of 311 records reference the deterioration and denigration of spaces and structures, including loose garbage, graffiti, and dilapidation. These reflect "physical disorder," a popular concept (or latent construct) often measured in urban science and of interest to sociologists, criminologists, public health researchers, and others. The 311 data set would then appear to provide many manifest variables for measuring the tendency of a community to generate or suffer from physical disorder. The challenge that faced our project, however, was twofold: identifying those case types and separating them out from all the other case types in the data set that have little bearing on physical disorder.

When we originally started working with the data in 2011, there were "only" 178 case types present in the data. I personally read through these one-by-one, identifying 33 that might be evidence of either denigration or neglect to spaces and structures. Importantly, we did not include instances of natural deterioration, like potholes or street light outages, or personal requests regarding things like street sweeping schedules or a special pick-up of a large trash item (e.g., couch). Many of the case types we identified were traditionally included in observational protocols for measuring physical disorder, such as graffiti and abandoned buildings. Others were unfamiliar but seemingly relevant, like cars illegally parked on a lawn. In these cases the data set's rich content made novel information available. Most notably, 311 was able to "see" the conditions inside private spaces, including failing utilities and rodent infestations. Though this diversity of manifest variables for measuring physical disorder was impressive, it is important to note that they constituted fewer than 20% of case types and about 50% of all reports. Thus, we had to discard over half of the database to isolate our desired construct. If we had simply summed cases across neighborhoods, we would have measured something very different and probably entirely uninterpretable.

Thirty-three is a large number of manifest variables. And as we looked closely at them, it seemed like there might be some important sub-categories, for instance, differentiating between disorder in private and public spaces. We tabulated the number of cases in each of these 33 cases for all census block groups and looked closely at their correlations using a technique called factor analysis (an advanced technique that goes beyond the scope of this book). Given the content of the cases and the results of this analysis, we confirmed that there were two distinct but related aspects of physical disorder: *private neglect* and *public denigration*. These further broke down into five more specific categories: housing, uncivil use of space, and big buildings for private neglect; and graffiti and trash for public denigration. (Note that these final measures only used 28 of the 33 case types that we originally identified as we determined that five of the case types were not as relevant as we thought.) This constituted a multi-dimensional measure of physical disorder that was more nuanced than any survey or observational protocol that preceded it. This is illustrated in Figure 6.5. That said, it still begged the question of whether the 311 system measured these conditions faithfully, or if they were biased in some way.

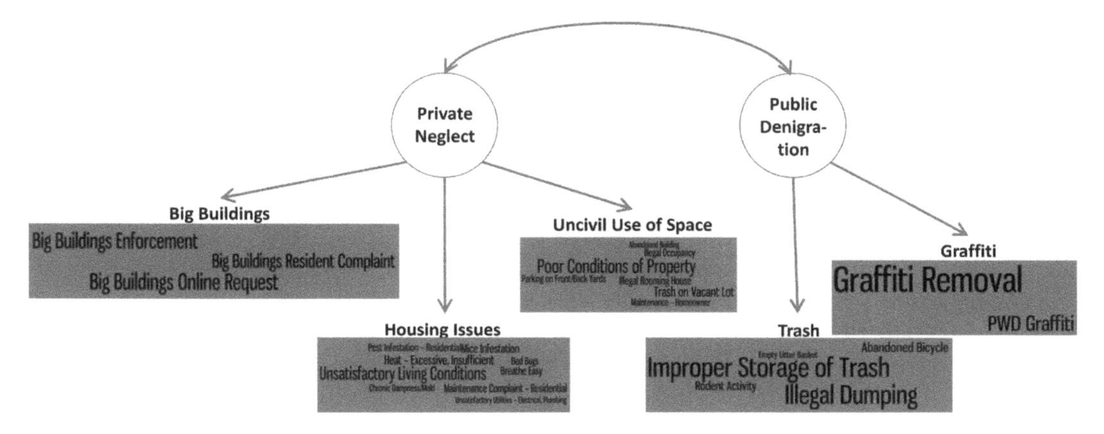

FIGURE 6.5
The case types indicating physical disorder could be combined into those that reflected issues in private spaces (or private neglect) and public spaces (or public denigration), and were further divided into five categories reflecting housing issues, uncivil use of space, big buildings, graffiti, and trash. (Credit: Author)

6.7 Biases and Validity

It is tempting to treat administrative records, social media posts, and other forms of naturally occurring data as faithful representations of the pulse of the city . Once one isolates the content that reflects a particular construct of interest, a new dimension of the urban landscape should be lain bare. But the data were not collected systematically with the goal of capturing objective information. Instead, aspects of the data-generation process may imbue the data with inherent biases, confounding any measures based on them. Especially in the case of crowdsourced content, like 311 reports and social media posts, there are questions about how often people of different backgrounds and perspectives are contributing content, and the kinds of things they see as worth contributing. This brings us back to the original challenge my colleagues and I faced at the beginning of this chapter: are 311 reports a reflection of objective issues or the tendency of community members to report issues? As such, analysts working with naturally occurring data must identify the potential sources of bias and account for them in some way.

6.7.1 Validity

Researchers are often concerned with the *validity* of their measures, defined as the extent to which it reflects the real-world features that they are intended to capture. The reason why should be apparent: a measure that does not actually reflect what it is supposed to is not especially useful. Survey measures of extraversion should accurately assess extraversion. Standardized tests should accurately evaluate ability in mathematics and language arts. Ecometrics should accurately quantify the physical and social characteristics of neighborhoods.

Evaluating validity might be split into statistical and non-statistical techniques. Non-statistical techniques are primarily conceptual and include *face validity*, or whether a measure appears on its surface to measure the construct of interest, and *content validity*, or whether a measure covers the range of information associated with a construct (but nothing more). In a sense, we have already engaged in an exercise of non-statistical validity in Section 6.6) as we defined the manifest variables that might capture a given latent construct. An extension of this reasoning is to consider whether one or more manifest variables might arise from more than one latent construct. For instance, as shown in Figure 6.6, 311 reports are a product of both objective conditions *and* the decision of local residents to report those conditions to the government, or what we might refer to as a "civic response rate." Similarly, a researcher with social media data often has to develop ways of removing posts made by automated "bots," or distinguishing posts made by individuals from those made by organizations.

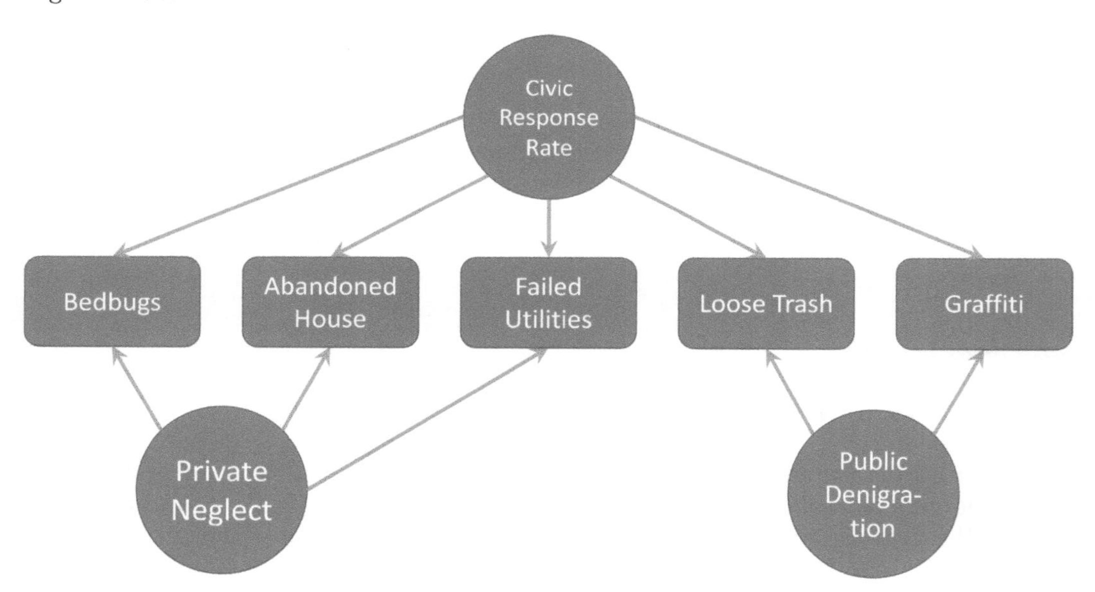

FIGURE 6.6
Various case types tracked by the 311 system could be manifest variables indicating disorder in private spaces (or private neglect) and public spaces (or public denigration). They also are theoretically all affected by the tendency of a neighborhood to report issues, or the civic response rate.

Statistical techniques for establishing validity are varied but often entail either *convergent validity* or *discriminant validity*. Both involve coordinating the measures of interest with external measures already believed to be valid. In the former, the analyst examines whether the newly developed measure agrees with one or more measures that capture similar constructs. For instance, does an IQ test predict academic performance? Does the volume of Twitter posts correlate with population numbers across neighborhoods? Does a measure of physical disorder derived from 311 records agree with observations made on the street? Discriminant validity is the complement to convergent validity, as it seeks to confirm that a measure is *not* measuring things it should not be. Are there sufficient differences between an IQ test and a personality test that they are not just measuring the ability to take a survey? Are all measures derived from 311 highly correlated because they reflect the underlying tendency of a neighborhood to report not just physical disorder but *anything*?

Both statistical and non-statistical techniques are necessary for developing valid measures. We have already begun to practice the non-statistical ones and will continue to do so. Statistical techniques for validity typically entail correlation tests and related analyses. Section 6.7.2 will reference these in a superficial way as we work through our validation of the 311 data, but we will learn more about conducting them in Chapter 10. For now, though, it is important to practice thinking about what a measure should correlate with as well as what it should not correlate with.

6.7.2 What We Did: Disentangling Physical Disorder from Reporting Tendencies

My colleagues and I faced an intriguing challenge. The 311 records were clearly a reflection of both physical disorder and the tendency of residents to report said physical disorder. To illustrate, in communities where residents were not inclined to make reports, an issue might sit unnoted for a lengthy period, or even indefinitely, creating a gap in the database. Meanwhile, neighborhoods with very vigilant residents might even generate multiple reports for a single issue, exaggerating the prevalence of disorder. Thus, we needed to find a way to separate physical disorder from the tendency to report issues, thereby measuring both things. Otherwise we would not be able to measure either of them.

This section describes the four steps my colleagues and I took to establish a set of validated, 311-based measures of physical disorder that account for the bias introduced by reporting tendencies. As you might imagine, this was a rather extensive effort, and if you are interested in a more in-depth description of all the methods used they are available in another book I wrote called *The Urban Commons* (O'Brien, 2018). Also, it is important to note that not all validation efforts need be this extensive, so please do not let it intimidate you from wanting to undertake your own validation exercises. Instead, look at the various steps taken here as parts of a menu of options for establishing validity.

6.7.2.1 Step 1: Assaying Disorder and Reporting

Our first step was to measure physical disorder and the tendency to report issues objectively. To accomplish this, I recruited a team of students to conduct audits in approximately half of Boston's neighborhoods. We identified 244 street light outages and quantified the level of loose garbage on every street. Meanwhile, the City of Boston's Public Works Department assessed the quality of all of the city's sidewalks. We used the level of garbage to map objective levels of physical disorder. We then cross-referenced the street light outages and sidewalk assessments with 311 reports. This gave us estimates of the likelihood of communities to report these types of issues. We found, indeed, that this tendency varied considerably across neighborhoods.

6.7.2.2 Step 2: Estimating Civic Response Rate from 311 Reports

We need the civic response rate to adjust for biases in any measure of physical disorder we might develop. Assessing the civic response rate through a neighborhood audit is useful in this process but using it to adjust our measures is only a temporary solution. Eventually variation in the civic response rate will shift, with some places becoming more responsive and others less so. We want to be able to adjust for the civic response rate repeatedly using only the information generated by the 311 system. We must then isolate relevant content from the database, this time reflective of civic response rate.

With that goal in mind, what goes into the civic response rate? We proposed two different elements. First was *engagement*, or knowledge of 311 and the willingness to use it. The second was *custodianship*, or the motivation to take responsibility for issues in the public space. Each of these is necessary for reporting a public issue: without the former, a custodian would not know what action to take; without the latter, someone with knowledge of the 311 system would have no motivation to use it.

This gave us two latent constructs for which we needed to develop manifest variables, as diagrammed in Figure 6.7. For engagement, these included: the proportion of neighborhood residents with an account; requests for the pickup of bulk trash items, the need for which would presumably be even across the city; and requests of snowplows during a snowstorm, being that a snowstorm hits the entire city (though controlling for certain infrastructural characteristics, like road length and dead ends). For custodianship, we identified a subset of 59 case types that reflect issues in the public domain, including many (but not all) of the case types regarding physical disorder as well as instances of natural deterioration, like street light outages. We then identified user accounts that had reported at least one of these cases. Further, we noted that a small fraction of these "custodians" made three or more reports in a year (~10%), whom we referred to as "exemplars." We then created two manifest variables for custodianship by tabulating the number of "typical custodians" and "exemplars" in a neighborhood. Presumably neighborhoods with more of either of these groups would report public issues more often and more quickly.

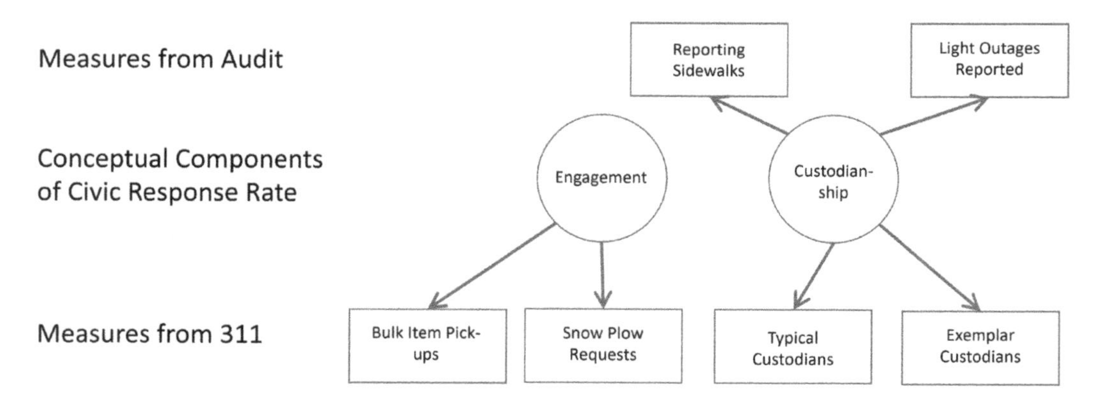

FIGURE 6.7
We broke the civic response rate into two components—engagement with the 311 system and custodianship for public spaces—each with its own manifest variables drawn from the 311 system. It turned out, though, that only the latent construct of custodianship was related to the manifest variables drawn from the objective audit. (Credit: Author)

We analyzed these metrics in combination with our objective measures of the civic response rate from the streetlight and sidewalk audits. This revealed that measures from the audits were only associated with our 311-derived measures of custodianship. That is to say, our objective measure of the civic response rate was more specifically an objective measure of custodianship, based on the convergent validity of these independent manifest variables.

6.7.2.3 Step 3: Calibrating the Adjustment Factor

At this point we have established that there is bias in reporting, created by differences in custodianship across neighborhoods. We might use our measures of custodianship then as a volume knob to balance out regional differences in custodianship. Where custodianship is lower, we can "turn up the volume" on the raw number of reports of physical disorder, assuming that some issues are going unreported. Where custodianship is higher, we need to turn down the volume to temper outstanding vigilance. Deciding just how much to adjust by, however, requires its own additional analysis. In short, we examined how well the measures of street garbage that we collected in-person predicted the prevalence of physical disorder reports made to 311 . We then assessed how well a consideration of the measures of custodianship could strengthen this relationship. The result told us just how much we needed to adjust. Interestingly, it turned out that disorder in public spaces needed substantial adjustment, but disorder in private spaces did not.

6.7.2.4 Step 4: Confirming Convergent Validity

We now have a measure of physical disorder that has been adjusted for each neighborhood's level of custodianship. But how do we know if it is correct? We decided to check its correlations with a handful of neighborhood indicators traditionally associated with physical disorder and blight, including: median income, homeownership, density of minority population, perceived physical disorder from a resident survey, and reports of gun violence made through 911. We found that our newly developed measures of private neglect and public denigration had many of the anticipated relationships. As noted, this act of checking correlations with external measures is often a key step in establishing convergent validity.

6.7.2.5 Summary of Establishing Validity with 311

We have now walked through the four steps that my colleagues and I went through to validate our measures of physical disorder. Again, this was an extensive effort and not every validity exercise requires quite so much work. That said, there are a number of tools and lessons that you might take away from it, including:

- Reasoning through content validity of both the desired measure and the sources of bias from which it suffers;
- Sourcing external data to assess convergent and discriminant validity;
- Identifying other variables that should correlate with your measure if it is valid;
- The potential to generate not one but *two* interesting sets of measures, one of which being the source of bias itself.

Regarding the last point, in the current case physical disorder was the well-established metric we wanted to replicate with big data, but custodianship turned out to be the original insight that revealed much about how 311 systems actually work as collaborations between government and the public. Consequently, the former has been used as a tracking metric and a variable in studies on criminology, which is how it has often been used in the past. The latter became the basis for an entire book titled *The Urban Commons* (O'Brien, 2018) and has added a new dimension to the study of neighborhoods.

6.8 Summary

This chapter has illustrated the challenges presented by naturally occurring data and the tools available for solving them by working through the development of measures of physical disorder (and custodianship) from 311 records. We have focused entirely on conceptual skills without even opening R or a single data set. These include being able to:

- Articulate the missing ingredients of naturally occurring data:
  - Identify the desired unit of analysis from the schema of a data set;
  - Isolate content relevant to a desired measure using the latent construct model;
  - Identify potential sources of bias in a data set and develop a strategy for addressing them to establish validity.

As noted at the outset, we will put these conceptual skills to good use in the next chapter as we gain the complementary technical skills. We will then be able to develop new aggregate measures for our desired unit of analysis.

6.9 Exercises

6.9.1 Problem Set

1. Define the following terms and their relevance to developing novel measures from naturally occurring data.
   a. Schema
   b. Content validity
   c. Convergent validity
   d. Latent construct
   e. Manifest variable
2. Propose one or more potential sources of bias you might expect for each of the following measures.
   a. Quality of sounds and smells street-by-street as drawn from Twitter postings.
   b. Political attitudes as drawn from Facebook postings.
   c. The median rent of apartments as drawn from Craigslist postings.
   d. The investment in a neighborhood by landowners as drawn from building permit applications.
   e. Social connectivity as drawn from cell phone records

6.9.2 Exploratory Data Assignment

Working with a data set of your choice:

- Propose at least one latent construct that you would like to measure with your data set and the unit of analysis for which you would want to measure it. Be sure to confirm that this unit of analysis is available from the schema of the data set.
- Describe how this construct is interesting, and at least one manifest variable for measuring it. Use the latent construct modeling notation from this week to bring it across.
- Suggest any potential sources of bias that might complicate the development and interpretation of the measure(s).
- *Note*: Your manifest variable can be an idea at this point. You will figure out how to code it in R in the coming chapters.

7

Making Measures from Records: Aggregating and Merging Data

Allegheny County, the home of Pittsburgh, Pennsylvania, had a problem with truancy. This is not an uncommon situation. A lot of urban areas have large populations of at-risk youth who are regularly absent from school. The County also recognized an institutional challenge to fixing the problem: the people responsible for supporting these youths were case workers, but case workers did not have access to school attendance records because they worked for the Department of Human Services (DHS), not the school district. To make things even more complicated, DHS was run at the county level and schools were run by each of Allegheny County's 43 school districts. Thus, case workers were not aware of when one of their cases was missing school. What to do?

Enter Erin Dalton, director of Allegheny County's Data Warehouse (and later Director of all of Health and Human Services). The Data Warehouse is an example of an *Integrated Data System* (IDS) that merges data from across the many Health and Human Services departments in a government for each individual. Through the Data Warehouse, Erin's team was able to link a case worker's profile with data on each of the kids they worked with from the appropriate school district. They were even able to do updates in real-time and created an alert system so that a case worker was notified if any of those children were absent for three consecutive days.

An IDS is a powerful infrastructure that takes advantage of the many different ways that people interact with government services. As Erin likes to quip, if the mayor (or county chair, in this case) has to sign both sides of a data-sharing agreement, you don't need a data agreement. Data-driven departments in over 35 states, counties, and cities have embraced this perspective (see Figure 7.1), enabling a variety of innovative approaches to practice, at least according to Actionable Intelligence for Social Policy, a national consortium of IDSes based at the University of Pennsylvania.

Let us consider for a moment how IDSes operate from a technical perspective. They are premised on the tools we discussed conceptually in the previous chapter. They first must create measures for individuals in the system by aggregating all records from a given data set referencing that individual. This is possible for any data set, whether it is generated by a hospital, human services, school districts, the criminal justice system, homelessness services, or otherwise. They can then merge these measures using unique identifiers (or key variables) made available by the schema (i.e., name, social security number, birthday, etc.). The technical process of aggregation and merger is the focus of this chapter.

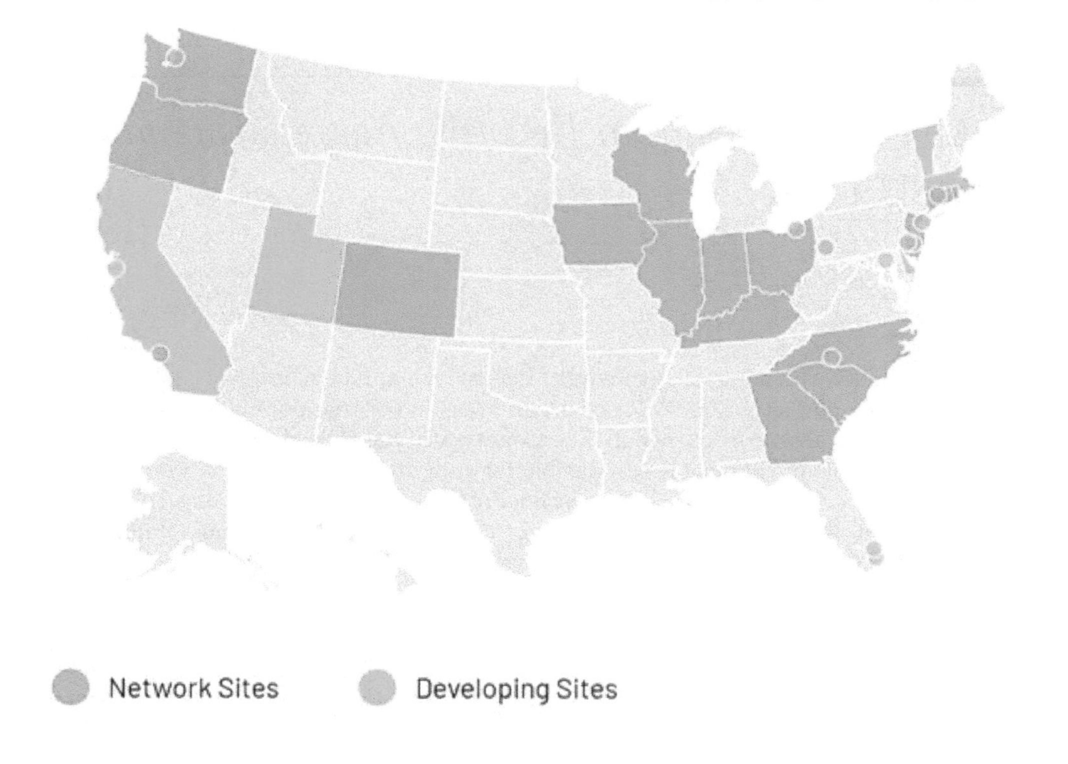

FIGURE 7.1
A map of the states, cities, and counties that are members of Actionable Intelligence for Social Policy, a network of IDSes. (Credit: Actionable Intelligence for Social Policy).

7.1 Worked Example and Learning Objectives

In this chapter, we will take the next step with the two tools fundamental to IDSes—aggregating and merging data sets—by learning the technical skills necessary to execute them. Of course, working with detailed human services data can require specialized access for privacy reasons. As such, in this chapter we are going to work through the original human services data: census data. And to illustrate how it can work at its most granular level, we will use census data for Boston from 1880, which is old enough that the raw data have been released: the gender, ethnicity, profession, and more are included for every individual. Working with these data we will learn to:

- Create aggregate measures in base R and `tidyverse`;
- Merge data sets on shared key variables in base R and `tidyverse`;
- Use SQL, another coding language common to database management that can be utilized in R through the `sqldf` package, to aggregate;
- Create bivariate (i.e., two-variable) visualizations in `ggplot2`, including dot plots.

Link: https://dataverse.harvard.edu/dataset.xhtml?persistentId=doi:10.7910/DVN/28677. You may also want to familiarize yourself with the data documentation posted there.

Data frame name: `census_1880`

Note: The data set contains census records from 1880-1930. We will limit the data to 1880 from the outset.

```
hist_census<-read.csv('Unit 2 - Measurement/Chapter 07 - Aggregation and
Combining Data/Example/Census Data 1880_1930.csv')

names(hist_census)[1]<-'year'
census_1880<-hist_census[hist_census$year==1880,]
```

7.2 Aggregating and Merging Data: How and Why?

Working with records is both challenging and empowering for an analyst. They are challenging because a record may reference one or more traditional units of analysis—a person, place, or thing—but it does not fully encapsulate that unit of analysis. A discrete unit might be referenced by 1, 10, 30, 100, or zero records. Often we are not interested in analyzing the records themselves but the units of analysis that they reference.

At the same time, records can be empowering. Their composite describes the desired units of analysis in highly nuanced, detailed ways. One might leverage a set of records to generate dozens of descriptors for those units of analysis. How many records for each individual unit appear in the data? What types of records? If the records have quantitative values associated with them, we can calculate the mean, the sum, the min, the max, or other statistical features of those values for each individual unit. This flexibility empowers the analyst to generate the precise measures needed to describe the objects of interest.

The key to going from records to units of analysis is *aggregation*, or the gathering of information at one level of organization to describe a higher one. For example, from records to people, or from people to neighborhoods, or from neighborhoods to cities, and so on. The other key tool here is merging, or the linkage of data sets on shared key variables. We can create aggregate measures for the same unit of analysis from multiple data sources and then link them by *unique identifiers*, or one or more variables that distinguish each unit in a set of units. And this is all made possible by a *schema*, or the organizational blueprint of how a data set's structure relates to other data sets at various levels of analysis.

7.3 Introducing the Schema for Historical Census Data

Before we get started, let's acquaint ourselves with the schema of the historical census data. Admittedly, we only have one data set, whereas a typical schema consists of multiple interlocking data sets at different units of analysis. Nonetheless, the census data does reference numerous nested units of analysis, making for a simple but illustrative schema for us to learn from.

By using the **names()** command we can take a look at the contents of the historical census data more closely. You might also want to download the data documentation and take a direct **View()** of the data as well. Each row is an individual with a variety of descriptors, including their gender, age, profession, and heritage (e.g., native-born vs. immigrant). There is also a series of variables that describe the broader organization of the data set and how it links to other units of analysis.

At the front end of the data set, we have the variable **serial**, which is a unique indicator for the household. Note that there are many cases in which multiple rows share the same value for **serial**, reflecting the fact that the data is a list of all individuals organized by household. There is also the variable **pernum**, which itemizes the individuals in each household (e.g., if there are two members of the household, they have the values 1 and 2). There are also enumeration districts (**enumdist**), or the rather small regions that the census used for organizing the geography of the city. Households (**serial**) are nested in enumeration districts. Further, enumeration districts are nested in census wards (**ward**), somewhat larger administrative regions. If we were working with data from across the country, we would also have wards nested in multiple cities (**city**). Instead, we just have all the wards that constitute Boston.

The historical census provides us with an ideal opportunity to aggregate data. We have individuals nested in households, households nested in enumeration districts, and enumeration districts nested in wards. We can aggregate up from our records to any of these other levels. (It is also worth comparing this structure with the example schema of the modern Geographical Infrastructure described briefly in Chapter 6. They are very similar!) If we happened to have outside data on any of these other levels—say, voting results for wards—we could link that information in as well. Here we will focus mainly on individuals, households, and wards.

7.4 Aggregation

Aggregation is the process of using cases from one unit of analysis to describe a higher order unit of analysis. Examples include using discrete credit card transactions to describe the purchasing patterns of individuals, using household characteristics to describe the demographics of a community, or using counts of 311 or 911 reports to measure disorder and crime on a street. Aggregation depends on three pieces of information: (1) the unit of analysis for the initial data set, (2) the unit of analysis to which we want to aggregate, and (3) the function by which we want to aggregate (e.g., the count of cases, the sum or mean of case values, etc.).

We have already learned the most basic way to aggregate, which is to use the **table()** command to generate counts for all values of a variable. For instance, to calculate the number of individuals in each household, we could do the following, using code originally seen in Chapter 4.

```
hh_people<-data.frame(table(census_1880$serial))
head(hh_people)
```

```
##      Var1 Freq
## 1 483155    2
## 2 483156    5
## 3 483157    6
## 4 483158    4
## 5 483159    6
## 6 483160    2
```

If you `View()` the result you will see a list of all households and the number of people living there. This technique is a little clunky, as the variable names (`Var1` and `Freq`) are artifacts of using the `table()` command. It also is limited to counts. In this section we will learn the commands that are designed to flexibly conduct all types of aggregations in base R and `tidyverse`.

7.4.1 aggregate()

In base R the primary command for creating aggregations is, fittingly, `aggregate()`. The `aggregate()` command can be specified in a number of ways, but its simplest form is `aggregate(x~y, data=df, FUN)` where `data=` indicates the data frame we want to aggregate from, `y` is the unit of analysis to which we want to aggregate, `x` is the variable we want to use to describe that level of analysis, and `FUN` is the function we will apply to `x`. The `~` symbol—which we also saw in Chapter 5 when learning how to make facet graphs—is used in R to indicate one variable (`x`) being analyzed as a function of another variable (`y`). We will encounter it with some frequency from here on, especially in Part III.

Let us begin by replicating the example we did above with the `table()` command. Oddly, base R does not have a good function for counting things. There are a couple of workarounds, one of which is to use `tidyverse`, as we will see in Section 7.4.2. For now, the historical census data offer us another option because it has already counted out the number of individuals in each household with the `pernum` variable. Thus we can take the maximum value on this variable for each household (`serial`).

```
hh_people2<-aggregate(pernum~serial, data=census_1880, max)
head(hh_people2)
```

```
##   serial pernum
## 1 483155      2
## 2 483156      5
## 3 483157      6
## 4 483158      4
## 5 483159      6
## 6 483160      2
```

If you compare this product side-by-side with `hh_people`, which we created using the `table()` command, you will note that they are identical except the variables are a bit more explanatory (`serial` and `pernum`).

What else might we do? We could use the `age` variable to calculate the oldest person in the house. Before we do so, let us take a quick look at this variable.

```
summary(census_1880$age)
```

```
##      Length       Class        Mode
##       37209 character character
```

We seem to have a problem. R thinks that this variable is a character variable. We need to fix this (if the following lines of code seem odd, check out `table(census_1880$age)` to get a sense of how (and why) we are modifying the variable).

```
census_1880$age<-ifelse(census_1880$age=='90 (90+ in 1980 and
1990)','90',as.character(census_1880$age))
```

```
census_1880$age<-ifelse(census_1880$age=='Less than 1 year
old','0',as.character(census_1880$age))
```

```
census_1880$age<-as.numeric(census_1880$age)
```

Now that we have replaced the text values with numerical ones and then converted the class of the variable to numeric, let's try again.

```
hh_oldest<-aggregate(age~serial,data=census_1880,max)
table(hh_oldest$age)
```

```
##
##   5   6   7   8   9  10  11  12  13  14  15  16  17  18  19  20
##   5   6   7  10  16  25  19  24  19  23  26  23  16  26  30  25
##  21  22  23  24  25  26  27  28  29  30  31  32  33  34  35  36
##  36  46  64  48 122 107 114 135 148 250 134 176 139 159 340 185
##  37  38  39  40  41  42  43  44  45  46  47  48  49  50  51  52
## 166 240 176 514 102 189 154 134 390 136 128 172 112 464  81 152
##  53  54  55  56  57  58  59  60  61  62  63  64  65  66  67  68
## 115 128 233 140  87 109  78 332  67 101  97  85 182  66  74  78
##  69  70  71  72  73  74  75  76  77  78  79  80  81  82  83  84
##  57 174  40  48  50  51  55  35  26  30  27  56  14  18  13  12
##  85  86  87  88  89  90  91  92  94  96  98  99 103 108
##   9   4  12   8   4   9   5   4   1   1   3   2   1   1
```

The table here is far more intelligible, ordered numerically rather than textually. The most common age for the oldest member of a household is now 40 (514 households). There are still some odd cases of minors being the oldest member of the household, but far fewer. This may tell us something about the completeness of the census data, the way households were defined for orphans, or typographical errors. We are not going to probe this oddity any further, however. That said, if we want to, we can now use similar code to calculate the youngest person for each household.

```
hh_youngest<-aggregate(age~serial,data=census_1880,min)
```

7.4.1.1 aggregate() with a Logical Statement

There are a variety of extensions of the use of `aggregate()` we might pursue. The first is to use a logical statement. Suppose we want to know how many minors are living in a house. We first need to define 'minor' in our records as individuals younger than 19, or `hh_youngest$age<19`.

We can incorporate this statement into an aggregate command, like so

```
hh_children<-aggregate(age<19~serial,data=census_1880,sum)
```

Our logical command has created a dichotomous variable based on whether age is less than 19, and then summed over this variable for every household. Minors are equal to 1 and adults are equal to 0 on this variable, so summing effectively counts the number of minors. If you look at the names of the resulting data frame, however, you will see that the second variable has an odd name because it is drawn from the logical statement.

```
names(hh_children)
```

```
## [1] "serial"   "age < 19"
```

We will need to fix this.

```
names(hh_children)[2]<-'children'
```

We might do something similar for immigration status. Looking at the codebook, any value greater than 2 on Nativity indicates a person who immigrated to the United States.

```
hh_immigrant<-aggregate(Nativity>2~serial,data=census_1880,max)
names(hh_immigrant)[2]<-'immigrant'
```

By using the maximum function, we have now determined whether each household has at least one immigrant living there (i.e., the maximum of the dichotomous variable of whether each resident is an immigrant or not; if anyone is an immigrant, the maximum is 1, otherwise it is 0).

7.4.1.2 `aggregate()` Multiple Variables at Once

You may have noticed that we are creating a new data frame each time we run an aggregation. We can bring them together later with merging. Even more inefficient, however, is that we have used the same function on multiple variables. We could instead apply the same logic of aggregation to multiple variables using the `cbind()` command, which stands for "column bind," or connecting multiple columns together.

```
hh_maxes<-aggregate(cbind(pernum, age)~serial,
                    data=census_1880,max)
head(hh_maxes)
```

```
##   serial pernum age
## 1 483155      2  55
## 2 483156      5  42
## 3 483157      6  41
## 4 483158      4  65
## 5 483159      6  50
## 6 483160      2  22
```

This command creates a data frame with three columns: **serial**, **pernum** as the number of people living in each household (i.e., the maximum value for pernum), and **age** as the age of the oldest person in the household. Importantly, we have to use `cbind()` for this, not `c()`. If we used `c()` it would combine all values from the two columns into a single list that would be twice the length of the list of households (`serial`). This would generate an error (go try!).

7.4.1.3 aggregate() with Two Grouping Variables

Suppose we want to create an aggregation based on two different variables. In past chapters, for example, we might have wanted to aggregate on the combination of month and year, or day of the week and month. Here, we might want to keep the information about the ward as we aggregate from the individual to the household. We then need both variables in our `aggregate()` command. This takes the following form:

```
hh_people<-aggregate(pernum~serial+ward,data=census_1880,max)
head(hh_people)
```

```
##   serial ward pernum
## 1 483155    1      2
## 2 483156    1      5
## 3 483157    1      6
## 4 483158    1      4
## 5 483159    1      6
## 6 483160    1      2
```

We now have a new data frame with three variables (note that if you are following the code faithfully, we have overwritten the previous `hh_people` that we created with `table()` earlier): `serial`, the ward of each household, and the number of people living there. This works because ~ indicates that one variable is being treated as a function of one or more other variables. Thus, the function applied to `pernum` is now organized by `serial` and `ward`, meaning across all of their intersections. This is something of a trivial case because households are nested in wards, meaning each `serial` value only shares rows with one `ward` value. A month-year example would be somewhat more complicated as each month occurs in each year, and the product would be the series of all months across multiple years of data.

7.4.2 Aggregation with tidyverse

Aggregation can also be done in tidyverse using the combination of the `group_by()` and `summarise()` functions, which we have already encountered in Chapter 4. These must be combined through piping. `group_by()` instructs R to organize any subsequent analyses by a variable (e.g., serial) and `summarise()` then creates variables according to specified equations. If we used the `summarise()` command without `group_by()` we would generate one or more measures describing the entire data set. With `group_by()`, however, we generate each of those measures for all values of the grouping variable. (Remember that you will need to `require(tidyverse)` to proceed with the worked example.)

Let us replicate our first aggregation, which was to count the number of people in a household using `pernum`.

```
hh_people<-aggregate(pernum~serial,data=census_1880,max)
```

In tidyverse, this becomes

```
hh_people<-census_1880 %>%
  group_by(serial) %>%
  summarise(fam_size=max(pernum))
```

We first indicate the initial data frame (census_1880) and then group_by() the variable serial and summarise() based on the equation max(pernum). The result is the same, though with the added benefit of being able to name our variable in advance (fam_size). Recall, however, that I said that tidyverse makes it easier for us to do counts of cases. This is the purpose of the n() function.

```
hh_people<-census_1880 %>%
  group_by(serial) %>%
  summarise(fam_size=n())
```

The group_by() and summarise() approach can easily be extended to each of the examples conducted above with aggregate() and more.

7.4.2.1 group_by() + summarise() with a Logical Statement

Above we tabulated the number of children in each household and the presence or absence of immigrants in each household by incorporating logical statements into our aggregate() commands. We can do the same by using the logical statements in the summarise() command.

```
hh_children<-aggregate(age<19~serial,data=census_1880,sum)
```

becomes

```
hh_children<-census_1880 %>%
  group_by(serial) %>%
  summarise(children=sum(age<19))
```

7.4.2.2 group_by() + summarise() Multiple Variables at Once

We can also calculate multiple variables at once with the summarise() command. What was once

```
hh_maxes<-aggregate(cbind(pernum, age)~serial,data=census_1880,max)
```

is now

```
hh_maxes<-census_1880 %>%
  group_by(serial) %>%
  summarise(fam_size=max(pernum), age_oldest=max(age))
```

But why stop there? summarise() is not constrained by a single function. We can calculate as many equations as we want at once, bringing together everything we have done so far:

```
hh_features<-census_1880 %>%
  group_by(serial) %>%
  summarise(fam_size=n(),
            age_oldest=max(age),
            age_youngest=min(age),
            children=sum(age<19),
            immigrant=max(Nativity>2))
```

Take a look at the output data frame to see what we have created.

7.4.2.3 group_by() + summarise() with Two Grouping Variables

The last thing we did with aggregate() was to have multiple grouping variables, thereby conducting the aggregation over all of their intersections. The same can be accomplished by adding variables to the group_by() command, as in:

```
hh_people<-census_1880 %>%
  group_by(serial,ward) %>%
  summarise(fam_size=n())
```

We could even combine everything we have done to this point with the following:

```
hh_features<-census_1880 %>%
  group_by(serial, ward) %>%
  summarise(fam_size=n(),
            age_oldest=max(age),
            age_youngest=min(age),
            children=sum(age<19),
            immigrant=max(Nativity>2))
```

And with that one multi-part command, we now have created a single data frame that contains the list of all households (serial) and the number of people living there, the ages of the oldest and youngest residents, the number of children, and whether any residents are immigrants, as well as the ward containing the household.

7.5 Merging

Merging data sets on key variables is often an important complement to aggregation. During the early stages of this analysis, we generated data frame after data frame, each with the serial variable and one other descriptor. Of course, the flexibility of tidyverse allowed us to overcome this inefficiency, but that was only possible because we were aggregating from a single data set. If, as in the case of an IDS, we had been working with multiple data sources to generate a variety of descriptors of households, they would have necessarily sat apart. But because they all described households, they would (at least in a well-constructed schema) share the same key variable. We can use such key variables to merge data set together to combine all variables in a single data frame.

7.5.1 merge()

Conveniently, the primary command for merging data sets in base R is `merge()`, which requires us to specify two data frames and the variable by which they will be linked. For example:

```
households<-merge(hh_people,hh_children,by='serial')
names(households)
```

```
## [1] "serial"   "ward"     "fam_size" "children"
```

This merges `hh_people` and `hh_children` using the key variable `serial`. Thus, as we see, every value of `serial` will now have all variables from `hh_people` and `hh_children` side-by-side.

One limitation of `merge()` is that it can only handle two data frames at a time. If you have three or more data frames, you need to merge them iteratively. Let's bring in the immigrant status variable next.

```
households2<-merge(households,hh_immigrant,by='serial')
```

Something funny happened here, though, as we can see with the `nrow()` command.

```
nrow(households)
```

```
## [1] 8555
```

```
nrow(households2)
```

```
## [1] 8544
```

We somehow dropped from 8,555 cases to 8,544 cases. Why? It appears that `hh_immigrant` had 11 missing cases. If we dig further, there were 130 NAs in the `Nativity` variable, which means that those 11 missing households likely had NA records.

`merge()` has the default of only keeping rows present in each data frame. We can override this using the `all=` argument. In this case, we want to keep every row in the `x` data frame, or the first one we specified, so we use the `all.x=` version of the argument and set it equal to `TRUE`.

```
households<-merge(households,hh_immigrant,by='serial',all.x=TRUE)
```

Depending on our goals, we can keep all rows in the `y` data frame, or the second one we specified (`all.y = TRUE`), or keep all rows in both data frames (`all.x.y = TRUE`).

Because `merge()` is sensitive to the particular composition of a data set and its values, there are a few other cautions and considerations that should be taken whenever merging. First, make sure that there are no duplicates on your key variables. If there are, the merge will keep these duplicates. If there are duplicates in both data frames, the final data frame will not only keep them but create new rows for all possible combinations (e.g., 3 duplicates of a single value in each data frame would generate 3*3 = 9 rows in the final data frame). With a large data set this can create an explosion of rows.

Second, R cannot handle duplicate variables. The command is designed to not duplicate the key variable in the final data frame, but it has to deal with any other variables repeated in both data sets in another way. It does so by adding the suffixes `.x` and `.y`. This can create issues with downstream code if you assume that all variable names remained the same.

Third, it is possible to merge data frames that have shared key variables but with different names. For instance, if we had renamed the variable serial to be household in one of our data frames, we could have still merged it with other data frames by specifying `by.x = 'serial'` and `by.y = 'household'`. In such situations, though, make sure that you are confident that the variables really are the same, or the merge will either not find any matches or run into the issue of duplication as noted above.

Fourth, it is also possible to merge on more than one key variable. This can come in especially handy if you have aggregated on multiple variables, as we did with serial and ward. This can be accomplished using `by = c('serial','ward')`.

7.5.2 `join` functions in `tidyverse`

`tidyverse` offers the family of `join` functions as an analog to `merge()`. There is in fact no `join` function, but multiple functions with different names that correspond to the various settings of the all argument.

- `inner_join()` reflects the default merge(), keeping only rows that occur in each data frame.
- `left_join()` corresponds to the all.x variant of merge(), keeping all rows in the first data frame specified.
- `right_join()` corresponds to the all.y variant of merge(), keeping all rows in the second data frame specified.
- `outer_join()` corresponds to the all.x.y variant of merge(), keeping all rows in both data frames.

In each case, the required arguments are the two data frames and the key variable. (`join` functions, like `merge()`, cannot handle more than one data frame at a time, in part because it would be logically difficult if not impossible to specify which rows to keep if each of three or more data frames had different sets of rows.) Thus,

```
households<-merge(hh_people,hh_children,by='serial')
```

becomes

```
households<-inner_join(hh_people,hh_children,by='serial')
```

and

```
households2<-merge(households,hh_immigrant,by.x='serial')
```

becomes

```
households2<-left_join(households,hh_immigrant,by='serial')
```

As you can see, `merge()` and the `join` family of commands are nearly identical and there are few if any objective advantages to one over the other. As such, it is simply a matter of preference and whether you tend to more easily remember the logic of the all argument or the different members of the `join` family of functions.

7.6 SQL and the sqldf Package

SQL is one of the most common softwares used for database management. Recall from the discussion of schemas that a database is consists of multiple interlocking data sets, also referred to in database terminology as "tables." SQL is a rather simple, elegant coding language used to "query" a database and extract the desired cases. This is because databases are often constructed to store very large amounts of data that you would not want to work with in their entirety—this was especially true decades ago when databases were first invented and hard drives were measured in megabytes instead of gigabytes.

SQL has been incorporated into R with the `sqldf` package (Grothendieck, 2017). A SQL statement is very simple, requiring only three components: `select` indicates the variables that you want; `from` indicates the table (or, in our case, data frame) you are drawing them from; and `where` articulates any criteria for the rows you want (like a subset in base R or `filter()` in tidyverse; technically, `where` is optional if you do not want to set any criteria). Importantly, there are no commas between commands, only between elements within commands (e.g., between multiple variables specified in the select command). We are going to learn these basic elements of SQL as well as a fourth (but optional) component of a SQL command that makes it possible to create aggregations: `group by`.

The main advantage of working with SQL to aggregate data rather than `aggregate()` is the ability to create multiple measures with different functions at once. However, with the advent of `tidyverse`, this is no longer as distinctive. I include it here, though, because it provides an excuse to introduce you to a skill that you might develop further if you want to go into database management at some point.

7.6.1 Aggregation with SQL

We will use SQL to illustrate a second level of aggregation from households to wards. To do so we will work from the `hh_features` data frame that we created in our last and most complete aggregation in the `tidyverse`. Before we can do that, though, we need to recode our integer variables to numeric variables because `sqldf` cannot handle integer variables.

```
hh_features$immigrant<-as.numeric(hh_features$immigrant)
hh_features$fam_size<-as.numeric(hh_features$fam_size)
hh_features$children<-as.numeric(hh_features$children)
```

`sqldf` uses a single command, `sqldf()`, into which a full SQL command as entered as text bounded by single quotation marks. To get started, let us first see how a simple SQL command works. We might, for example, want to look quickly at a list of only household serial numbers and their wards.

```
require(sqldf)
options(scipen=100)
head(sqldf("select ward, serial
            from hh_features
            "))
```

```
##   ward serial
## 1    1 483155
## 2    1 483156
## 3    1 483157
## 4    1 483158
## 5    1 483159
## 6    1 483160
```

Note that we have entered this without storing it in a new data frame.

We could also look at all variables for a single ward:

```
head(sqldf("select *
            from hh_features
            where ward = 5
            "))
```

```
##   serial ward fam_size age_oldest age_youngest children
## 1 484456    5        4         51            1        1
## 2 484457    5        3         46           10        1
## 3 484458    5        5         61           12        1
## 4 484459    5        5         68           25        0
## 5 484460    5        3         53            9        1
## 6 484461    5        4         41            3        2
##   immigrant
## 1         0
## 2         1
## 3         0
## 4         1
## 5         1
## 6         0
```

Note that entering * after select chooses all variables, in this case for all households in ward 5. Now we are ready to use SQL to create an aggregation. Instead of printing out the result we will now create a new data frame containing all of the aggregations for our wards.

```
wards<-sqldf("select ward,
            sum(fam_size) as population,
            count(serial) as households,
            sum(immigrant)/count(immigrant) as prop_imm_hh,
            sum(fam_size)/count(fam_size) as avg_hh_size,
            sum(children)/count(children) as children_per_hh
            from hh_features
            where immigrant != 'NA'
            group by ward")
```

Let's break this down. First, we selected ward followed by a series of newly calculated variables. This highlights two special features of SQL: (1) you can calculate new variables on the fly, whether for aggregations or otherwise; and (2) you can name them flexibly using `as`. Thus, `sum(fam_size) as population` is identical to `summarise(population=sum(fam_size))` in `tidyverse`. Also, SQL, unlike base R, has a `count()` function we can call upon for counting. We have thus calculated for each ward: the total population; the number of households; the proportion of households with at least one immigrant resident (`prop_imm_hh`) as the sum of our dichotomous immigrant variable divided by all households (calculated with `count()`); the average household size (`avg_hh_size`), again calculated with a `sum()` divided by a `count()` because SQL has no internal command for calculating means; and the average number of children per household (`children_per_hh`). If you are feeling up to it, see if you can replicate these equations in `tidyverse`.

After `select`-ing all of our variables, we had to indicate that we were generating them from the `hh_features` data frame. We also limited to cases in which immigrant was not equal to NA, as this would have created issues for SQL. We then grouped the newly calculated variables (`group by`) by ward. Importantly, I have taken a lazy shortcut here for the sake of illustration. It is, of course, necessary to remove households with no clear immigration status if we are calculating variables related to immigration status; this is indeed the purpose of `na.rm =` in most statistical functions. However, there is no reason we need to drop these 11 households when calculating things like average household size or average number of children. We have done so here only because it is far more efficient to demonstrate the tool in one fell swoop rather than rerun multiple aggregations. That said, you are welcome to do this the right way on your own.

7.6.2 Merging with SQL

SQL is a comprehensive language for managing the tables contained in one or more databases. It of course can do much more than just select variables and rows and aggregate them. One of its other capacities is merging. However, merging in SQL is a little bit more complicated than in R, and `sqldf` does not support it. We will pause in our learning of `sqldf` here, but there are plenty of resources for going further if you would like to do so.

7.7 Creating Custom Functions

As described numerous times, R's flexibility and open-source nature allow analysts to easily develop new functionality and packages. We have not yet discussed how this is done, however. The fundamental building block of development in R is the creation of custom functions, and it is quite straightforward. This can come in quite handy when generating aggregations because they depend on the precise function that we want to apply to our records to then describe the higher unit of analysis.

We can illustrate by creating a function for identifying the mode of the distribution (i.e., the most common value). Interestingly, R does not have a function for this despite it being commonplace in statistics. We can create such a function ourselves.

```
mode <- function(x) {
  ux <- unique(x)
  ux[which.max(tabulate(match(x, ux)))]
}
```

We have now made an object (you can see it in your Environment) called `mode`. That object is a function as defined in the two lines of code contained in `{}`. `function(x)` indicates that we are creating a function that operates on a single input, `x`. The function itself then creates a temporary vector, `ux`, that is a list of all unique values in `x`. The following line tabulates the number of matches between `x` and each value in `ux` and selects the value with the most such matches (`which.max()`), which is the functional definition of a mode.

Let's see how it works in practice. In the historical census data, birthplace is recorded for each individual. It might be interesting to know what the most common birthplace is for each household and see how it is distributed across the city.

```
hh_mode_bpl<-aggregate(bpl~serial,data=census_1880,mode)

data.frame(table(hh_mode_bpl$bpl)) %>%
  arrange(desc(Freq)) %>%
  head()
```

```
##            Var1 Freq
## 1 Massachusetts 4636
## 2       Ireland 1496
## 3         Maine  517
## 4        Canada  454
## 5 New Hampshire  251
## 6       England  207
```

A quick look at the values across households reveals that for the majority the most common birthplace of household members is Massachusetts, but that for nearly 20% the most common birthplace is Ireland. This is then followed by Maine, Canada, New Hampshire, and England. This is consistent with the large influx of Irish immigrants to Boston in the second half of the 19th century.

7.8 Bivariate Visualizations

We now have a data frame with a series of numerical variables describing the population of each of Boston's wards in 1880. This makes for a convenient excuse to expand our knowledge of visualization in `ggplot2` to work with bivariate plots. We have featured more than one variable in a graph previously, but it has always involved categorical variables that split up a univariate graph in one way or another (e.g., stacked or faceted histograms). For the first time we will work with the full variation of multiple numerical variables with the `geom_point()` and `geom_density2d()` commands.

7.8.1 geom_point()

Recall that our data frame of wards contains population, number of households, average household size, average number of children, and proportion of households with immigrant residents. We can examine the association of any two of these variables with a dot plot, created with the `geom_point()` command. For example, does the number of children per household increase with average household size? First we establish our base `ggplot()` command.

```
size_w_child<-ggplot(wards,
              aes(x=children_per_hh, y=avg_hh_size))
```

and then we can add on the `geom_point()` command.

```
size_w_child+geom_point()
```

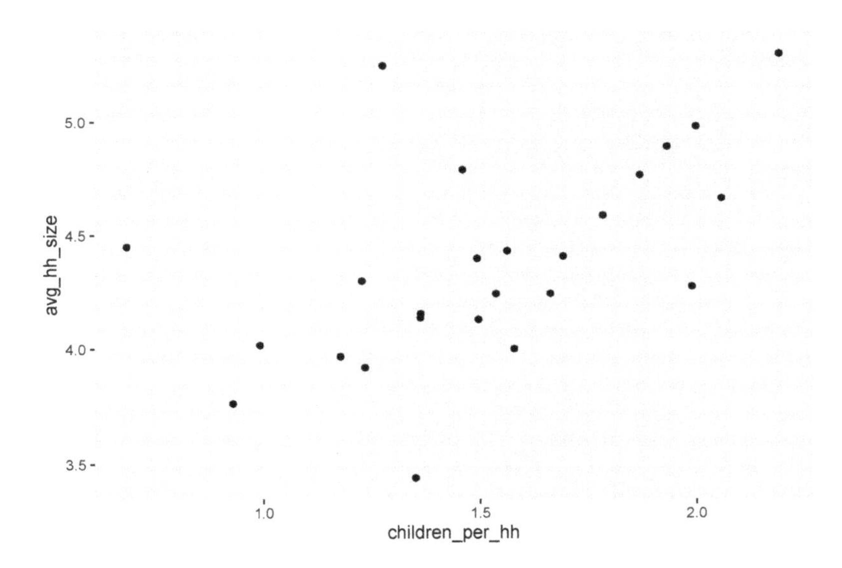

Unsurprisingly, there is a very strong correlation. As a community's households have more children on average, the average household size goes up. There are a few outliers that violate this relationship, likely because of different norms of how many adults live together (or can afford not to). Of course, we can use `ggplot`'s broad functionality to make this a little prettier.

```
size_w_child+geom_point(colour = 'blue', shape = 0, size = 5) +
  labs(x='Children per Household',y='Avg. Family Size')
```

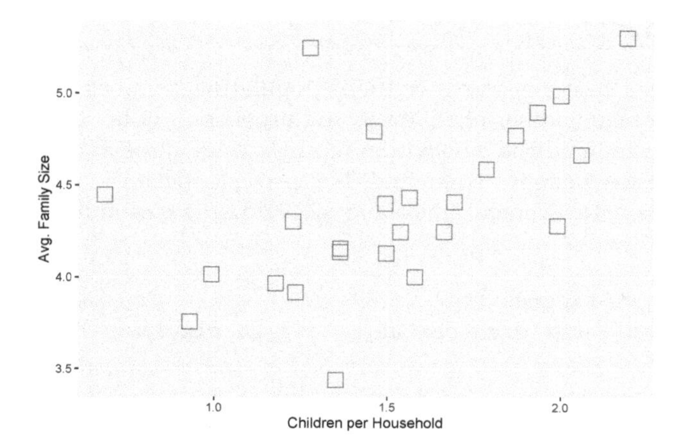

Possibly more interesting would be whether communities with more immigrant households have more children on average, suggesting that immigrant households tended to have more children in 1880 Boston.

```
child_by_imm<-ggplot(wards,
              aes(x=prop_imm_hh, y=children_per_hh))
child_by_imm+geom_point(color = 'blue', shape = 0, size = 5) +
  labs(x='Proportion of Immigrant Households',
  y='Children per Household')
```

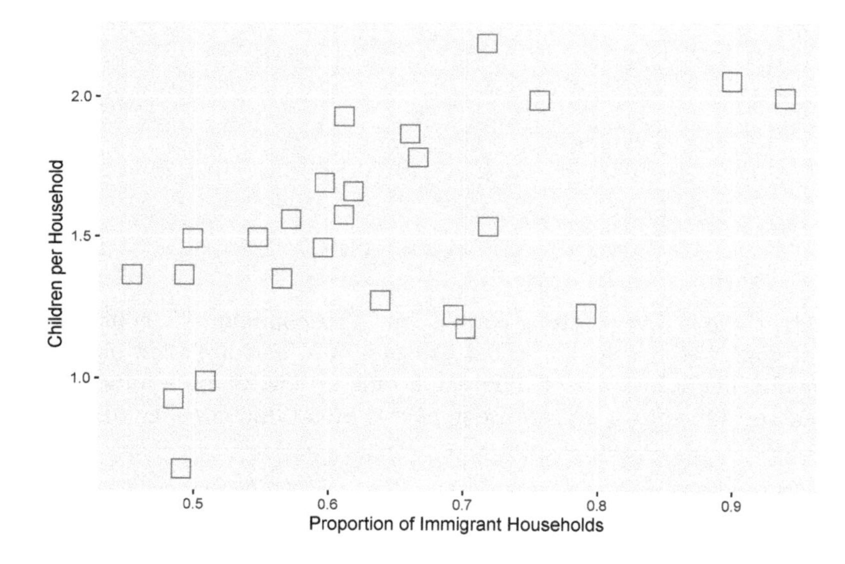

Again, we see a substantial correlation, suggesting that immigrant households tended to have more children than non-immigrant households.

7.8.2 geom_density2d()

In Chapter 4 we learned about the command geom_density() to make density plots as an alternative to histograms. Density plots offer smoother estimates of the density of data across the distribution. The same logic can be applied to bivariate distributions as an alternative to dot plots.

```
size_w_child+geom_density2d(color = 'purple') +
  labs(x='Children per Household',y='Avg. Family Size')
```

One nice aspect of a density plot is that it can make the correlation (or lack thereof) between two variables clearer. As we see here, the orientation of the gradient is diagonal, an observation that can sometimes be difficult to identify when distracted by outliers.

We can do the same for our second bivariate comparison as well.

```
child_by_imm+geom_density_2d(color = 'purple') +
  labs(x='Proportion of Immigrant Households',
       y='Children per Household')
```

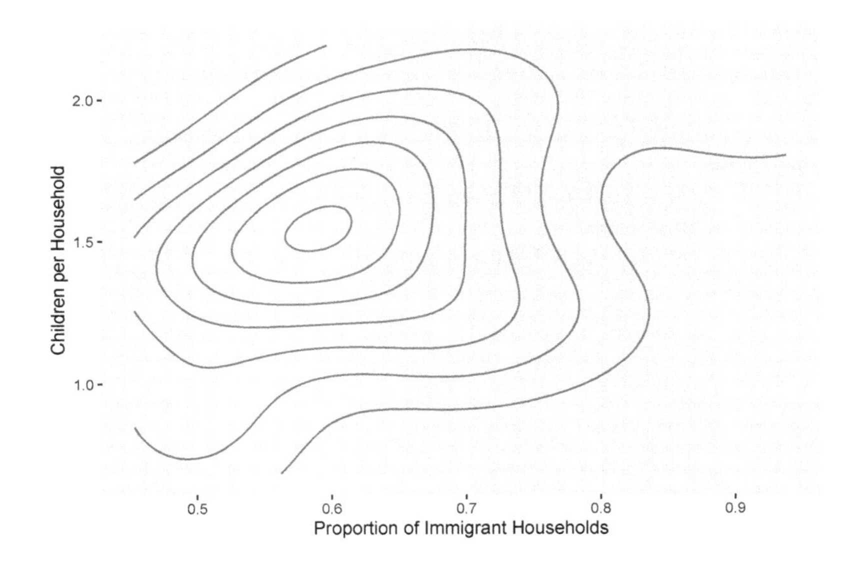

Again, the gradient is on a diagonal, though we can see the curve on the far right accommodating the outliers with extremely high proportions of immigrant households.

7.9 Summary

By aggregating records and merging the products, we created a data frame with a series of custom-designed quantitative measures describing a given unit of analysis, in this case the wards of Boston in 1880. While this might seem a somewhat obvious statement, it is actually the first time in this book that it is true! Previously, we have worked with data at the record level dominated by categorical variables and the occasional quantitative descriptor, be it temperature or precipitation during bike collisions or the price of apartments posted on Craigslist. Here instead we took census records and created measures to describe the population of each of Boston's wards. The measures were of our design and reflected things that we wanted to know-—and therefore that we could easily interpret and describe. This is an important step in the ability to work with and communicate the content of any set of records.

In the process, we have developed the following skills, all of which are applicable to any aggregation effort, be it from records to individuals, individuals to households and neighborhoods, or otherwise:

- *Aggregate records to higher units of analysis* by using the `aggregate()` command in base R, including:
  - *Incorporating logical statements*,
  - *Aggregating multiple variables* with the same function,
  - *Aggregating on multiple key variables at once*;
- *Aggregate records to higher units of analysis* by using the `group_by()` and `summarise()` commands in `tidyverse`, including:
  - *Incorporating logical statements*,
  - *Aggregating multiple variables* using a variety of functions and equations,
  - *Aggregating on multiple key variables at once*;
- *Merge separate data frames on shared key variables* using the `merge()` function in base R;
- *Merge separate data frames on shared key variables* using the family of `join` functions in `tidyverse`;
- *Query and aggregate records* from a data frame (or database) with SQL language, through the `sqldf` package;
- *Create custom functions* that could then be called in aggregations or other commands requiring statistical functions;
- *Visualize bivariate relationships* between numerical variables using `geom_point()` and `geom_density2d()`.

7.10 Exercises

7.10.1 Problem Set

1. Suppose you are working with the data set used for the worked example in this chapter. Write code for creating each of the following aggregations in base R, `tidyverse`, and SQL. Briefly explain the logic for each.
   a. Proportion of immigrants in ward.
   b. Female-headed household.
   c. Proportion of female-headed households in a ward.
2. You have two data frames of census blocks: `blocks` and `cblocks`. The first has the unique identifier for each block as `Blk_ID`. The second has the same information in `FIPS`, which is a common label in census data.
   a. Write code to merge these data frames. Briefly explain your reasoning.
   b. The first has blocks for the whole state, the other is just greater Boston. How does your merge generate each of the following outcomes:
      i. All census blocks.
      ii. Only the census blocks in greater Boston.
3. You have two data frames with `Blk_ID`, each with 100 rows. You merge them and get 178 rows. How might this have happened? How would you attempt to fix it?

7.10.2 Exploratory Data Assignment

Working with a data set of your choice:

1. Create new measures based on aggregations of your data. If you are working with the same data set across chapters, you can use this as an opportunity to begin calculating the latent constructs that you proposed in the Exploratory Data Assignment in Chapter 6.
2. Be sure to describe why each measure is interesting and how the specific calculations you are using capture it. In addition, if these are intended as manifest variables of one or more latent constructs, describe how they are reflective of this underlying construct.
3. Describe the content of these new variables statistically and graphically using tools we have learned throughout the book. Be sure to include at least one dot plot that shows the relationships between two variables at the aggregate level.

8

Mapping Communities

In 1854, there was a cholera outbreak in London. At the time, most doctors believed that cholera (and other diseases) were spread through a "miasma," or a noxious bad air. John Snow, an obstetrician and anesthesiologist, thought otherwise, reasoning that (a) there was no evidence for miasma, and (b) contaminated water made a much simpler explanation that was consistent with his anecdotal observations. The outbreak gave him an opportunity to make his case in a more systematic fashion.

Snow mapped the home address of each of the recorded deaths stemming from the cholera outbreak, as seen in Figure 8.1. In doing so he revealed a clear pattern. The deaths were spatially clustered around a single water pump on Broad Street. Further, when he drew a line (faintly visible in the map) around the households that were nearer to that water pump than any other—thereby making them most likely to access water from it—it contained almost all of the deaths. This was strong evidence that the water pump was the source of the cholera outbreak and, in turn, for Snow's theory that cholera was transmitted by contaminated water.

The story of Snow's map, the contaminated Broad St. water pump, and the transmission of cholera is one of the iconic illustrations of the power of spatial data. Sometimes, simply plotting points on a map can communicate intricate patterns and reveal important relationships and mechanisms. And we have come a long way since 1854, with sophisticated file formats, software, and analytic tools for examining and communicating spatial data. These Geographical Information Systems (GIS) can be applied to nearly any question. Researchers at Columbia University mapped incarcerations in New York City, identifying concentrations that were literally costing the state millions of dollars, or what they referred to as "million dollar blocks" (Kurgan et al., 2012). Mike Batty and his team at the University College London's Centre for Advanced Spatial Analysis have digitized the road network of London going back to 1786, thereby modeling how the extension of main roads and the development of side roads resemble the growth of vein networks in leaves (Masucci et al., 2013). Recent advances in participatory GIS have enabled community organizations to crowdsource the ways that residents perceive their space, reconfiguring maps to reflect the experiences of those who live there. And NASA and others have used sensors on planes and satellites to map ground conditions, from estimated population to surface temperature to the density of trees.

Working with GIS is a distinctive skill because the data have a different structure. Think about it. Instead of just a list of objects (i.e., rows) and their attributes (i.e., columns), we now need all of the information necessary to place those objects on a map. For reasons we will get into, that can be rather complicated to do, and thus GIS comprises a variety of specialized tools for solving these challenges. In this chapter, we will learn a bit more about the data structures used to organize spatial information, known as *shapefiles*, and how to work with them.

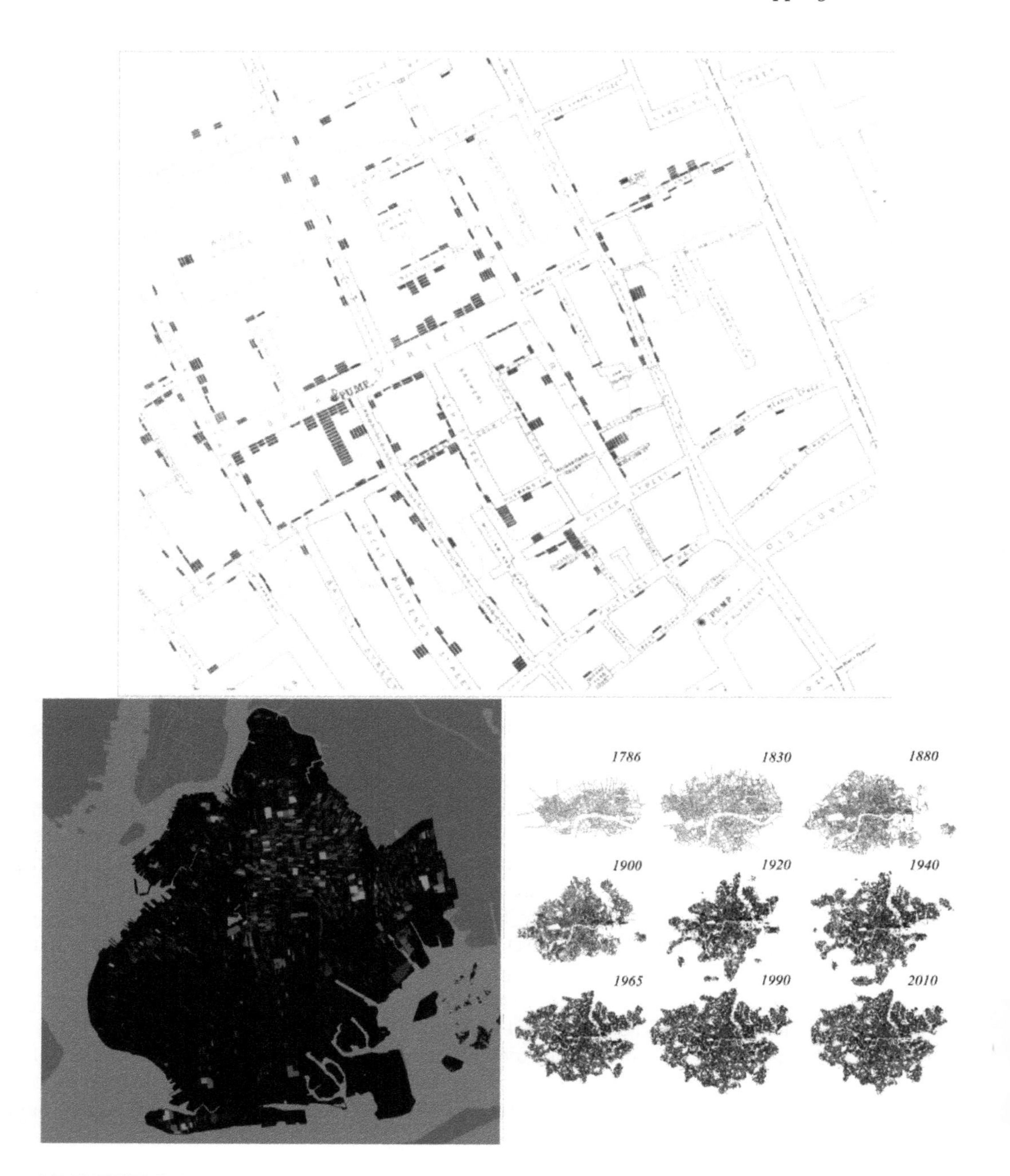

FIGURE 8.1
Mapping spatial data can illustrate a wide range of phenomena, from John Snow's demonstration that a contaminated water pump was responsible for a cholera outbreak in London in 1854 (top), to more modern applications, like the identification of blocks in Brooklyn, New York where more than 1 million dollars were spent on the incarceration of residents (bottom left), to the examination of the evolution of London's road network over the centuries (bottom right). (Credit: John Snow, Columbia University's Center for Spatial Research's, University College London's Bartlett Centre for Advanced Spatial Analysis)

8.1 Worked Example and Learning Objectives

In this chapter, we will learn about *shapefiles*, the primary data structure for organizing spatial information, and the skills necessary to work with them in R. To do so, we will leap from last chapter's example of historical census data to modern census data, mapping the demographics of Boston today. This will leverage the Boston Area Research Initiative's Geographical Infrastructure for the City of Boston (see Chapter 6 for more on the schema), specifically the scale of census tracts (approximations of neighborhoods with ~2,500 residents). We will learn to:

- Describe a shapefile and why spatial data requires a unique structure;
- Identify the multiple types of spatial data;
- Import and manipulate a shapefile in R using the `sf` package;
- Make a map using `ggmap`, a companion package to `ggplot2`;
- Utilize point data, whether in a spatial form to start with or not;
- Connect to other tools for spatial analysis outside of R.

Links

Shapefile for tracts: https://dataverse.harvard.edu/dataset.xhtml?persistentId=doi:10.7910/DVN/SQ6BT4

Census indicators: https://dataverse.harvard.edu/file.xhtml?persistentId=doi:10.7910/DVN/XZXAUP/IDL8OA&version=3.0

You may also want to familiarize yourself with the data documentation posted alongside each of these datasets.

Data frame name: `tracts_geo`

8.2 Intro to Geographical Information Systems

Spatial data are unique for the very reason that they contain spatial information. This creates a variety of complications that Geographical Information Systems (GIS) software need to account for. In this section we will learn more about what these complications are and their implications, the types of spatial data that we typically encounter, and `sf`, the R package designed to work with them (Pebesma, 2021).

8.2.1 Structure of Data

Compare, for a moment, the two images in Figure 8.2. On the left side, we have a spreadsheet with a list of the states in the United States, including their names and attributes like size, population, etc. Alongside it, we have a simple map of the 48 contiguous states. Visually, the image on the right is easier to digest. We see all of the states at once, their relative sizes are more or less apparent, and we could even recolor it to reflect differences in population, or any other attribute that our hearts desire. But a lot goes into making that map. How, for instance, does the software know all of the detailed curves that define the shape of each state? How does it know where to place them? And how is it still capable of knowing all of the attributes contained in the spreadsheet?

STUSPS	NAME	LSAD	ALAND	AWATER
DC	District of Columbia	00	1.583404e+08	18687198
NC	North Carolina	00	1.259237e+11	13466071395
UT	Utah	00	2.128862e+11	6998824394
ND	North Dakota	00	1.787075e+11	4403267548
SC	South Carolina	00	7.786492e+10	5075218778
MS	Mississippi	00	1.215335e+11	3926919758
CO	Colorado	00	2.684229e+11	1181621593
SD	South Dakota	00	1.963470e+11	3382720225
OK	Oklahoma	00	1.776629e+11	3374587997
WY	Wyoming	00	2.514585e+11	1867670745
WV	West Virginia	00	6.226647e+10	489028543
ME	Maine	00	7.988743e+10	11746549764

FIGURE 8.2
The information contained in a spreadsheet describing the states of the United States (left) can be more easily communicated through a simple map of them (right).

Describing spatial objects requires the coordination of two sets of information. First, it relies on the traditional spreadsheet with rows (cases) and columns (variables). The variables in this case are called the "attributes" of the spatial objects. Second, it requires information on the spatial placement and shape of those objects. As we will see, this information can be more or less complicated depending on the type of spatial information. For an example like the states, plotting their shapes entails the encoding of lots of irregular edges, which consists of lots of detail that would be difficult to put in a traditional spreadsheet. For instance, suppose the border of one state is defined by a line that connects 25 different points (or vertices) and another requires 40 different points (or vertices). Would the spreadsheet have 40 additional columns for spatial information, with some columns empty for certain rows? If so, how would we know when creating the file how many columns were necessary? Would all vertices be contained as a list in a single column? If so, how would we be able to access each point individually?

This complexity is solved by the *shapefile*, which has the suffix *.shp*. In fact, a shapefile is not a single file, but a composite of complementary files describing the same data, including:

- *.shp* catalogs all of the objects and their vertices.
- *.dbf* is a specialized form of the traditional spreadsheet or data frame, with all of the objects and their attributes.
- *.shx* is an index of all the objects, permitting coordination between the .shp and .dbf.

Each of the files must have the same name, differing only in the file type (e.g., *Tracts.shp*, *Tracts.dbf*, and *Tracts.shx*); otherwise the software will not be able to coordinate them. Often there are numerous other files in addition to these three required elements, many of which provide extra detail and enhance the processing of the data by GIS software and the precision with which they can be analyzed and mapped. One of these is the *.prj*, which encodes the coordinate projection, something we are about to learn more about.

8.2.2 Coordinate Projection Systems

The world is round. Maps, at least ones that we print on paper or create on our computer screens, are typically flat. How do we deal with that? This is the job of the coordinate projection system (CPS), which uses a fixed equation to translate the curvature of the Earth to a flattened approximation. This generally works pretty well, though it has its weaknesses. First, the further we zoom out, the more problematic the curvature becomes. To illustrate, we can generally see about 3 miles into the distance before things fall beneath the horizon—that is to say, things more than 3 miles away are not in our line of sight because the curve of the Earth's surface hides them from our view. Thus, CPSes are always going to be an approximation.

Take, for instance, what happens when we try to generalize a CPS to the entire planet. This can be seen when we apply different CPSes to maps of the entire world, as in Figure 8.3. The Mercator Projection with which we are most familiar tends to exaggerate land near the poles and shrink land near the equator, making Antarctica look gigantic and Africa undersized relative to their actual land masses. Part of the issue here is that, as we zoom out, we are forced to make more and more of an approximation that violates the three-dimensional reality of how places are situated along the Earth's curved surface.

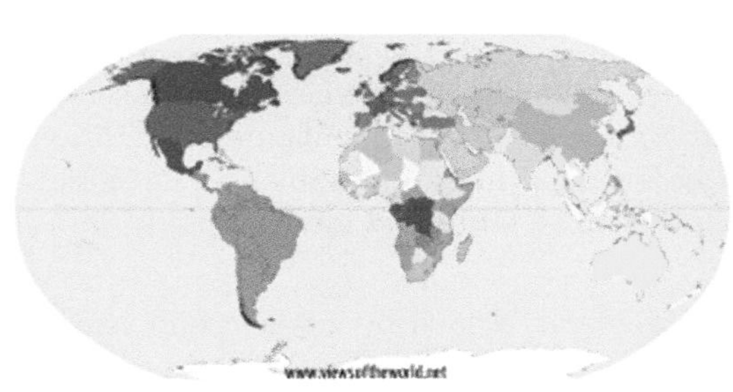

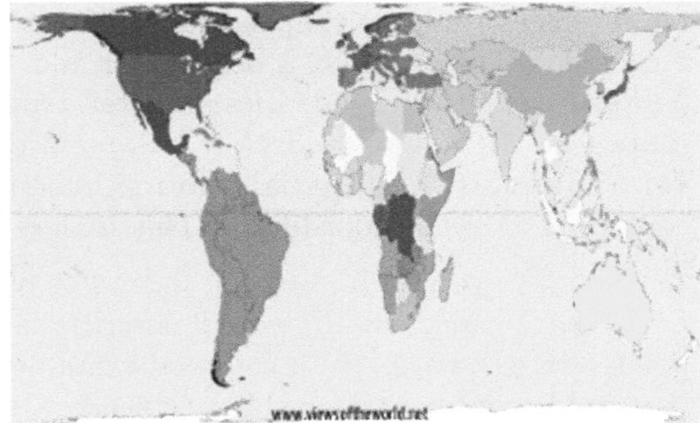

FIGURE 8.3
Most people are familiar with the Robinson projection for representing the world (left), which exaggerates the size of land near the poles and makes land near the equator look smaller. New techniques have been used to better represent the true land area of countries (right). (Credit: www.viewsoftheworld.net)

The existence of curvature alone is not the entirety of the problem. A further complication is that the curvature of the earth is not consistent across places but varies by where you are. The curve of the ground follows a gradient from horizontally flat at the poles and perfectly vertical at the equator. Consequently, a single CPS is never going to provide a perfect translation between the three-dimensional reality of space into a two-dimensional approximation for all places. We need different CPSs for each region. There are CPSs maintained by the United States Geological Survey that are reasonable approximations for mapping the entire United States. There are also distinct CPSs maintained for each state that are more precise for where they are located. Other countries maintain their own CPSs.

In nearly all cases, spatial data that you access will come with a standardized CPS. Rather than create one yourself, you can lean on the hard work of others before you. What can be tricky, though, is that not all generators of spatial data use the same projection system. For instance, you might find one shapefile of data for Boston uses a Massachusetts-specific CPS and another uses a national CPS. This will often create minor discrepancies that seem trivial, but will make maps look ugly (e.g., streets being 20 feet out of line and covering images of buildings) and analyses imprecise. It can also create rather bemusing (or confounding) mismatches, like the time I accidentally mapped a bunch of 311 reports occurring in Boston to the middle of Angola. As we will see, though, because each CPS is simply a set of equations translating longitude and latitude to a two-dimensional plane, GIS software can translate between them seamlessly.

8.2.3 Types of Spatial Data

There are four main types of spatial data that you are likely to encounter: points, lines, polygons, and rasters. As summarized in Figure 8.4, these are categorized by the amount of information needed to define them and, consequently, how GIS software needs to organize that information. All are based on one or more coordinates, each one including a latitude and longitude or other way of placing an item on a two-dimensional, x-y plane. A given *.shp* can only contain objects of one of these types.

Points. Each object in a point .shp describes a discrete location with a single coordinate (x, y). Examples include events that are mapped to a single address (311, 911 calls). Also, many places that are shapes in real life can be abstracted as points if one is zoomed out far enough. For example, a building or address can often be represented as a point, and on national maps towns and cities are often represented as points, despite having more complex shapes. As we will see in Section 8.5, point data are the easiest to reconfigure as traditional data frames as their spatial information can be encoded in two variables—-the x and y components of the coordinate. This makes it possible to store and analyze point data in both forms.

Lines. Each object in a line .shp describes a line segment defined by a start point and end point $\{(x_1,y_1), (x_2,y_2)\}$. A common example of this is a street map. It is important to note that in GIS, a long street is not typically a single object. It is a series of line segments that run from intersection to intersection (or from intersection to dead end), defined by each of those end points. It is possible, though, to use attributes like a name column to identify every street segment that composes, say, Main Street.

Polygons. Each object in a polygon .shp describes a two-dimensional shape with both length and width. These are defined by a series of three or more vertices, placed in the order in which they are connected, much like a dot-to-dot activity. This can be represented as $\{(x_1,y_1), \ldots, (x_1,y_1)\}$. An example of this is the states in Figure 8.2. We will also work primarily with this type of shapefile in the worked example of census tracts here.

Raster. Rasters are a special type of spatial data that organizes a landscape into grid cells, giving a value to each of one or more attributes. Rasters are popular for data collected by sensors on satellites or planes that try to describe conditions continuously across space. Interestingly, they do not actually need to be in the *.shp* format but can be represented in *.jpg* or *.gif* formats. This is because the grid is all the spatial information that needs to be described, and a graphic format is simply a grid. We will not work much with raster data in this book except in the form of "base maps," which are pre-constructed maps that have been exported as rasters that we can then place our own information on top of. For example, if we use a base map provided by Google Maps to put context around the locations of Craigslist postings, the Google Map is a raster.

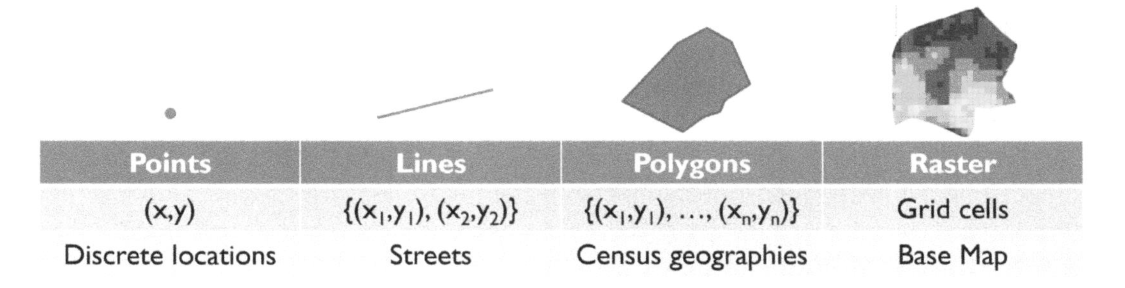

FIGURE 8.4
A brief summary of the four main types of spatial data, including the quantity of geographic points needed to draw them and illustrative examples.

8.2.4 Making a Map: Layering *.shps*

Each shapefile contains the spatial and attribute information for a clearly defined set of objects. For instance, returning to BARI's Geographical Infrastructure for the City of Boston, there are *.shps* for addresses, streets, multiple census geographies, and multiple administrative boundaries. Most often, we want to create a map of the city that includes two or more of these. For instance, in a standard map of the United States, we want to see the states as polygons, but also capital cities as points, major highways, and large rivers and lakes as lines. Each of these elements is stored in its own .shp and can be visualized in a variety of ways, depending on its attribute data. How we choose to visualize each is called a *layer*, and a map consists of multiple layers. In the sections that follow, we will see how we import *.shps* into R and then layer them to create what we would recognize as a map.

8.3 Working with Spatial Data in R

8.3.1 The `sf` Package

GIS software entails a series of specialized functions for working with shapefiles, including coordinating the multiple files that compose them, representing them graphically, and analyzing their content. The most common software for doing so is ArcGIS, which is built by the company ESRI. ESRI is one of the pioneers in spatial data analysis and is responsible for designing the current *.shp* file format. Most municipalities, for instance, have a license for ESRI software. ESRI provides powerful tools for working with spatial data. In recent years, the opensource movement has given rise to Quantum GIS (or QGIS), a freeware alternative to ArcGIS that has many (but not quite all) of the same tools and functions—it is even designed to look nearly identical. Unless you have specific, advanced needs, like the creation of three-dimensional maps and animations, you could probably do your work successfully in QGIS.

R also has the capacity to work with spatial data thanks to multiple packages designed for that purpose. The most popular currently is `sf`, which stands for "simple features." The `sf` package, as its name implies, has streamlined the way R handles spatial data, making the code for working with spatial data more straightforward and the computational tasks more efficient than previous packages. Both of these advances are good for analysis and that is why we are going to focus on `sf` in this chapter. To be certain, ArcGIS and QGIS have more tools and are better equipped to generate highly customized, attractive maps. `sf` can do the job effectively, however, and it allows you to keep your data analysis and spatial visualization in a single platform and syntax. You may find as you develop more skills, however, that you pick and choose between these softwares depending on the circumstance.

8.3.2 Importing Spatial Data into R

To get started with a spatial data set in R, we first need to import it, which requires a specialized function for reading in *.shps*. The `st_read()` function in sf is able to import *.shp* files into R, analogous to `read.csv()`. When we run:

```
require(sf)
```

```
tracts_geo<-st_read("Unit 2 - Measurement/Chapter 08 - Mapping
Communities/Example/Tracts_Boston_2010_BARI.shp")
```

we are notified that this is an ESRI shapefile with 178 features (observations) and 16 fields (though it is listed in the Environment as having 17 variables, the last of which is the spatial information), and its features are polygons. R also tells us what its "bounding box" is, or the coordinates that, if used to create a rectangle, would contain all of the objects, and that the CRS, or coordinate reference system (a synonym for CPS), is NAD83, which was designed for data in North America in 1983.

If we want to get to know the data further, we can `View()` it or examine the `head()`. If you scroll over to the far right, you will find the last column, geometry, contains a list of coordinates for each row. These are the vertices describing the polygon. We can use the `str()` command, as we did in Chapter 3 when looking for initial information about a data set and its variables. We also can learn more about the projection with `st_crs()`.

```
st_crs(tracts_geo)
```

```
## Coordinate Reference System:
##   User input: NAD83
##   wkt:
## GEOGCRS["NAD83",
##     DATUM["North American Datum 1983",
##         ELLIPSOID["GRS 1980",6378137,298.257222101,
##             LENGTHUNIT["metre",1]]],
##     PRIMEM["Greenwich",0,
##         ANGLEUNIT["degree",0.0174532925199433]],
##     CS[ellipsoidal,2],
##         AXIS["latitude",north,
##             ORDER[1],
##             ANGLEUNIT["degree",0.0174532925199433]],
##         AXIS["longitude",east,
##             ORDER[2],
##             ANGLEUNIT["degree",0.0174532925199433]],
##     ID["EPSG",4269]]
```

Most of this is unintelligible to the average user, but it is good to know it is there. The key is that this shapefile is in the NAD83 projection and will align cleanly with others on the same projection. If two shapefiles have different projections, we may need to convert the projection of one or the other, although often `sf` is able to solve that problem for us on the fly because it is aware of these differences and has built-in transformations.

8.3.3 Plotting a Shapefile

Let's get to know our shapefile better in the way that it was meant to be viewed—as a map! This is done with the `plot()` function. `plot()` stops a step short of truly making a map with multiple layers and full customization, but it does give us a first view into a shapefile the way `summary()` or `str()` might for a data frame.

```
plot(tracts_geo)
```

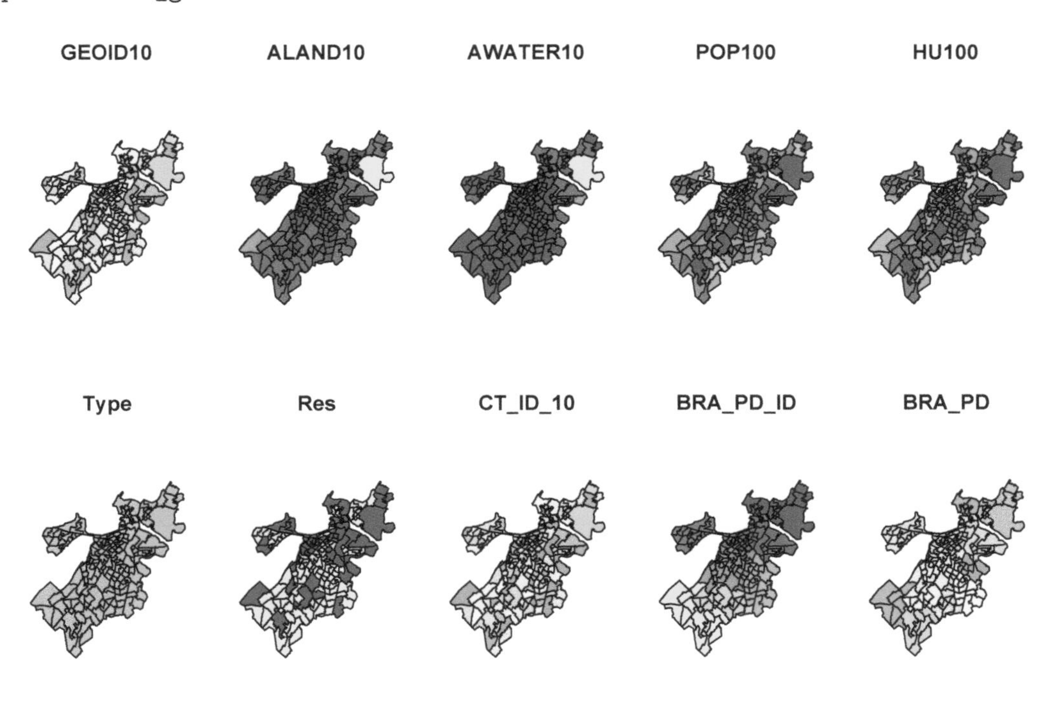

Note that we receive an error warning that it will only plot 10 of the attributes for us. You can read more about these variables in the data documentation, but many of the variables and their coloration probably make immediate sense. `ALAND10` and `AWATER10` are the land and water area contained in each tract, and they are largest at the edges of the city and on the coast, respectively (brighter colors). Population (`POP100`) and households (`HU100`) also vary across communities.

What if we want something simpler though, like just seeing the polygons themselves. We can do this with the `st_geometry()` command within the `plot()`.

```
plot(st_geometry(tracts_geo))
```

One fun feature of `plot()` is that it can layer plots on top of each other using the `add=TRUE` argument. So, if we want to add more information, like, say, whether a census tract is predominantly residential or not, we can do the following.

```
plot(st_geometry(tracts_geo))
plot(st_geometry(tracts_geo[tracts_geo$Res==1,]),
     add=TRUE, col='red')
```

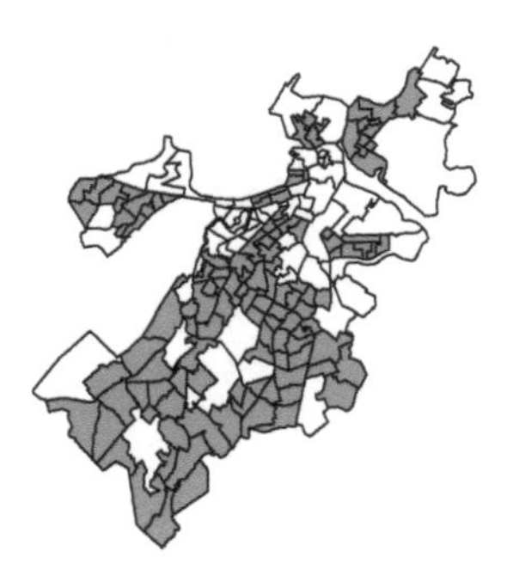

What if we want to visualize a continuous variable? Our instincts might suggest the following code:

```
plot(tracts_geo$POP100, main = 'Population in 2010',
     breaks = 'quantile')
```

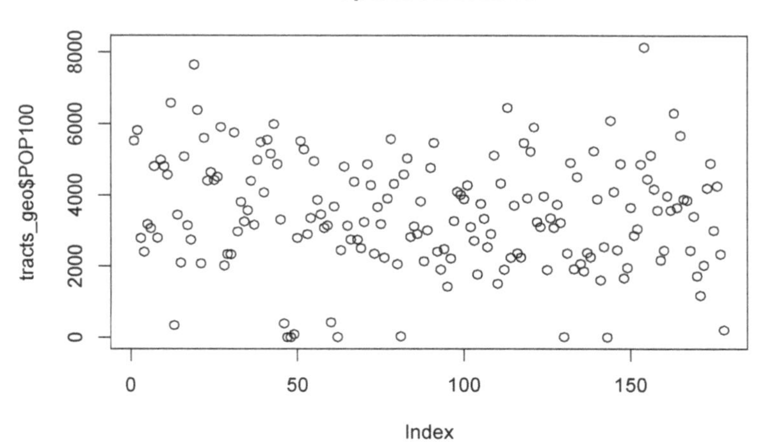

Unfortunately, this produces a dot plot that inexplicably has an index for the census tracts on the x-axis and the population on the y-axis.

This is because $ notation in this case creates a numeric vector that has been separated out from the shapefile. Instead, we need [] notation.

```
plot(tracts_geo['POP100'], main = 'Population in 2010',
     breaks = 'quantile')
```

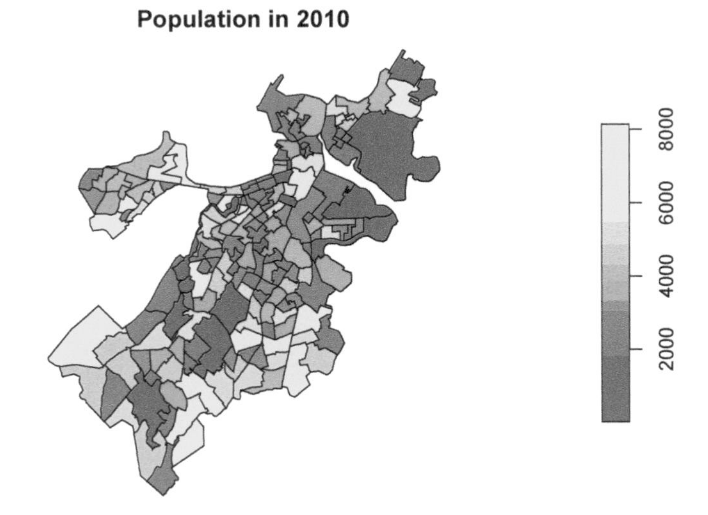

Now we have a plot of a single variable that clearly communicates the number of people living in each census tract. (Note that I have added `breaks='quantiles'` to the command to make it more readable as some tracts are outliers with much more population than others. This is not entirely necessary. Feel free to see what the map would look like without this additional argument, or other options for `breaks=`.)

8.4 Making a Map

Let's make a map for real, one that capitalizes on additional information from our schema, has multiple layers, and that we can fully customize. This is going to consist of four steps: (1) importing and merging additional data, (2) creating a base map, (3) making a multi-layer map, and (4) customization.

8.4.1 Importing and Merging Additional Data

Our shapefile contains all census tracts for the City of Boston. Census tracts are good approximations of neighborhoods and many institutions release data describing them, all of which could easily be merged with our shapefile and incorporated into a map. Most notably, the U.S. Census Bureau releases a plethora of variables describing their population, households, and other characteristics. BARI curates a subset of these and publishes them with each update of the American Community Survey. We will incorporate these for the 2014-2018 estimates by importing them and merging them with our shapefile.

```
demographics<-read.csv("Unit 2 - Measurement/Chapter 08 - Mapping
Communities/Example/ACS_1418_TRACT.csv")

tracts_geo<-merge(tracts_geo,demographics,by='CT_ID_10',
                  all.x=TRUE)
```

Note at this point a few details. First, the `merge()` command was able to work with `tracts_geo` as if it were a standard shapefile, merging the columns from `demographics` based on the shared `CT_ID_10` variable. `tracts_geo` now has 74 variables instead of 17. Also, BARI's census data are for all of Massachusetts and have 1,478 census tracts, so we use the `all.x=TRUE` command to account for the differences between the files and to safeguard the rows that we know we want to keep. Last, if you would like, take a look at `names()`. You will notice that `geometry` remains the last variable, as that is required for `sf` to be able to continue working with it as a shapefile.

8.4.2 Creating a Base Map

Before we map our census tracts, we need a base map layer to place them on. Otherwise, as in the plots above, we will see the outline of Boston but without any geographic context. Incorporating a base map will require `ggplot2` and its companion package `ggmap`. The latter provides us with the command `get_map()` for accessing publicly available base maps. The default is an opensource map from Stamen, which we will use here.

```
require(ggplot2)
require(ggmap)
Boston<-get_map(location=c(left = -71.193799,
                           bottom = 42.22,
                           right = -70.985746,
                           top = 42.43))
Bostonmap<-ggmap(Boston)
Bostonmap
```

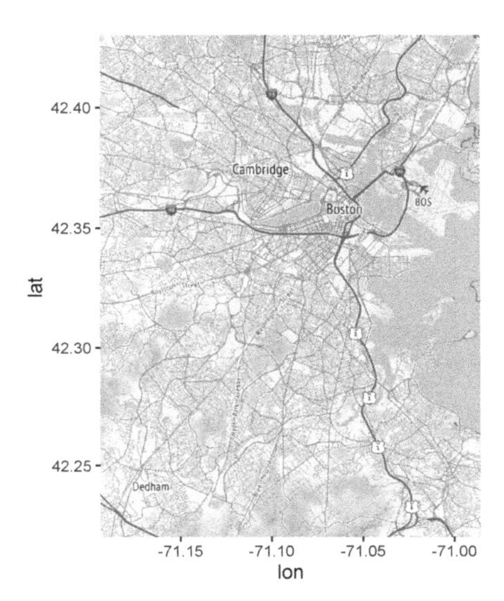

Let's walk through what happened in those three lines of code and why we now have a map of Boston. First, we specified the bounding box for our map (which if you compare to above is quite similar to the bounding box for our census tracts). Another way to do this is to specify a single point and a zoom level; see the **ggmap** documentation for details. We stored the product in the object Boston. Note that the console generated a series of lines of the form: `Source : http://tile.stamen.com/terrain/12/1240/1517.png`

These are the many tiles or grid cells that constitute the base map that we downloaded, which is actually a raster.

Now that we have a base map object, we need to convert it into a `ggplot2` compatible object. The `ggmap()` command does this, acting as an equivalent to the `ggplot()` command, which creates the basis for a non-spatial graphic. And with that, we have a map of Boston.

8.4.3 Making a Multi-Layer Map

We are now ready to make a multi-layer map that combines our base map with a visualization of the census tract data. We can start simple, replicating our plot distinguishing between neighborhoods that are and are not predominantly residential.

```
Bostonmap + geom_sf(data=tracts_geo,
                    aes(fill=as.factor(Res)),
                    inherit.aes = FALSE)
```

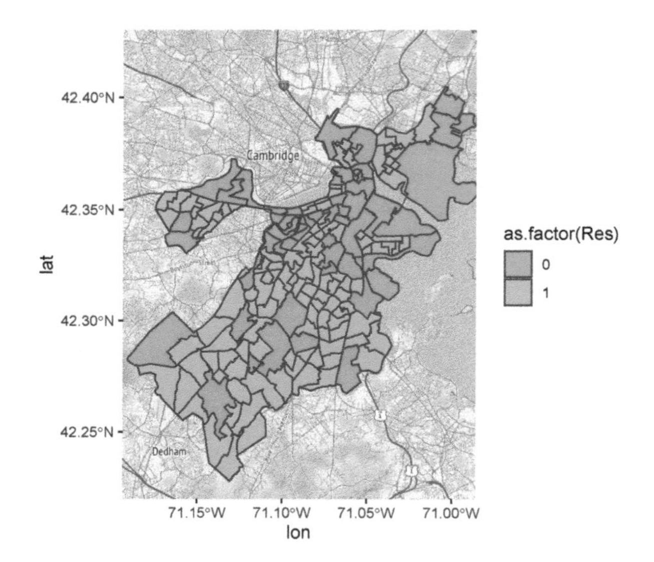

That's it! One line of code, thanks to the elegance of **ggplot2**. Note that we already have the base map stored as **Bostonmap**, so this is equivalent to:

```
ggmap(Boston) + geom_sf(data=tracts_geo, aes(fill=as.factor(Res)),
                        inherit.aes = FALSE)
```

What we have done new is incorporate the **geom_sf()** function, which follows the standard structure of **ggplot2** commands, specifying a geometry, in this case of the form sf. It includes, as is often the case, a **data=** argument, an **aes()** argument that can be used to specify the logic for **fill=** or the color of each object. We also need to specify **inherit.aes=FALSE** to avoid any disagreements between our **ggmap()** command and the new data. This proves useful because the two have different coordinate projections. As the console tells us, R detected this and converted all layers to match the NAD83 projection of our shapefile.

We could color the tracts according to the values of a continuous variable, like income, which will be the basis for the rest of this example. The technical term for this is a *chloropleth*, or the use of a color gradient to describe the distribution of a variable across spaces.

```
Bostonmap + geom_sf(data=tracts_geo, aes(fill=MedHouseIncome),
                    inherit.aes = FALSE)
```

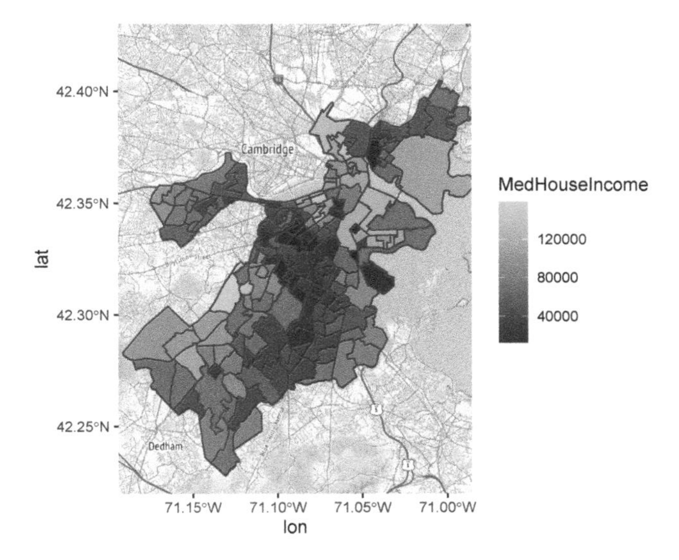

We can now see that the south-central neighborhoods have the lowest incomes in the city, and peripheral areas, which are more suburban, are more affluent. There are certainly ways that we can customize this map to be a bit easier to read, though.

8.4.4 Customization

Let's finish by customizing our map of income in Boston. As we go through, we will iteratively add new elements and new arguments to existing elements in our `ggplot2` command. First, it is hard to know how to interpret the median income of places with very few residents, so we will exclude tracts with a population lower than 500.

```
Bostonmap+ geom_sf(data=tracts_geo[tracts_geo$POP100>500,],
                   aes(fill=MedHouseIncome),inherit.aes = FALSE)
```

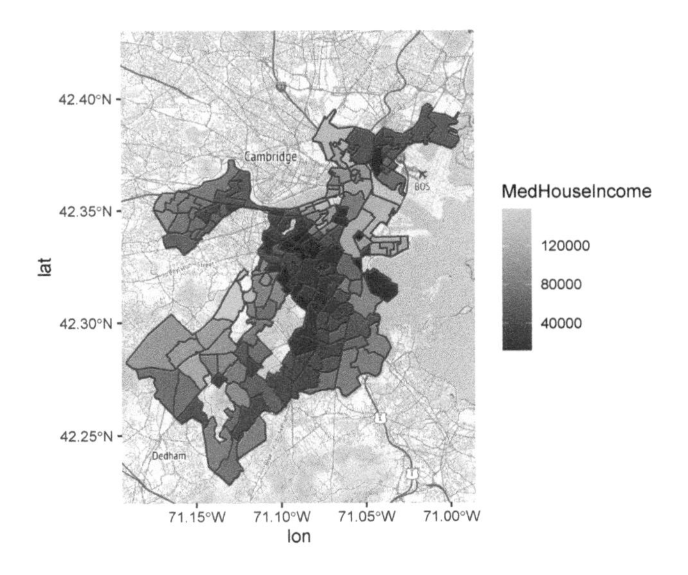

Maybe we also want to change the color gradient, say, running from red to green, changing color at the midpoint.

```
Bostonmap+ geom_sf(data=tracts_geo[tracts_geo$POP100>500,],
                   aes(fill=MedHouseIncome),inherit.aes = FALSE)+
  scale_fill_gradient(high = "green", low = "red")
```

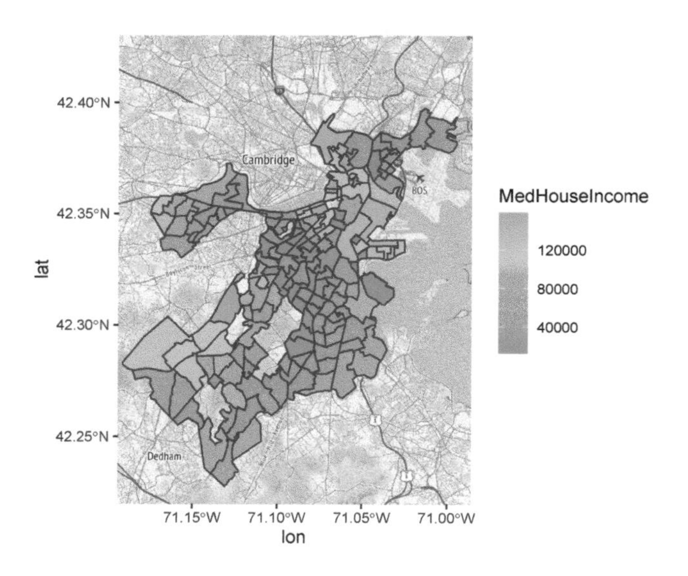

Last, a better legend label would be helpful.

```
Bostonmap+ geom_sf(data=tracts_geo[tracts_geo$POP100>500,],
                   aes(fill=MedHouseIncome),inherit.aes = FALSE)+
  scale_fill_gradient(high = "red", low = "green")+
  labs(fill='Median Household \nIncome')
```

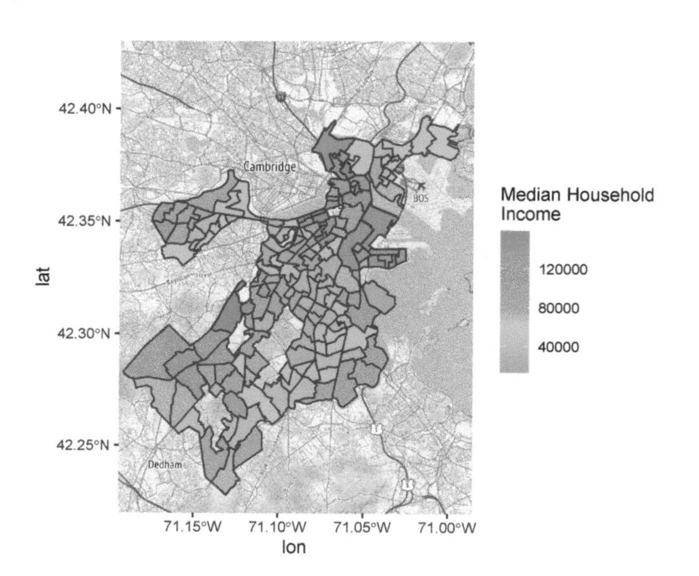

8.4.5 Summary and Extensions

We have learned how to create a multi-layer map. This is a very generalizable skill. You can now create a chloropleth for any variable in the demographic indicators that we just merged with the shapefile. You could also do the same for any other data set at the census tract level—or any other shapefile you might encounter. Though the precise representation of points and lines will be slightly different, many of the same conventions apply. Also, this map only has two layers on it, but you could easily add more, including points or a road map. The world is your oyster.

8.5 Working with Points

A very common type of spatial data is points. Points are special for a couple of reasons. First, they are frequently the way that we represent the events and transactions that are captured by administrative and internet systems. Each record references the discrete location where said event occurred, and that is represented with a latitude and longitude. Second, latitude and longitude can easily be organized into two columns. Thus, there are a lot of data sets out there that are currently in a .csv or data frame format that can be coerced into a mappable form. We are going to illustrate this with an old friend from Chapter 3, the reports received by Boston's 311 system . We will specifically analyze cases in 2020 referencing graffiti. Note that we will use tidyverse commands in the example that follows.

```
CRM<-read.csv("Unit 1 - Information/Chapter 3 - Telling a Data
Story/Example/Boston 311 2020.csv")

require(tidyverse)
CRM<-CRM[str_detect(CRM$type,'Graffiti'),]
```

8.5.1 "Map" Records with *lat-long*

If we look at the `names()` in CRM, we will note that two of the columns are titled `latitude` and `longitude`, which, respectively, can be treated as y and x coordinates (oddly, latitude comes first but is north-south, meaning a y-coordinate, and longitude is east-west, meaning an x-coordinate). We can leverage these to treat them as point data.

```
Bostonmap +  geom_point(data=CRM, aes(x=longitude, y=latitude))
```

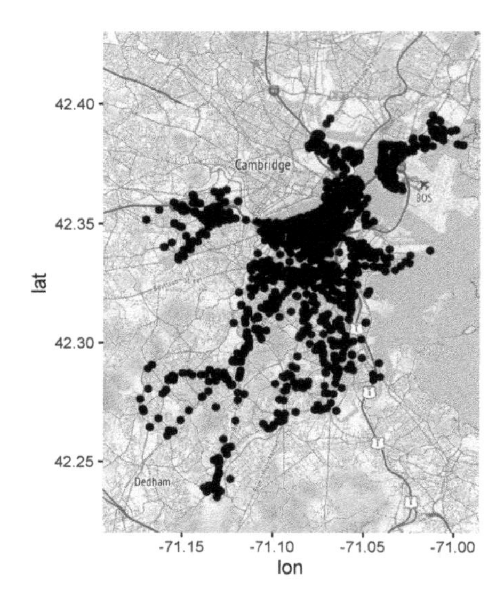

We now have a map of all the complaints about graffiti received by Boston's 311 system in 2020. Note that the command for doing this was `geom_point()`, which we used to create dot plots in Chapter 7. It is not really a spatial command, but we were able to exploit the fact that a map is simply a two-dimensional plane with an x- and y-axis. Luckily, our points were drawn from a coordinate system that was consistent with that of our base map, making this possible. Otherwise, R would have no information to transform the two to match each other.

We can draw some simple inferences from this map, the most obvious being that most graffiti reports come from the downtown business district and few occur in the outlying neighborhoods. But the points are so numerous and dense that it is hard to say much more. We might get more out of a density plot, though, another skill we learned in Chapter 7.

```
Bostonmap + geom_density2d(aes(x=longitude, y=latitude),
                          bins=200, data = CRM)
```

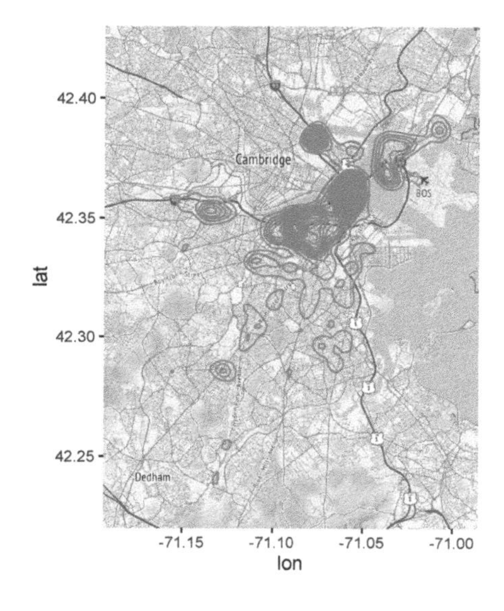

Now we see something that looks akin to a topographical map for the density of graffiti complaints. We note that, as the points suggested, there are a lot more graffiti complaints in the downtown area near the harbor and along the river. As far as the plot is concerned, there might as well not be any reports in the outlying neighborhoods. The same thing is apparent if we use the aesthetically fancier `stat_density2d()`.

```
Bostonmap + stat_density2d(aes(x=longitude, y=latitude,
                               fill = ..level.., alpha = .5),
                           size = .5, bins = 200,
                           data = CRM, geom = "polygon")
```

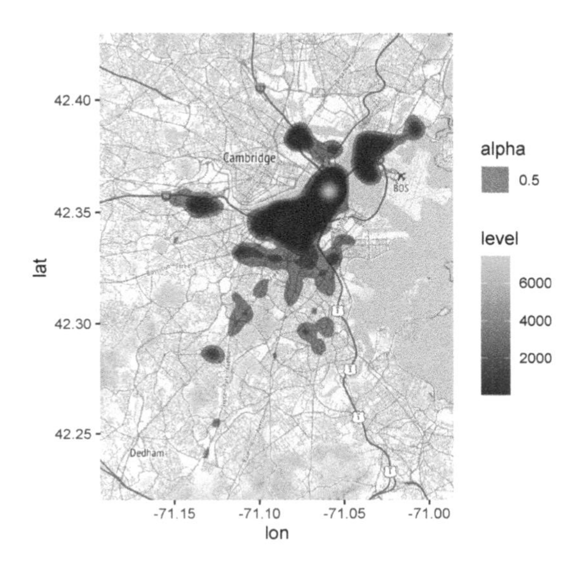

This tells the same story but with a bit more color. Clearly, the downtown area is where graffiti—or at least complaints about graffiti—concentrates.

8.5.2 Converting Records into sf Points

Thus far we have leveraged the `latitude` and `longitude` columns to visualize our 311 records as if they were spatial, but we have not actually converted them to spatial data as recognized by `sf`. This means we cannot leverage their spatial information for specialized tools or relate them to other spatial information, like our census tracts. Because the events occurred within census tracts, we might want the two data sets to speak to each other.

We can convert records with latitude and longitude to sf points using the `st_as_sf()` command.

```
CRM_geo<-st_as_sf(CRM, coords=c('longitude','latitude'),
                  crs=4269)
```

This command required us to indicate our data frame, input coordinate variables, in the order of x and y (`cords=c('longitude','latitude')`), and the CRS. Note that CRS 4269 is the numeric code for NAD83, which is how these points were generated.

Before we move on to use these points as spatial features, take a quick look at the new object.

```
class(CRM_geo)
```

```
## [1] "sf"         "data.frame"
```

Its class is now "sf" "data.frame".

Notably, it has one variable fewer than the original data frame. Why? (If you want to take a moment to solve this conundrum, please do.)

Answer: `st_as_sf()` took the two coordinates and combined them into a single geometry variable, which, as we have seen, contains the spatial information for all `sf` objects.

8.5.3 Spatial Joining Points with Polygons

Now that our graffiti complaints are an `sf` object, we can relate them to our census tracts. The easiest way to do this is with the `plot()` command.

```
plot(st_geometry(tracts_geo))
plot(st_geometry(CRM_geo),col='blue',add=TRUE)
```

By using the `add=TRUE` argument, we were able to build a multi-layer plot of points on polygons. This illustrates that the two overlap as we might expect. What if we want to analyze their relationship? Say, for example, we want to know which census tract each point falls within. This is exactly the kind of thing that sf is designed to accomplish.

The `st_join()` function conducts what is referred to as a *spatial join*. Just as we learned in Chapter 7 about joins linking records from different data sets based on shared variables, a spatial join links them based on shared spatial information.

```
CRM_tract<-st_join(CRM_geo,tracts_geo,left=TRUE)
```

We have now created a new version of our `CRM_geo` object that has 101 variables, which makes sense because we had 28 variables in the previous version of this object and there were 73 non-geometry variables in `tracts_geo`. If we look at the names, we now have attached to every record the `CT_ID_10` of the tract containing it and all of its associated information.

8.5.4 Aggregating Spatially Joined Data

Last, now that our points are linked to our tract polygons, we can use the former to describe the latter. To do so we will return to the tools for aggregation that we learned in Chapter 7. Suppose we want to map how many graffiti complaints come from each census tract. We can do this by:

```
tracts_graf<-CRM_tract %>%
  group_by(CT_ID_10) %>%
  summarise(graffiti=n()) %>%
  st_drop_geometry()
tracts_geo<-merge(tracts_geo,tracts_graf,
                  by='CT_ID_10',all.x=TRUE)
```

Note that we had to use `st_drop_geometry()` before merging because `sf` assumes that any manipulation of an `sf` object will still be an `sf` object, and you cannot merge two `sf` objects because of the geometries.

Now we can create a map of graffiti complaints by census tract.

```
Bostonmap+ geom_sf(data=tracts_geo[tracts_geo$POP100>500,],
                   aes(fill=graffiti),inherit.aes = FALSE)+
  scale_fill_gradient(high = "red", low = "green")+
  labs(fill='Graffiti \nComplaints')
```

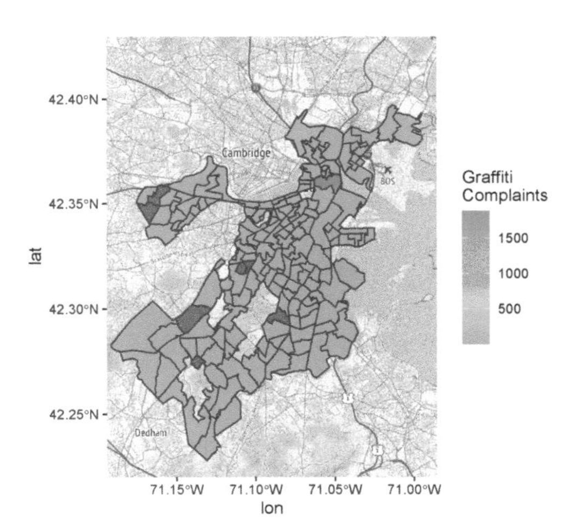

We do seem to have a problem, though, as there is a single outlier dominating the map. This also explains the extremities of the density maps we saw above. Why might this be? It turns out that that census tract contains City Hall, which is a default for mapping cases without location information. We should probably do it again without that tract.

```
Bostonmap+ geom_sf(data=tracts_geo[tracts_geo$POP100>500 &
                                   tracts_geo$graffiti<1000,],
                 aes(fill=graffiti),inherit.aes = FALSE)+
  scale_fill_gradient(high = "red", low = "green")+
  labs(fill='Graffiti \nComplaints')
```

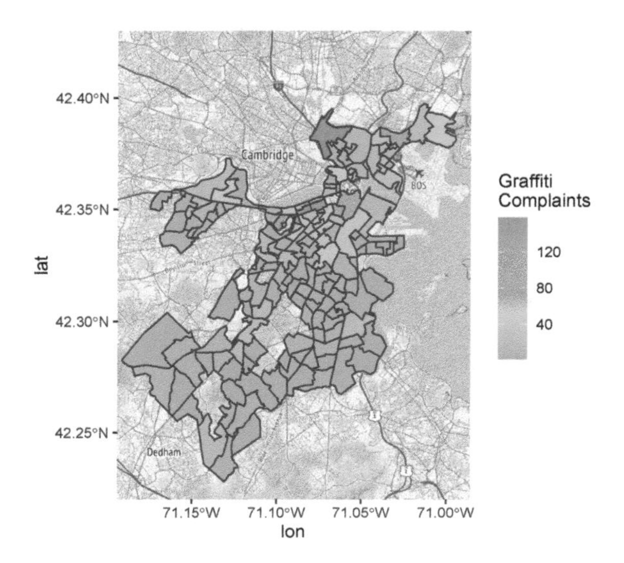

This is a bit more interpretable—still with a clear concentration of cases in the downtown area, but with at least some variation in other parts of the city.

8.5.5 Summary and Extensions

We have seen here the power of coercing records with latitude and longitude data into sf objects. In this case we have spatially joined them to census tracts and then created aggregate measures. This could be extended in multiple ways.

- We might create other aggregate measures based on any number of variables or any of the logics developed in Chapter 7.
- We can do other types of spatial joins. Here we did the default "intersect" or "over" join, where each point falls in a polygon. There are other logics, though, including nearest neighbor, which is popular for joining points to lines (e.g., events to the nearest street), or "majority contained within," which can be used to join lines to polygons, especially when they cross over polygon borders.
- We could extend this to applying any spatial analytic tool we might like to the points. Learning more of these is beyond the scope of this book, but there are many resources out there for, including the `sf` documentation, for learning more about the vast scope and capabilities of GIS.

8.6 Connecting R to Other GIS Software

R is just one software package capable of working with spatial data, and it is not even the most popular. We have learned it here because it is consistent with the curriculum of this book and because it is useful to be able to process and analyze data and maps in a single interface. Other options, though, include ArcGIS and QGIS, each of which is designed specifically to work with spatial data and are probably preferable if your goal is to make professional maps with detailed customization.

There are ways to integrate R into ArcGIS and QGIS, as both have plug-ins that can accommodate R syntax. It is logical for the connection to go in this direction (as opposed to working in R and linking to ArcGIS or QGIS from there) because R is essentially a syntax that is then incorporated into a user interface (like RStudio). The power of ArcGIS and QGIS is to be able to do hands-on work with spatial data, so you would likely want to work in that context and leverage R syntax to serve up the content you need.

One exception, though, is Leaflet, which has become popular for creating interactive online maps. Leaflet is code-based itself with a structure similar to `ggplot2`, with commands stacked upon each other to specify the layers of a map and customize their visualization. As you might expect, there is a `leaflet` package that introduces this capacity for R (Cheng et al., 2021). We will work through an example of creating a Leaflet map. This is possibly a bit tangential for some readers of this book. If you choose to skip this, it will not affect your ability to engage with any of the material in subsequent chapters.

8.6.1 Making an Interactive Leaflet Map

Commands in the `leaflet` package are much like `ggplot2`, wherein we start with a `leaflet()` function that is often left empty, followed by a series of additional instructions. These are organized here through piping. We will proceed by gradually elaborating on a single map in eight steps.

8.6.1.1 Step 1: Creating a Base Map

First, as before, we need to establish a base map. The code for doing so is a bit different but should be interpretable. We have to indicate the source of our base map with `addProviderTiles()` (which we get in this case from CartoDB) and we set the extent of our base map with `setView()`, this time specifying a center point and a zoom level.

```
require(leaflet)
mymap <- leaflet() %>%
  addProviderTiles("CartoDB.Positron") %>%
  setView(-71.089792, 42.311866, zoom = 11)
mymap
```

8.6.1.2 Step 2: Adding a Layer

We now can add our own layer to the base map, in this case with the **addPolygons()** function (there is also the **addPoints()** command, etc.).

```
mymap <- leaflet() %>%
  addProviderTiles("CartoDB.Positron") %>%
  setView(-71.089792, 42.311866, zoom = 11) %>%
  addPolygons(data = tracts_geo)
mymap
```

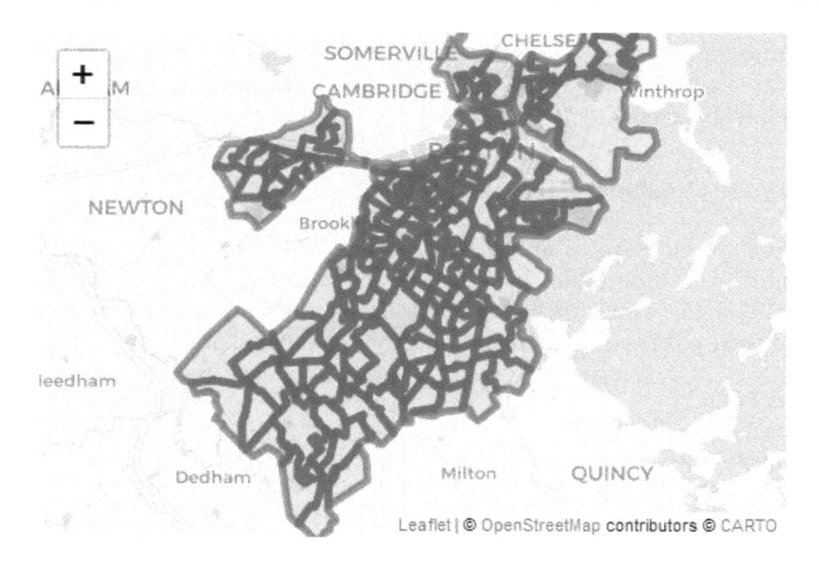

8.6.1.3 Step 3: Making a Chloropleth

One thing that can be a little challenging about Leaflet is that it does not make any assumptions about what colors you would like a chloropleth to have. As such, it does not actually create a chloropleth unless you tell it how to. So, we need to first create a coloration with the `colorNumeric()` function and then pass that to the `color=` argument in the `leaflet()` command.

```
pal <- colorNumeric(palette='Blues',
      domain=tracts_geo$graffiti[tracts_geo$POP100>500 &
                                tracts_geo$graffiti<1000])

mymap <- leaflet() %>%
  addProviderTiles("CartoDB.Positron") %>%
  setView(-71.089792, 42.311866, zoom = 11) %>%
  addPolygons(data = tracts_geo[tracts_geo$POP100>500 &
                                  tracts_geo$graffiti<1000,],
              color = ~pal(tracts_geo$graffiti))
mymap
```

Here we have created a rule set for coloring our census tracts based on the color blue and the distribution of the graffiti variable. This then determines the coloration in the following map.

8.6.1.4 Step 4: Adjusting Opacity

In the previous map, the tracts were completely opaque. This may not be what we want if we want to maintain a sense of the spaces underneath them that they are describing. We can adjust the opacity with the `fillOpacity=` argument.

```
pal <- colorNumeric(palette='Blues',
      domain=tracts_geo$graffiti[tracts_geo$POP100>500 &
                                tracts_geo$graffiti<1000])

mymap <- leaflet() %>%
  addProviderTiles("CartoDB.Positron") %>%
  setView(-71.089792, 42.311866, zoom = 11) %>%
```

```
  addPolygons(data = tracts_geo, fillOpacity = .5,
              color = ~pal(tracts_geo$graffiti)
              )
mymap
```

8.6.1.5 Step 5: Introducing Interactivity with Highlights

One of the main attractions of Leaflet is that it creates interactive maps. We are now ready to introduce this capacity by creating "highlights." Highlights are when objects on a map light up when the mouse cursor is placed over it. This is specific to a given layer on the map, so the **highlight=** argument goes within the **addPolygons()** command. We set the weight (thickness of the line) to 3, the color to red, and specify that it should bring the object to the front—which would be relevant if there were multiple overlapping layers. Note that this and the following step only alter interactivity and not appearance, so I will only show the code and not print the result, which will look the same as that created in Step 4.

```
pal <- colorNumeric(palette='Blues',
       domain=tracts_geo$graffiti[tracts_geo$POP100>500 &
                                  tracts_geo$graffiti<1000])

mymap <- leaflet() %>%
  addProviderTiles("CartoDB.Positron") %>%
  setView(-71.089792, 42.311866, zoom = 11) %>%
  addPolygons(data = tracts_geo, fillOpacity = .5,
              color = ~pal(tracts_geo$graffiti),
              highlight = highlightOptions(weight = 3,
                                           color = "red",
                                           bringToFront = TRUE)
  )
mymap
```

8.6.1.6 Step 6: More Interactivity with Pop-Ups

Another interactive feature available from Leaflet is "pop-ups," in which a user can click on an object and have information about it appear. This is again specific to a given layer, so the popup= argument is within the `addPolygons()` command. We have used the `paste()` command to indicate that, when a tract is clicked on the map, it should print the `BRA_PD`, which is the name of the larger neighborhood containing that tract, and the number of graffiti complaints, with a colon in between.

```
pal <- colorNumeric(palette='Blues',
      domain=tracts_geo$graffiti[tracts_geo$POP100>500 &
                                  tracts_geo$graffiti<1000])

mymap <- leaflet() %>%
  addProviderTiles("CartoDB.Positron") %>%
  setView(-71.089792, 42.311866, zoom = 11) %>%
  addPolygons(data = tracts_geo, fillOpacity = .5,
              color = ~pal(tracts_geo$graffiti),
              highlight = highlightOptions(weight = 3,
                                           color = "red",
                                           bringToFront = TRUE),
              popup = paste(tracts_geo$BRA_PD,
                            ":",tracts_geo$graffiti))
mymap
```

8.6.1.7 Step 7: Add a Legend

Of course, no map is complete with an interpretable legend. We can do this with the `addLegend()` command, whose arguments should be largely self-explanatory.

```
mymap <- leaflet() %>%
  addProviderTiles("CartoDB.Positron") %>%
  setView(-71.089792, 42.311866, zoom = 11) %>%
  addPolygons(data = tracts_geo, fillOpacity = .5,
              color = ~pal(tracts_geo$graffiti),
              highlight = highlightOptions(weight = 3,
                                           color = "red",
                                           bringToFront = TRUE),
              popup = paste(tracts_geo$BRA_PD,
                            ":",tracts_geo$graffiti))%>%
   addLegend(pal = pal,
             values = tracts_geo$graffiti[tracts_geo$POP100>500 &
                                          tracts_geo$graffiti<1000],
                        position = "bottomright",
                        title = "Graffiti \nComplaints",
                        opacity = 1
              )
mymap
```

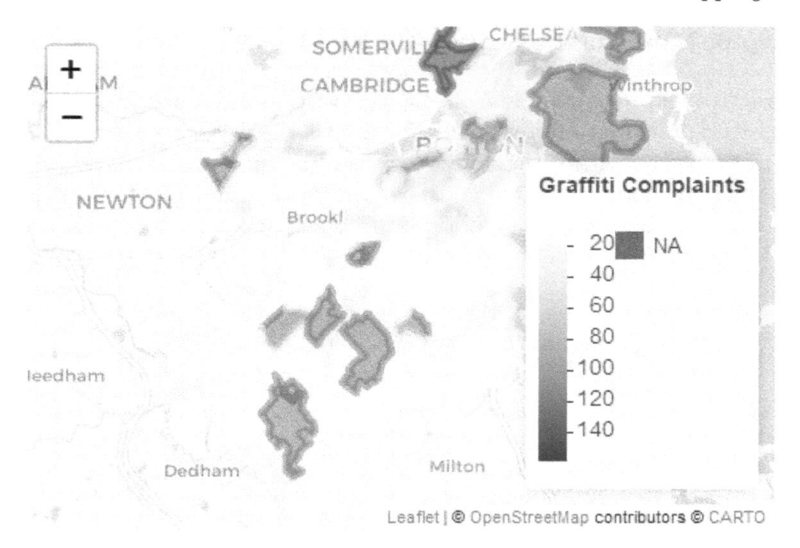

8.6.1.8 Step 8: What Next?

At this point, our map feels arguably complete. But there is always the opportunity to add more. For example, you could put more points, maybe of the locations of parks or government buildings. You might want the interactive pop-ups to offer more information. There might be other customizations you would like to pursue. Whatever it is you are looking to do, you can likely accomplish it with the code we have learned or by looking through the documentation of the `leaflet` package.

8.7 Summary

In this chapter we have learned how to work with spatial data. This is no easy feat! As discussed at the outset, organizing and representing spatial information is complicated and can be quirky, idiosyncratic, and error-prone. That said, advances over the years have made the process increasingly streamlined and accessible. The skills you have learned here are the tip of the iceberg, but they are both extensible to any other spatial data you might encounter and can act as a gateway to a variety of more complex spatial transformations, analyses, and visualizations that await. As you go forth, you can already do the following:

- *Describe what makes spatial data distinct* and the challenges that these distinctions solve;
- *Identify the different types of spatial data*;
- *Use the `sf` package* to:
  - *Import shapefiles (.shps) into R*,
  - *Plot shapefiles in R* using `plot()`, and
  - *Merge additional data with shapefiles* on shared key variables;
- *Create a base map* with `ggmap`;
- *Make and customize a multi-layer map* with `ggmap` and `ggplot2`;
- *Leverage latitude-longitude data in records* to incorporate them into maps and spatial analyses;

- *Spatial join one layer with another* to merge their information together based on their intersections (including creating aggregate measures from those spatial joins);
- *Make an interactive map* with the `leaflet` package.

8.8 Exercises

8.8.1 Problem Set

1. Define each of the following terms. How does it address one or more of the specific challenges or opportunities created by spatial data?
   a. Shapefile
   b. Coordinate Projection (or Reference) System
   c. Chloropleth
   d. Spatial join
2. Visit the open data portal of your choice. See if you can find an example of each of the four types of spatial data. How are you confident that it is of that type?
3. Using the merged data set of tracts and census indicators from the worked example, make three additional chloropleths (you may want to read the census indicators documentation to better understand all of the measures). Describe what you observe. Do you notice any interesting patterns or similarities between the spatial distributions of these variables?
4. Visit the Boston Data Portal[1] and download another component of the Geographical Infrastructure for the City of Boston.
   a. Make at least one map visualizing a variable already in the data frame associated with that shapefile.
   b. Find *another* data set on the Boston Data Portal that references the same geographic scale (Hint: use keywords associated with the scale of the shapefile, like "streets" or "census block groups"). Merge its variables into your first shapefile and make an additional visualization.
   c. Incorporate an additional data set of a different type (i.e., if you have polygons, incorporate lines or points; if you have lines, incorporate polygons or points, etc.), visualized in the way that makes the most sense for communicating a meaningful map.

8.8.2 Exploratory Data Assignment

Working with a data set of your choice. (Note: you will need access to at least one shapefile to complete this assignment. Many open data portals for cities have shapefiles. You can also download shapefiles from the U.S. Census or create shapefiles from records with latitude and longitude.)

1. Create new measures based on aggregations of your data to a geographic scale. It is possible that you have already done this for the Exploratory Data Assignment in Chapter 7. Merge these measures to the shapefile for that geographic scale.

[1]https://dataverse.harvard.edu/dataverse/BARI

2. Create a chloropleth for at least one of these variables.
3. Describe the spatial distribution of the variable and how you believe this furthers our understanding of the landscape of the communities you are seeking to better understand.
4. Feel free to incorporate other shapefiles, whether points, lines, polygons, or rasters, that you believe help illustrate the patterns you see in the spatial distribution of that variable. If you do so, please explain their contributions to the map.

9

Advanced Visual Techniques

The Kantar Information is Beautiful awards recognize the most impressive data visualizations of each year. The submissions cover a broad range of subjects, from arts and entertainment to maps and places, from news and current affairs to science and technology, from climate and environment to sports statistics. There is even a category called "Unusual." The true versatility of data visualization is especially apparent, however, when one looks at specific submissions and awardees, including those pictured in Figure 9.1: a map of where houses in Houston damaged by Hurricane Henry have been scooped up by investors—-from mom-and-pop landlords to billion-dollar Wall Street funds; the history of U.S. immigration by continent conceived of as rings in a tree trunk (also known as "dendrochronology"; (Cruz et al., 2019)); a multimedia visualization of how American municipalities bus homeless individuals to other places, often big cities; and even quirky visualizations like a swarm plot of the vocabulary size of rappers, as measured by the number of unique words in their first 35,000 lyrics.

The Kantar Information is Beautiful awards illustrate just how powerful, prominent, and sophisticated the world of data visualization has become. In fact, if you browse through the submitters, you will notice that many are from major media outlets, like *The New York Times*, *The Guardian*, and *The Washington Post*, while others are from advocates, scientists, and independent analysts (a few, like MIT's Senseable City Lab, have been mentioned previously in this book). And, if we are being honest, these visualizations really are beautiful! There are dozens of creative ways to visualize data, but to be certain this is not an "automatic" exercise. As with everything we have learned and will learn, the creation of data visualizations requires careful thought and design to ensure that the final product communicates the story clearly without distorting the meaning.

In this chapter, we are going to learn a handful of advanced data visualization techniques that, if used properly, are sure to help communicate your discoveries and impress your audience. In many cases, they are not all that difficult to execute, they just require forethought about how they are best designed to reveal the most important information. You could skip this chapter and have no difficulty completing the material that follows, but if you want to expand your toolbox for visualization beyond `ggplot2`, you are ready!

9.1 Worked Example and Learning Objectives

In this chapter we will learn five advanced visualization techniques. Using these tools requires a bit more thought and consideration about how to best represent your data, but most only call for a few lines of code. Conveniently, nearly all of them use additional packages that build on the syntax of `ggplot2`. We will learn about:

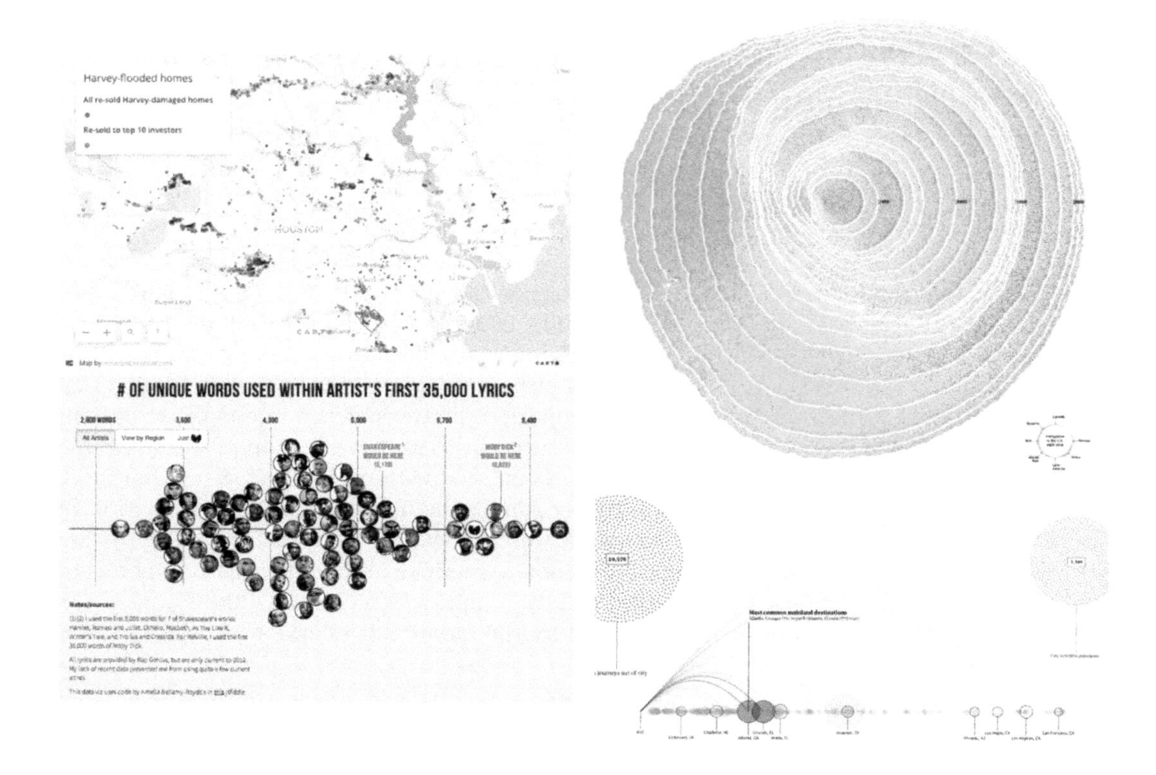

FIGURE 9.1
Submissions to the Kantar Information is Beautiful awards are diverse, including topics as wide-ranging as how the houses in Houston damaged by Hurricane Henry have been scooped up by investors (top left), the history of U.S. immigration by continent represented as rings in a tree (top right), the number of unique words in each rapper's first 35,000 lyrics (bottom left), and how American municipalities bus homeless individuals to other places (bottom right). (Credit: *Houston Chronicle*, Cruz et al., *The Guardian*, Matthew Daniels)

- *Multiplots*, which combine multiple graphical components into one image;
- *Streamgraphs*, which are interactive representations of the frequency of categories over time;
- *Heat maps*, which are visually appealing representations of the relationships between two sets of categories;
- *Correlograms*, which are visual representations of the strength of correlations between multiple variables;
- *Animations*, which place multiple visualizations in a sequential order to show change, often over time;
- and, of course, we will consider conceptually when it is appropriate and necessary to use these sorts of tools and when something simpler might suffice.

We will again illustrate these techniques with 311 case records as we are already familiar with some of the nuances of these data, though this time we will use the version curated by BARI, which has additional geographic information. Also, to take advantage of the time-stamps in the data, we will do a small amount of pre-processing and `require(lubridate)` and `require(tidyverse)`.

Link: https://dataverse.harvard.edu/dataset.xhtml?persistentId=doi:10.7910/DVN/CVKM87.

You may also want to familiarize yourself with the data documentation posted there or the worked example in Chapter 3.

Data frame name: `CRM`

```
require(lubridate)
require(tidyverse)

CRM<-read.csv("Unit 2 - Measurement/Chapter 09 - Advanced
Visualization/Example/311 Cases 2015_2019 Unrestricted.csv")

CRM$OPEN_DT <- as_datetime(CRM$OPEN_DT)
CRM$year <- year(CRM$OPEN_DT)
CRM$month <- month(CRM$OPEN_DT)
CRM$day <- day(CRM$OPEN_DT)
CRM$Snow<-str_detect(CRM$TYPE,'Snow')
```

9.2 Data Visualization: How and Why?

They say that a picture is worth a thousand words, and good data visualization can be worth thousands and thousands of data points. Effective data visualizations are capable of thoroughly communicating information in captivating ways that are accessible to audiences of all levels of data literacy. This has inspired visualizers—who in some ways are their own entire sub-class of analyst—have developed innumerable new and compelling ways to represent data graphically, and these tools have rapidly become widespread. Many organizations have also turned to Tableau as a straightforward cookbook for making data visualizations. Those with the capacity for custom analysis in R leverage `ggplot2`, which as we have seen includes dozens of different techniques. And there are many, many additional packages for R that enable more sophisticated visualizations, most of which capitalize on the syntax of `ggplot2` . The five that we will learn about in this chapter are but a small subset of what is possible. You will note throughout, however, that these tools are very specific about the data structure needed to inform the construction of the visualization, which means we will often need to create custom aggregations as we go (see Chapter 7 for a refresher, as these bits of code will only be referenced in brief).

When I think about advanced visualization, I am reminded of a quip that one of our more illustrious, senior professors made to us young quantitative types when I was in graduate school. "If you need an overly complex statistical tool to analyze your data, your question is too complicated to be worth asking." We all laughed, but there was an important lesson in here: focus your attention on the insights you want to pursue, and the methods will follow. If the question requires a simple set of tools, then that is what you should use. Do not prioritize sophisticated techniques for their own sake. The same is true for advanced visualizations. Animations, streamgraphs, and multiplots are great tools, but they have to be aligned to the question. Sometimes they will be flashy—little more than technical showboating—without making the information any more accessible. Sometimes, they will even obscure or distract from what you are trying to accomplish. For this reason, when describing each of the five techniques we will start by identifying when it is most appropriate, and maybe even certain times when you might be better off using a different technique.

9.3 Multiplots

Sometimes a story calls for more than one graphic, but we want those graphics coordinated in a single visual. For example, one of the visualizations in Figure 9.1 was a combination of three different graphics describing the tendency of municipalities to bus homeless individuals to larger cities (in fact, what you see there was a subset of even more visualizations in that graphical collage). This is called a multiplot, or a single visualization that coordinates two or more graphics. This is a relatively simple extension of standard visualizations, but one should always consider whether multiple graphics help add detail to the story or can either distract or overwhelm the audience.

The process of making a multiplot has two steps. First, we need to make the individual graphics that we want to compose the multiplot. Second, we need the `gridExtra` package, which can organize those graphics into a grid (Auguie, 2017).

9.3.1 Making Individual Graphics

First, making our individual graphics will require skills we have learned in previous chapters. Suppose we want to visualize the quantity of 311 cases by neighborhood, using `BRA_PD` ("planning districts" defined by the Boston Planning Redevelopment Agency, formerly known as the Boston Redevelopment Authority (BRA), which largely conform to the historical neighborhoods of the city), and see how these counts evolve over time. To make this more tractable, we are going to subset to 10 major neighborhoods that capture the demographic and economic diversity of the city.

```
require(gridExtra)

nbhd_data <- CRM %>%
  filter(BRA_PD %in% c('Back Bay/Beacon Hill',
              'Allston/Brighton','Central', 'Roxbury',
              'North Dorchester', 'South Dorchester',
              'South End', 'West Roxbury','East Boston',
              'Charlestown'), LocationID!=302615000) %>%
  group_by(BRA_PD, year) %>%
  summarise(count_per_nhood = n())
```

We have now created `nbhd_data`, which consists of counts of 311 reports by neighborhood for each year. We might then create a bar plot that represents total cases by neighborhood.

```
p1 <- ggplot(nbhd_data, aes(x=BRA_PD,
                    y = count_per_nhood, fill = BRA_PD)) +
  geom_bar(stat="identity") + xlab("") + ylab("") +
  theme(legend.position = "none",
        axis.text.x = element_text(angle = 45, hjust=1))
p1
```

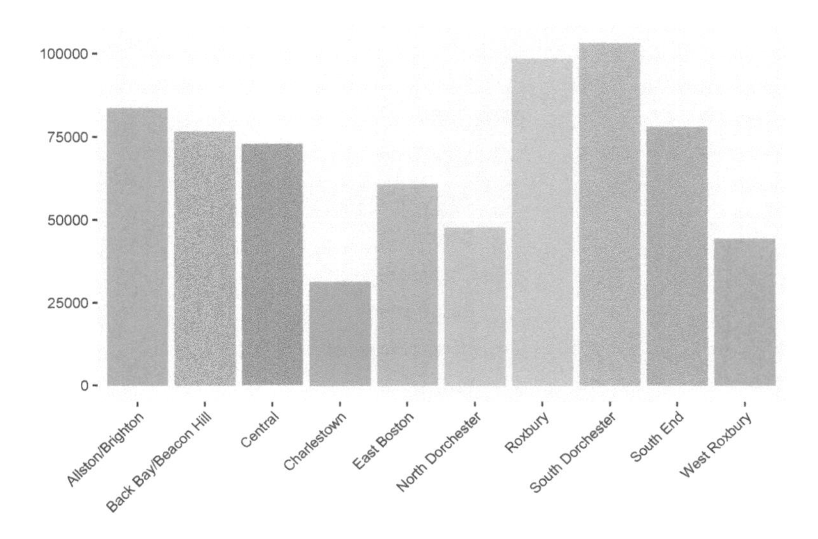

We can also create a line plot with change over time for each neighborhood.

```
p2 <- nbhd_data %>%
  ggplot( aes(x=year, y=count_per_nhood, group=BRA_PD,
              color=BRA_PD)) +
  geom_line() + xlab("") + ylab("") +
  theme(axis.text.x = element_text(angle = 45, hjust=1))
p2
```

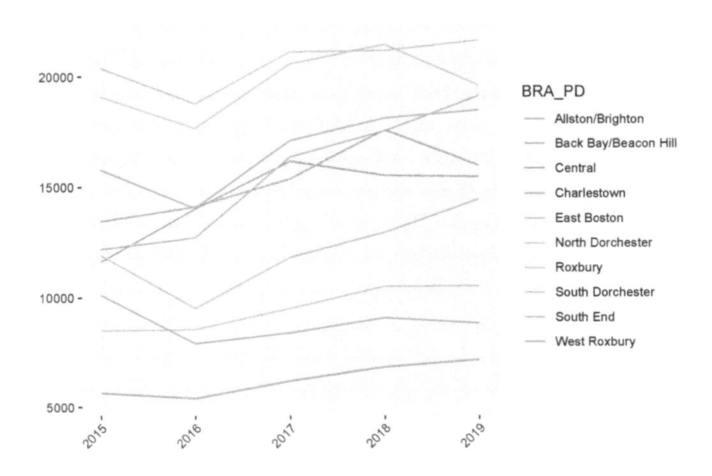

9.3.2 Coordinating Individual Graphics in a Multiplot

Now suppose we want to put those two graphics together in a single graphic. This can be done with the **grid.arrange()** command from **gridExtra**.

```
grid.arrange(p2, p1, nrow = 1, widths=c(250,150))
```

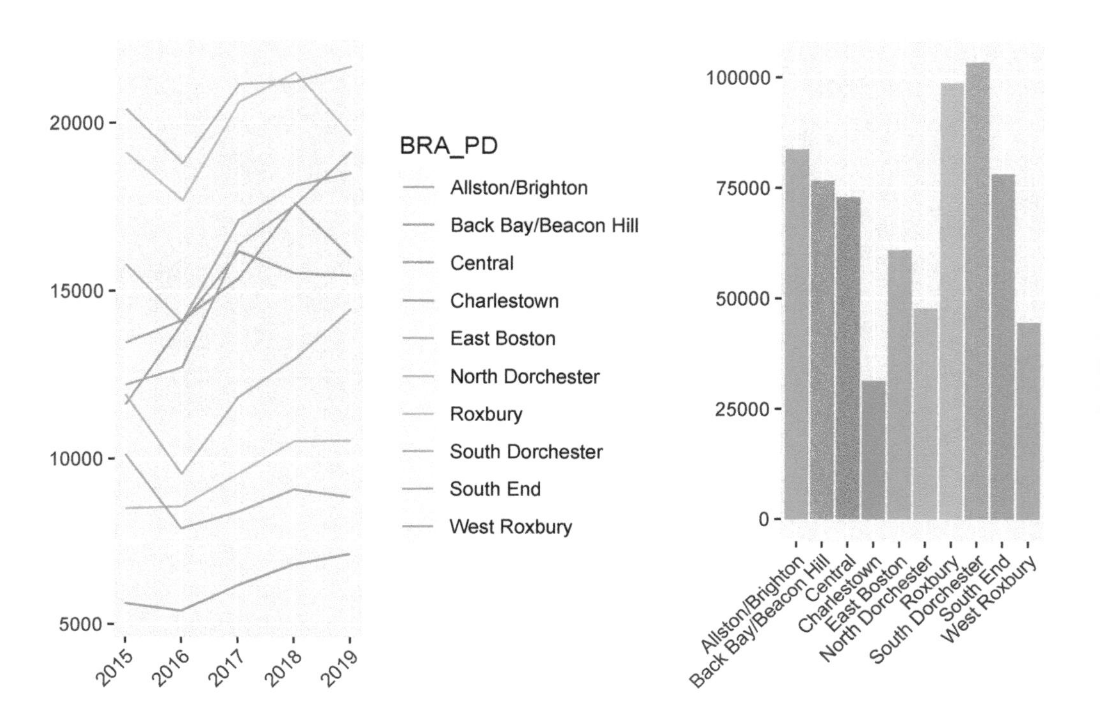

How did that work? All we had to do was tell **grid.arrange()** the graphics we wanted to bring together in the multiplot. In this way it capitalizes on the capacity of **ggplot2** to create graphics as objects. We told it how many rows we wanted, which is more relevant if we have more than two graphics. There are multiple ways to customize the organization of a multiplot. Here I have used the **widths=** argument to accommodate the different widths of the two original graphics. (Try without this command and see what happens.)

Substantively, seeing these two graphs side-by-side helps us understand the spatial and temporal trends of 311 reports. On the right, we see that some neighborhoods produce many requests for services (e.g., South Dorchester) and others generate relatively few (e.g., Charlestown). It is not immediately clear whether this is a function of population, area, or need, however. We see on the left, though, that over time the demand for 311 services has climbed across the board. Nonetheless, these rises have been steadier in some areas (e.g., South End) than others (e.g., Charlestown).

9.4 Streamgraphs

Suppose we want to go beyond the superficial coordination of the multiplot and combine time and space simultaneously. This can be accomplished with a *streamgraph*, which layers counts by categories across time (or some other numeric variable, but most often time). A streamgraph is especially fun because it can be interactive, with the viewer being able to click on layers of the stream and turn on and off specific categories to see how they contribute to the overall pattern. The downside of streamgraphs is when there are so many categories as to overwhelm the graphic, or either the categories or the timespan lacks variation. Streamgraphs can be made with the fittingly titled package **streamgraph** (Rudis, 2019), though it has been made by a developer who has not put the package on CRAN. Thus, we have to use the following code to access it from GitHub.

```
devtools::install_github("hrbrmstr/streamgraph")

require(streamgraph)
```

9.4.1 Executing a Streamgraph

Returning to our previous example, it might be interesting to look at something more specific than the total volume of 311 requests. A given case type, like snow removal requests, might tell a more detailed story about variations in space and time. We then need to recreate our aggregate data set, this time filtering for cases whose `TYPE` references snow.

```
stream_data <- CRM %>%
    filter(BRA_PD %in% c('Back Bay/Beacon Hill',
              'Allston/Brighton','Central', 'Roxbury',
              'North Dorchester', 'South Dorchester',
              'South End', 'West Roxbury','East Boston',
              'Charlestown'), LocationID!=302615000) %>%
  group_by(BRA_PD, year) %>%
  summarise(Snow = sum(str_detect(TYPE,'Snow')))
```

Note that this data frame contains the three basic elements required by a streamgraph: (1) time or other continuous variable that we want to be the basis of the x-axis, or the length of the stream, as it were (i.e., `year`); (2) a numeric variable for the y-axis, which reflects the width of the stream (i.e., `snow`); and (3) a categorical variable that will differentiate the layers of the stream (i.e., `BRA_PD`). We are now ready to run the streamgraph command, whose customizations are constructed with piping.

```
pp <- streamgraph(stream_data, key="BRA_PD", value="Snow",
                  date="year", interactive = TRUE,
                  height="300px", width="1000px") %>%
  sg_legend(show=TRUE, label="names: ") %>%
  sg_axis_x("%Y")
pp
```

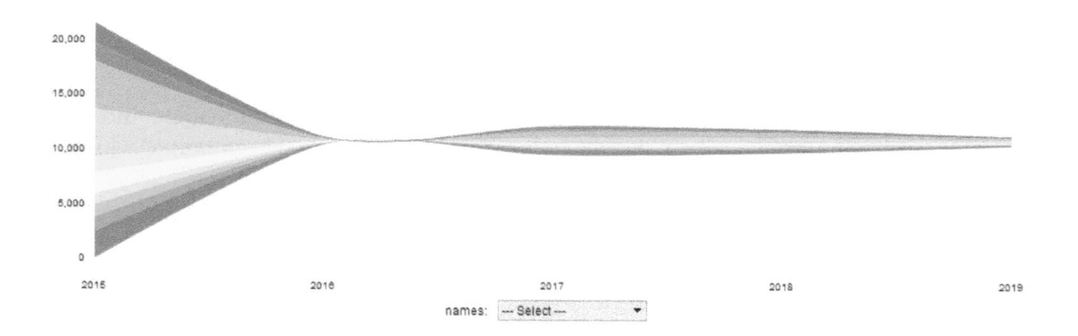

The result is a rather colorful interactive graph. We can see multiple layers, each of which corresponds to one of our ten neighborhoods. Notably, some are thicker than others, reflecting greater need in some neighborhoods than others. Also, see how the graph is really thick in 2015, nearly absent in 2016, and then rebounds more modestly in the years that follow. This reflects the annual patterns of snow in Boston. 2015 saw a record-breaking amount of snow; 2016, not so much.

Streamgraphs are interactive, as well. By mousing over them we can see the counts of snow removal requests for each neighborhood each year. We can scroll left to right to see the full timespan. Last, we can use the drop-down menu to highlight particular categories (i.e., neighborhoods) in the data. Again, this can be generalized to any count variable for any category. Technically we can also make the x-axis represent any continuous numeric variable we want, though it would have to, like time, make sense to represent as the stages of a stream.

9.5 Heat Maps

Sometimes we want to look at the distribution of two variables together. We have seen this previously with two-variable tables in Chapter 4, which contain counts for the intersection of each value of the two variables. This is also known as a crosstab. Technically, this is exactly what we have done in this chapter so far, as we have created data frames with counts of cases by neighborhood in each year. It may be more engaging and immediately interpretable, however, to represent this graphically. This is what a *heat map* does by translating the numbers in a crosstab into colors proportional to their size. Heat maps, though, may not be as effective when one of the variables is dominated by one or two values, as these will obscure any of the richness of the crosstabs—which is the whole reason we would want to use a heat map in the first place. Heat maps are made with `ggplot2`'s `geom_tile()` command .

9.5.1 A Heat Map for Two Categorical Variables

Building further on our example from the streamgraph, let us consider the distribution of multiple case types, including snow removal, graffiti removal, and some other key issues, across neighborhoods. We will first need to create a data frame that has counts organized by these two variables (BRA_PD and TYPE). We will again subset to 10 key neighborhoods and only 6 key case types.

```
heat_data <- CRM %>%
      filter(BRA_PD %in% c('Back Bay/Beacon Hill',
                  'Allston/Brighton','Central','Roxbury',
                  'North Dorchester','South Dorchester',
                  'South End','West Roxbury','East Boston',
                  'Charlestown'),LocationID!=302615000,
             TYPE %in% c("Graffiti Removal",
                  "Poor Conditions of Property", 'Bed Bugs',
                  'Snow Removal', 'Rodent Activity',
                  'Improper Storage of Trash (Barrels)')) %>%
  group_by(TYPE, BRA_PD) %>%
  summarise(Frequency = n())
```

We can then create a heat map using `geom_tile()` with our two main variables as `x` and `y` and `Frequency`, or our count variable, as the `fill=` argument.

```
p <- ggplot(heat_data, aes(x=BRA_PD, y=TYPE,
                    fill= Frequency)) + geom_tile() +
  theme(axis.text.x  = element_text(angle=45, hjust = 1,
                         vjust=1, size=8),
                 axis.text.y  = element_text(size=8)) +
  ylab("") + xlab("")
p
```

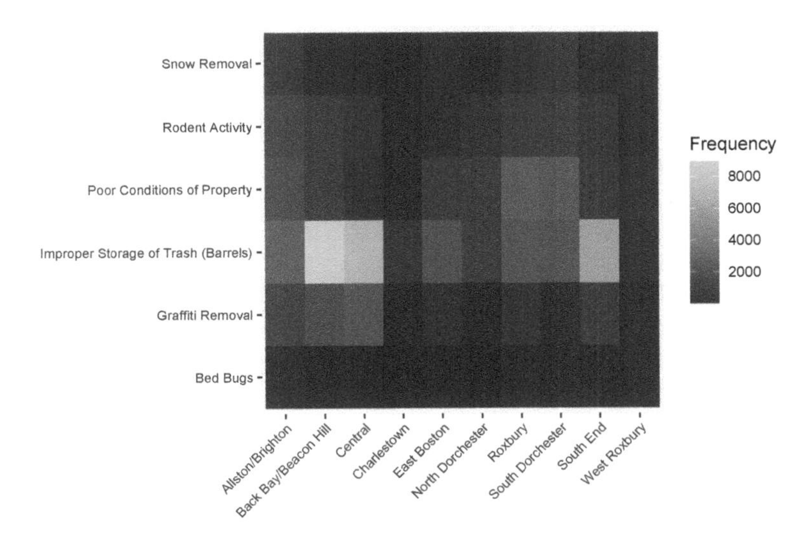

We can see here a few interesting trends. First, improper storage of trash is the most frequent complaint in most if not all neighborhoods. This is especially true for Back Bay/Beacon Hill and South End, which are high-density downtown neighborhoods. But these neighborhoods do not necessarily have the most cases for all types. We see that Roxbury and South

Dorchester, which are outside of downtown but have large populations and high levels of economic disadvantage, have the most complaints regarding poor conditions of property.

9.6 Correlograms

Our heat map revealed an interesting pattern. It appears that issues with trash disposal were not entirely correlated with those reflecting blight and dilapidation of properties. Is it possible that different case types might cluster across neighborhoods in different ways? A correlogram will help us to zoom out and view the strengths of the relationships between multiple variables, quantified as correlation coefficients. We will learn more about correlation coefficients and what they mean in Chapter 10, but for now just know that they range from -1 to 1; a positive value indicates that two variables rise and fall together and a negative value indicates that as one goes up the other goes down; and larger absolute values mean the positive or negative relationship is stronger (i.e., more consistent). Correlograms can be made using the `ggcorrplot` package (Kassambara, 2019). They are generally a strong tool provided the question is, "How do a bunch of variables relate to each other?", but they can be difficult to interpret if there are too many variables.

9.6.1 Creating a Correlogram

To create a correlogram, we first need a set of variables that we want to correlate for a given unit of analysis. In this case, our variables are counts of different case types and our unit of analysis is the neighborhood. Because we no longer need to limit to a subset of neighborhoods to make the graphics readable, we will expand to all places.

```
require(ggcorrplot)
correl_data <- CRM %>%
  filter(!is.na(BRA_PD)) %>%
   group_by(BRA_PD, year) %>%
   summarise(graffiti = sum(TYPE == "Graffiti Removal"),
    poor_cond = sum(TYPE == "Poor Conditions of Property"),
    bedbugs = sum(TYPE == 'Bed Bugs'),
    snow = sum(TYPE == "Snow Removal"),
    trash = sum(TYPE == "Improper Storage of Trash (Barrels)"),
    rodents = sum(TYPE == "Rodent Activity"))
```

We then need to create a correlation matrix, which quantifies the strength of relationship between all variables as correlations. This is done by using the `cor()` command on all variables except the first (which is the `BRA_PD` identifier).

```
corr <- cor(correl_data[,-1])
```

We are now ready to create a correlogram based on this correlation matrix.

```
ggcorrplot(corr, hc.order = TRUE,
  type = "lower",
```

```
    lab = TRUE,
    lab_size = 3,
    method="circle",
    title="Correlations between 311 Case Types by Neighborhood")
```

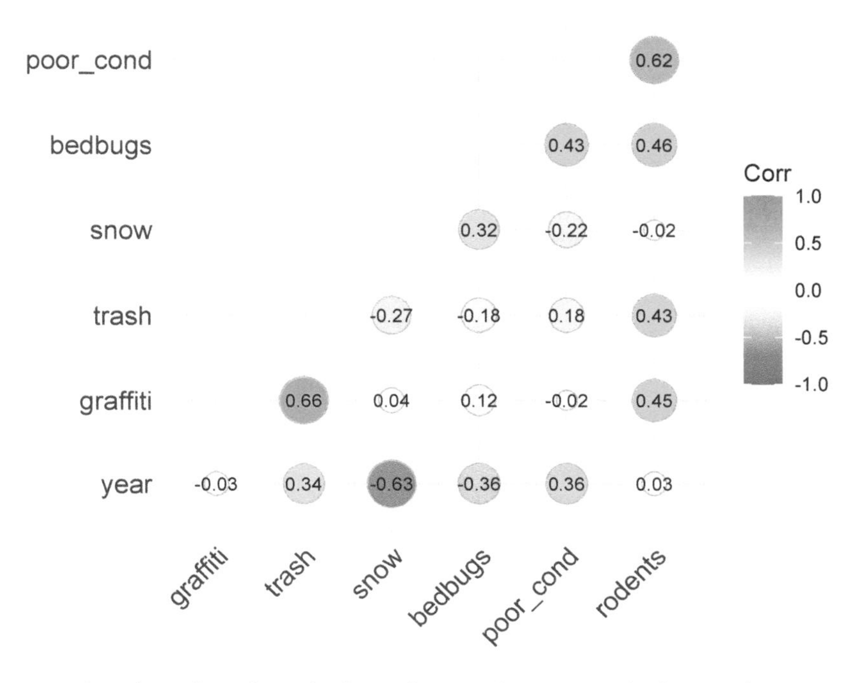

The function takes the values directly from the correlation matrix, but we have customized in a few ways. We have specified that the variables be ordered in a way that highlights groups of variables that correlate strongly with each other (`hc.order=TRUE`), that the correlations be represented below the diagonal (`type="lower"`; because we do not need them reproduced in both halves), that the values be printed on the graph (`lab=TRUE`) at a size of 3, and that the magnitude of the correlations be reflected in the size of circles (`method="circle"`).

Looking at this graph confirms our suspicion that the poor disposal of trash and the prevalence of poorly maintained properties do not coincide much at all, featuring one of the smallest values in the chart. There are also some interesting relationships that tell a fuller story. The prevalence of poorly maintained properties in a neighborhood coincide with more rodent activity and bed bugs, which stands to reason. Meanwhile, poor trash disposal correlates strongly with graffiti. This appears to reflect the two constructs we met in Chapter 6 of neglect of private spaces and denigration of public spaces. Last, snow tends to correlate with the first group, though it is not clear why this would be the case, except if those areas also have more people and roads (and more residential roads, which are often plowed after main arteries and thereby elicit more calls for plowing).

9.7 Animations

Some might say that animations are the holy grail of data visualization. Instead of a single, static graphic, they move! They change, showing shifts in the data over time (or some other variable). Think about how impressed your audience will be! To the uninitiated, animations seem really complex, but they are not actually that difficult to create. To understand why, it helps to think back to those little flipbooks you might have had as a kid for animating a short scene, or to documentaries about how cartoons are made. What we see as continuous movement is actually many individual images viewed one-at-a-time in a high-speed sequence. In the same way, data animations are multiple graphics shown one-at-a-time according to a logical sequence (most often a timeline). We can do this in R with the **gganimate** package (Pedersen and Robinson, 2020). We will also need the **gifski** package to render the graphics into an animated *.gif* (Ooms, 2021).

While animations are one of the most sophisticated visualization techniques, they also call for the greatest caution. Analysts must always ask themselves if an animation is making the patterns in the data more accessible, or if it is just superfluous window-dressing that may do more to distract than to enhance. We will consider this as we walk through two examples.

9.7.1 Animating a Bar Graph

Keeping in mind that an animation is essentially stapling together a series of related graphics, let us start simply and create a single, static graph upon which we might want to build. We will again subset to a handful of neighborhoods and case types, though we do not need to aggregate this time around.

```
require(gganimate)
require(gifski)
options(scipen=10000)

data_animate <- CRM %>%
      filter(BRA_PD %in% c('Back Bay/Beacon Hill',
                  'Allston/Brighton','Central','Roxbury',
                  'North Dorchester','South Dorchester',
                  'South End','West Roxbury','East Boston',
                  'Charlestown'),LocationID!=302615000,
            TYPE %in% c("Graffiti Removal",
                  "Poor Conditions of Property", 'Bed Bugs',
                  'Snow Removal', 'Rodent Activity',
                  'Improper Storage of Trash (Barrels)'))
```

We can now create a bar chart of the number of records per case type.

```
ggplot(data_animate, aes(x=TYPE)) +
  geom_bar(stat='count') + coord_flip()
```

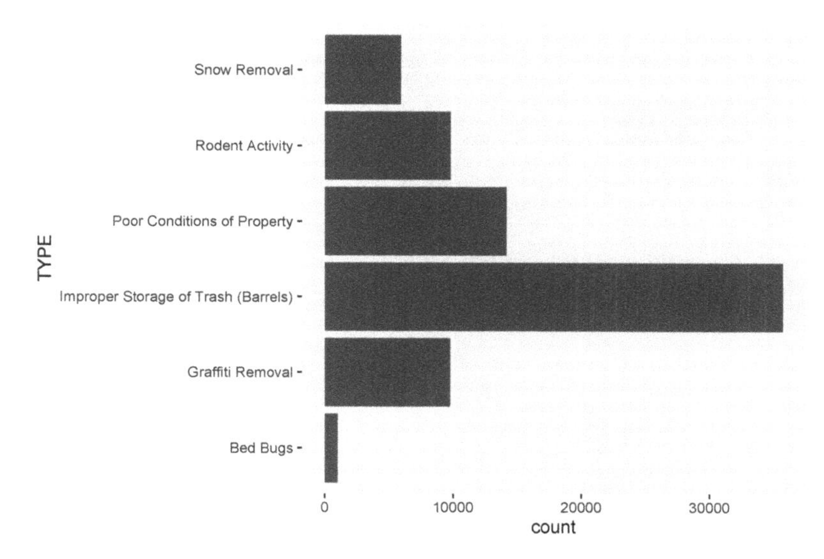

This describes a pattern that we surmised from the heat map in Section 9.5: trash disposal issues are the most common of the case types we have chosen, followed by requests for snow plowing and poor conditions of properties. Bed bugs are the least common. But has this distribution shifted over time?

```
p <- ggplot(data_animate, aes(x=TYPE)) +
  geom_bar(stat='count') + coord_flip() +
  transition_states(
    year,
    transition_length = 2,
    state_length = 1
  ) +
  ease_aes('sine-in-out') +
  labs(title = 'Number of cases: {closest_state}')

animate(p, duration = 10, fps = 20, width = 400, height = 400,
        renderer = gifski_renderer())
```

This code has a lot of components, so let us walk through it. We start with the same commands for the bar graph through `coord_flip()`. We then have a command for `transition_states()`, which is from the `gganimate` package. The arguments here are for: (1) the variable over whose values we want to repeatedly remake our graph (i.e., `year`); (2) how long we want transitions between images to take (`transition_length=`); and (3) the length that each image should stay stable (`state_length=`). The `ease_aes()` command specified a mathematical logic for making the movement between states smooth (here we use sine curves, which make change the slowest at the beginning and end of a transition). The next line of code features the `animate()` function, which specifies the object we just created (`p`), the duration, the frames per second (remember the analog to cartoons?), the width, height, and a renderer for stitching all of the pieces together. Here we use `gifski_renderer()` for that purpose. Note that we also enter a dynamic field as part of the text in the label so that the title changes to reflect the data currently visible in the graph (`{closest_state}`).

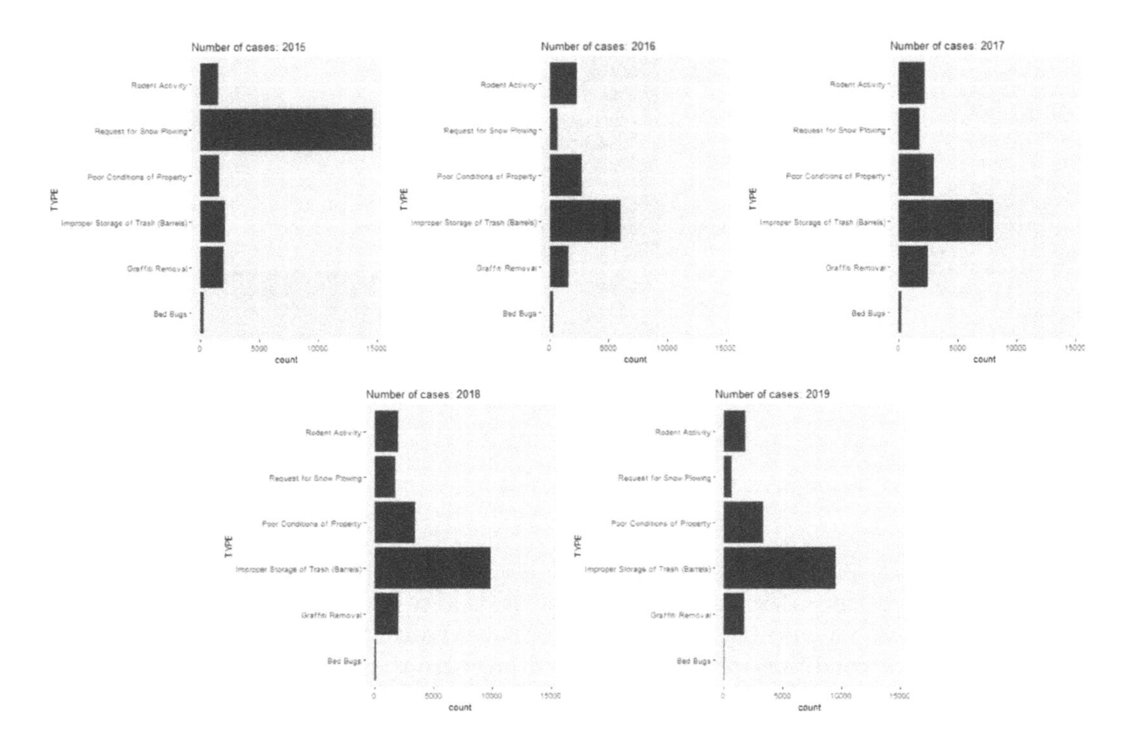

FIGURE 9.2
The frequency of different types of 311 requests across the months of the year. An animation would transition smoothly between these images.

Because this is a book, we cannot watch the animation live (though see the HTML version at ui.danourban.com if you would like to see it). But we can see the multiple states that were stitched together to create the animation, lain out in Figure 9.2. Here we see that a lot of case types stay relatively stable between years. The most striking changes are in requests for snowplows and improper storage of trash. For the former, we have already seen that 2015 had way more requests for snowplows because of a record-breaking amount of snowfall that winter. For the latter, it is not clear why this might be the case. It could be that trash storage was less of a problem in 2015, but it is more likely that something changed in how the 311 system coded these types of issues at that time. We can also save this as a *.gif* that you can insert into a presentation or other medium.

```
anim_save("Unit 2 - Measurement/Chapter 09 - Advanced Visualization/Example/ty
barplot x year.gif")
```

While we are at it, what about animating the same graphic across neighborhoods?

```
p <- ggplot(data_animate, aes(x=TYPE)) +
  geom_bar(stat='count') + coord_flip() +
  transition_states(
    BRA_PD,
    transition_length = 2,
    state_length = 1
  ) +
  ease_aes('sine-in-out') +
```

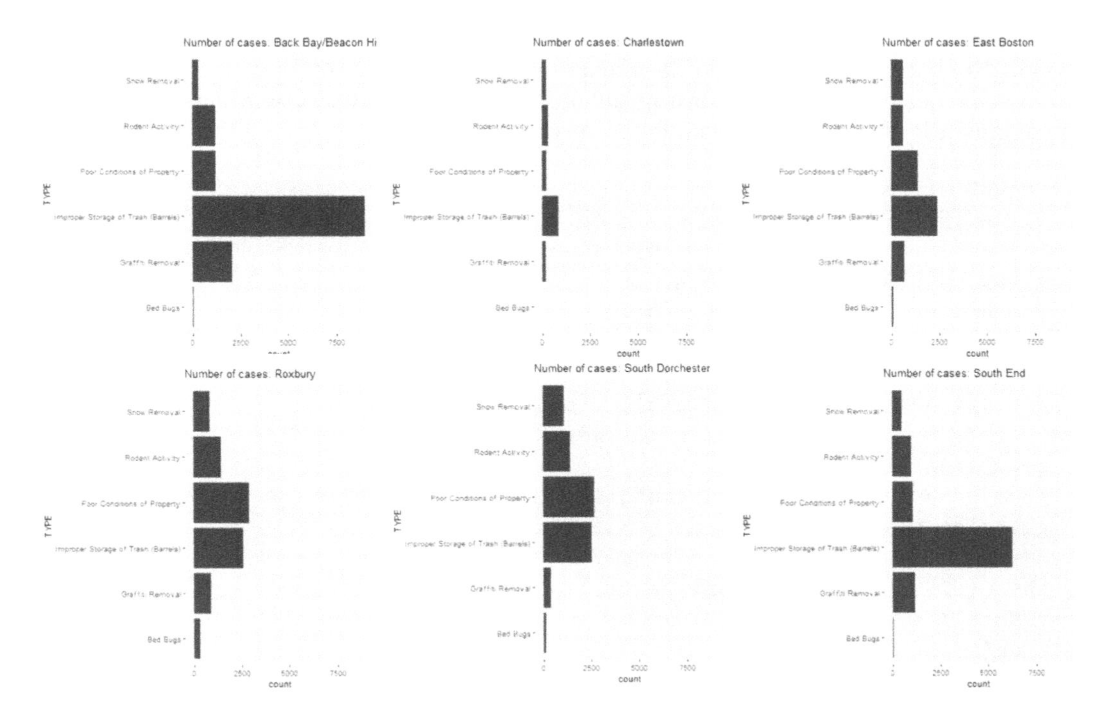

FIGURE 9.3
The frequency of different types of 311 requests across a subset of neighborhoods. An animation would transition smoothly between these images.

```
  labs(title = 'Number of cases: {closest_state}')

animate(p, duration = 10, fps = 20, width = 400, height = 400,
        renderer = gifski_renderer())

anim_save("Unit 2 - Measurement/Chapter 09 - Advanced Visualization/Example/
type barplot x nbhd.gif")
```

In Figure 9.3 we see a lot more action, and it corresponds to the relationships we saw previously between case types in the heat map and correlogram. Reports of improper storage of trash are remarkably high in Back Bay and Beacon Hall, but this is tempered in places like Roxbury and South Dorchester, where complaints about the upkeep of properties and snowplows become more prevalent, both in a relative and absolute sense.

9.7.2 Animating a Line Graph

Let us make one more animation, one that illustrates a different `gganimate()` command. Whereas we previously used the `transition_states()` command to animate a series of replications of the same graph, the `transition_reveal()` command allows us to reveal the elements of a graph sequentially. This is often used for line graphs, as it can "draw" the lines step-by-step.

We can start by creating a line graph that shows the monthly levels of calls by three neighborhoods: Roxbury, which is a majority-minority residential neighborhood; Back Bay/Beacon Hill, which is an affluent downtown neighborhood; and West Roxbury, which is an affluent, predominantly White neighborhood with a suburban feel.

```
data_git <- CRM %>%
  filter(BRA_PD %in% c("Roxbury", "Back Bay/Beacon Hill",
                       "West Roxbury")) %>%
  group_by(BRA_PD, month) %>%
  summarise(count_per_nhood = n())

data_git %>%
  ggplot( aes(x=month.abb[month], y=count_per_nhood,
              group=BRA_PD, color=BRA_PD)) +
  geom_line() +
  geom_point() + scale_x_discrete(limits = month.abb) +
  ggtitle("Changes in the number of reported requests
in three neighborhoods during the year") +
  theme(plot.title = element_text(size=9)) +
  ylab("") +
  xlab("Month")
```

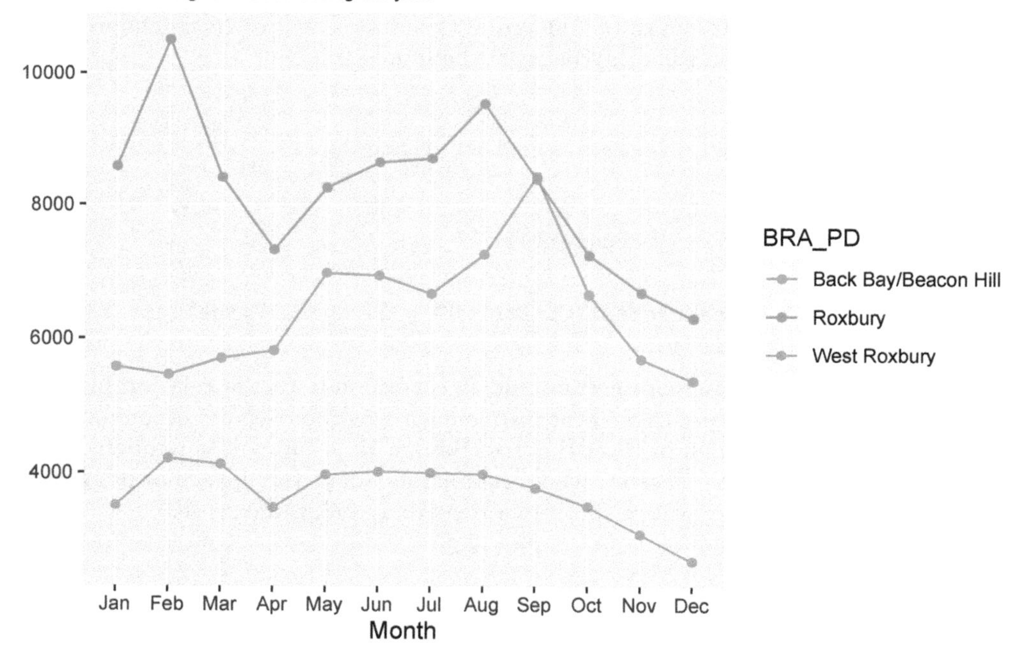

We see here that Roxbury generally has more requests, especially in February, likely owing to the uptick of snowplow requests in residential neighborhoods at that time. Back Bay/Beacon Hill sees a sharp rise in September, probably in response to the influx of college renters and the "move-in" season. Meanwhile, West Roxbury is relatively quiet, with just a moderate uptick in the snowy parts of the winter. Now, let's turn this into an animation using the `transition_reveal()` command, with the month variable as our order. The example images in Figure 9.4 illustrate how the "revealing" unfolds.

```
p2<-data_git %>%
  ggplot( aes(x=month.abb[month], y=count_per_nhood,
              group=BRA_PD, color=BRA_PD)) +
  geom_line() +
  geom_point() +
  scale_x_discrete(limits = month.abb) +
  ggtitle("Changes in the number of reported requests
          in three neighborhoods during the year") +
  theme(plot.title = element_text(size=9)) +
  ylab("") +
  xlab("Month") +
  transition_reveal(month)

animate(p2, duration = 10, fps = 20, width = 400, height = 400,
        renderer = gifski_renderer())
```

Again, we can save this as a .gif file.

```
anim_save("neighborhood cases x month.gif")
```

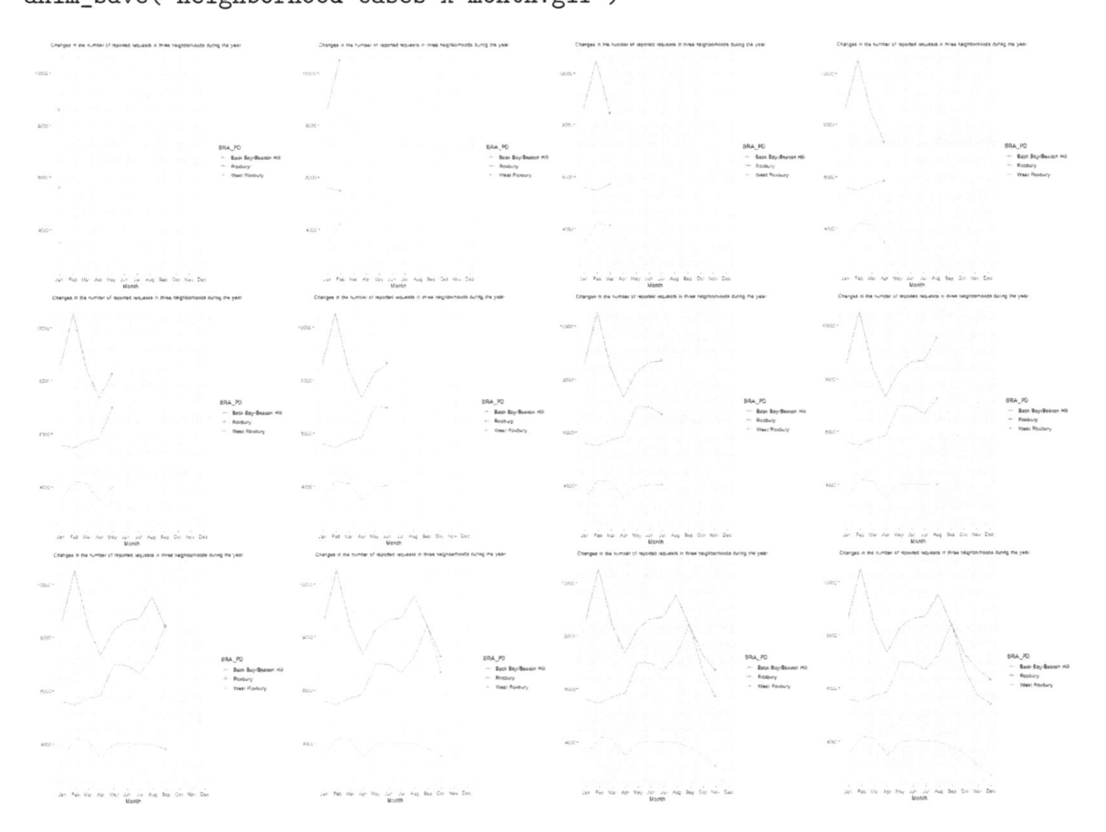

FIGURE 9.4
How the frequency 311 requests unfolds across the year for three neighborhoods. An animation would transition smoothly between these images.

9.7.3 Animations Redux

We have now animated bar graphs and line graphs. Fun and fancy, right? It is worth scrutinizing this process to think about what we learned about animations in general. I would suggest three main takeaways. One is that animations are simpler than they seem, both conceptually and technically (they can be finicky, though; you are bound to run into errors as you try this on your own). They are simply the act of stitching together a series of graphs based on a template, and they only require a few specialized commands. This is extensible to all sorts of `ggplot()` objects, including maps. Second is that they, like many graphics, suffer when there is too much complexity. If we had too many stages, categories, or other elements, they would lose their interpretability. This is especially true here given that the audience is trying to capture information as the graphic moves between states. Unlike a static graph, what you see now will be replaced by different information in a moment. Third is the related question of whether an animation in fact helps to communicate the information. I would argue that the animation of the bar graph over neighborhoods was striking, potentially making it worth the extra sophistication. I am not entirely convinced that revealing the lines sequentially by month in the line graph was necessarily the best way to present that information, especially because the lines disappeared at the end. A talented visualizer must always consider this potential weakness of animation to be little more than a fancy technique for its own sake.

9.8 Summary

We have now learned how to do a series of advanced visualizations that you can apply to any data set you might like. These are also just a start—there are so many other techniques out there that you might use. As you seek these out, you will be able to incorporate many of them seamlessly into your toolbox as they build on the syntax of `ggplot2`. Just keep in mind two things. First, these tools are very particular about how data translates into a visual and require you to reconfigure your data to meet those expectations, often with aggregations. Second, like my graduate school professor would happily remind you (and me, for that matter), always think about what information you are trying to communicate and how the visualization you create will help you to do so. There is no need to make fancy visuals for their own sake if they do not accomplish that fundamental goal. To summarize, you have learned in this chapter how to create:

- *Multiplots*, which combine multiple graphical components into one image;
- *Streamgraphs*, which are interactive representations of the frequency of categories over time;
- *Heat maps*, which are visually appealing representations of the relationships between two sets of categories;
- *Correlograms*, which are visual representations of the strength of correlations between multiple variables;
- *Animations*, which place multiple visualizations in a sequential order to show change, often over time.

You have also learned the importance of determining when it is appropriate and necessary to use these sorts of tools and when something simpler might suffice.

9.9 Exercises

9.9.1 Problem Set

1. For each of the following visualizations, describe (a) the type of relationship it is best equipped to communicate, (b) when you want to avoid it, (c) the package and function(s) needed to create it, and (d) the data structure or variables needed.
    a. Animation
    b. Correlogram
    c. Heat map
    d. Multiplot
    e. Streamgraph
2. For each of the following, use the 311 records to make a new version of graphics from the worked example using new variables and/or subsets.
    a. Animation
    b. Correlogram
    c. Heat map
    d. Multiplot
    e. Streamgraph
3. Visit the Kantar Information is Beautiful web site (https://www.informationisbeautifulawards.com/) and browse some of the submissions. Identify three that you find interesting and believe we have not yet learned the techniques for making. See if you can determine the name of that type of graphic and what package or command might make the same graphic in R (Hint: some submissions will have GitHub pages or other resources that make their code public).
    a. Extra: See if you can execute that type of graphic on the 311 records or another data set.

9.9.2 Exploratory Data Assignment

Complete the following working with a data set of your choice.

1. Choose at least three of the visualization techniques presented in this chapter and execute them on your data set.
2. Be sure to interpret the results. In doing so, justify for the reader why this advanced technique helps us to better understand the data.
3. Extra: Browse the submissions on the Kantar Information is Beautiful web site https://www.informationisbeautifulawards.com/ and identify one type of visualization which we have not yet learned how to make. See if you can determine the name of that type of graphic and what package or command might make the same graphic in R (Hint: some will have GitHub pages or other resources that make their code public). Apply that technique to your data set, interpret the results, and justify that technique's value for revealing that particular piece of information.

Measurement: *Unit II Summary and Major Assignments*

Summary and Learning Objectives

Unit II has focused on the technical and interpretive skills necessary to translate records into custom measures describing units of analysis referenced in the data, like people, places and things, and how to communicate them both statistically and visually.

Technical Skills

- Creating aggregate measures;
- Merging data sets on shared key variables;
- Importing, manipulating, and mapping spatial data in the form of shapefiles (.shps);
- Creating various advanced visualizations for two or more variables.

The packages we have learned include:

- `sqldf` for querying data frames and creating aggregations;
- `sf` for working with spatial data as "simple features";
- `ggmap` for creating maps of spatial data with ggplot2;
- `gridExtra` for creating visualizations that include two or more individual graphics;
- `streamgraph` for creating graphs of changes in counts across categories over time;
- `ggcorrplot` for creating correlograms of multiple correlations between variables;
- `gganimate` and `gifski` for making data animations.

Interpretive Skills

We have learned how to address the "missing ingredients" of a naturally occurring data set, including:

- Specifying the unit of analysis available from a schema;
- Defining constructs of interest and isolating the information needed to measure them;
- Identifying sources of bias and establishing validity;
- Describing the structure and organization of spatial data;
- Evaluating the best visualization technique for communicating a particular piece of information.

Unit-Level Assignments

Community Experience Assignment

The community exploration assignments in this book are designed to align skills you have learning with real-world contexts. They are most useful in conjunction with the Exploratory Data Assignments at the end of each chapter, especially when you have been working through them with a single data set. They provide an opportunity to "ground truth," or really evaluate the assumptions and objectives that have guided your analysis thus far. There will be one in each unit. These can also be combined with a service-learning or capstone oriented course.

This second community exploration assignment will focus on what we can learn about a neighborhood through the lens of public media. Please:

1. Select a neighborhood based on something notable about the measure(s) that you are developing, e.g., the highest or lowest value, an interesting combination of values, etc.
2. Explore how this neighborhood is represented and portrayed through news articles, web sites, blogs, community organizations, social media, and other online resources.
3. Write a 3-5 page virtual walk memo describing the logic for why you visited this place, what you discovered, and what this tells you about the interpretation of your data. It will need to include illustrative images from the media you utilized.

Post-Unit Assignment: Constructing a Novel Metric from Raw Data

This unit of the book has focused on using a theoretical concept to guide the construction of one or more novel metrics from our raw data sets, describing a particular aggregate unit of analysis (e.g., parcel, restaurant, neighborhood), and then communicating the distribution of those metrics. This will culminate in a paper that will be organized in a series of short sections:

- An Overview of the measure and why it is interesting. About one-two paragraphs.
- A textual description of how the measure was constructed, justifying any specific decisions that were made (for example, categorization of case types). This will include:
  - A summary of new variables at the record level (i.e., the original database) that were constructed first as part of the overall calculation of your aggregate measure(s).
  - A summary of the new aggregate measure(s) that you've calculated and how you have done so.
- A short description of the new variable's distribution and/or values. What's the mean? Where is it highest? Anything else fun or interesting? etc. This should include at least one tabular visualization (e.g., histogram) and a map, if appropriate.
- An appendix with an annotated R syntax articulating all steps required to create the measure(s) (should be your portion of the R syntax copy-and-pasted; see below).

Suggested Rubric (Total 10 pts.)

Measurement-Concept: 3 pts.

Measurement-Execution: 2.5 pts.

Visualization: 2.5 pts.

Details (Grammar, etc.): 2 pts.

Unit III

Discovery

10

Beyond Measurement: Inferential Statistics (and Correlations)

One of the oldest observations in the study of cities is that bad things—crime, medical emergencies, low academic and career achievement, poor upkeep of buildings—tend to cluster together in certain communities. Possibly the first person to systematically examine this clustering was Charles Booth, whose 1889 maps of London color-coded each street according to its level of "poverty" (Booth, 1889). Notably, poverty was measured as a collection of issues, including low income, unpleasant jobs (including prostitution), crime, poor sanitation (or physical disorder, in modern parlance), public drinking and fighting, and so on. As shown in Figure 10.1, streets colored black were home to the "lowest class" and red and yellow reflected the "well-to-do" and "wealthy."

Not long after Booth, two independent set of scholars also recognized that certain communities suffered from many issues, whereas other communities saw few to no such issues. Interestingly, they both proposed similar explanations for this outcome. W.E.B. DuBois, studying Black communities in Philadelphia, argued that certain social challenges, like weak community institutions (e.g., churches, community centers), lack of opportunity, and limited relationships between neighbors, were responsible for these ills (DuBois, 1899). Likewise, members of the University of Chicago's new Department of Sociology (affectionately referred to now as the Chicago School of Sociology) , led by Robert E. Park and Ernest W. Burgess, linked the clustering of maladies in certain communities to the *social organization*, or strength of relationships between neighbors and local institutions, which could be called upon to manage behavior and maintenance in public spaces and advocate for local priorities (Park et al., 1984). They believed that the social organization was undermined by demographic characteristics that limited the ability of a community to build such relationships, including poverty, residential turnover (i.e., households moving in and out), and immigrant concentration. This perspective has been one of the most influential theories in urban science, inspiring lines of research in sociology, criminology, public health, developmental psychology, and community development, among others. Both DuBois and the Chicago School noted that the strength of the social organization was weaker in communities populated by ethnic minorities, especially Black Americans. (Note: The long-term impact of the Chicago School's version of the theory is not necessarily evidence that it was "better" than DuBois'. It is probably more indicative of the immediate influence that White scholars in the early 20th century were able to have with their peers in the academy relative to a preeminent Black scholar.)

Analytically speaking, the concept that bad things cluster together relies on *correlations*—that is, whether some variables tend to go up and down together across a set of objects. Does crime tend to be higher in neighborhoods with lower income? Are medical emergencies higher in communities with dilapidated housing? Are all of these issues higher in communities with more non-White residents and lower in communities with more White residents? This chapter begins the unit on *Discovery*, which focuses on *inferential statistics* and the use of various analytic techniques to evaluate relationships among variables. This includes, but is certainly not limited to, correlations. We will learn about the conceptual basis for inferential statistics and demonstrate them for the first time with correlation tests, setting the stage for additional statistical tools we will learn in the proceeding chapters.

10.1 Worked Example and Learning Objectives

In this chapter we will introduce the fundamental concepts underlying inferential statistics. We will then illustrate them using correlations, specifically evaluating whether "bad things cluster together" in the neighborhoods of modern Boston . We will also look at how these things correlate with the demographic factors that W.E.B. DuBois and the Chicago School of Sociology believed were relevant to a community's social organization and, in turn, the various outcomes occurring there. Along the way, we will learn to:

- Differentiate between descriptive and inferential statistics, including the logic of studying samples;
- Describe the distributions of numeric variables and the assumptions that statistical tests make about them;
- Interpret and report the magnitude of a relationship between variables (i.e., effect size) and whether it is likely to be observed by chance (i.e., significance);
- Run and interpret correlations between pairs of variables using multiple commands;
- Visualize a series of correlations with the `GGally` package.

For this chapter we will return to census indicators of Boston's census tracts. To expand the story, however, we will leverage two other data sets that BARI publishes annually: indicators (or *ecometrics*, a term we learned in Chapter 6) of crime and medical emergencies from 911 dispatches. Before getting started we will download, import, and merge these and then subset to a series of variables that will be the focus of our analysis. We will also use `tidyverse` throughout.

Links:

Demographics: https://dataverse.harvard.edu/dataset.xhtml?persistentId=doi:10.7910/DVN/XZXAUP

911: https://dataverse.harvard.edu/dataset.xhtml?persistentId=doi:10.7910/DVN/XTEJRE

You may also want to familiarize yourself with the data documentation posted alongside each.

Data frame name: `tracts`

```
demographics<-read.csv('Unit 3 - Discovery/Chapter 10 - Inferential
Statistics and Correlations/Example/ACS_1418_TRACT.csv')
eco_311<-read.csv('Unit 3 - Discovery/Chapter 10 - Inferential Statistics
and Correlations/Example/311 Ecometrics Tracts Longitudinal.csv')
eco_911<-read.csv('Unit 3 - Discovery/Chapter 10 - Inferential Statistics
and Correlations/Example/911 Ecometrics CT Longitudinal, Yearly.csv')
```

FIGURE 10.1
Various urban scientists have sought to demonstrate that and explain why lots of unwanted outcomes, like crime, poor health, low academic and career achievement, and low income, cluster in certain communities. This includes Charles Booth, whose maps of London in 1889 quantified "poverty" street-by-street (top), W.E.B. DuBois, who studied Black communities in Philadelphia in the 1890s (bottom left), and the early Chicago School of Sociology, who studied the emerging immigrant neighborhoods of industrial Chicago (bottom right). (Credit: booth.lse.ac.uk; Wikimedia commons, Public Domain; ludilozezanje at deviantart.com)

```
tracts<-merge(demographics,eco_311,by='CT_ID_10',all.y=TRUE)
tracts<-merge(tracts,eco_911,by='CT_ID_10',all.y=TRUE)

tracts<-tracts %>%
select('CT_ID_10','White','Black','Hispanic',
       'MedHouseIncome','Guns_2020','MajorMed_2014')
```

10.2 Foundations of Inferential Statistics

10.2.1 Inferential Statistics and Samples

The two main types of statistics are descriptive statistics and inferential statistics, and each entails the analysis of a given set of items (i.e., people, places, things) known as a *sample*. *Descriptive statistics* describe, summarize, and communicate the features of one or more variables for the sample, typically using a small set of indicators or representations. We have learned and practiced the tools of descriptive statistics numerous times in this book by calculating means, medians, minima, maxima, and modes, and by creating frequency tables, histograms, and other visualizations. *Inferential statistics*, on the other hand, uses the characteristics of the sample to attempt to generalize (or make inferences, as the name implies) about the full population from which the contents of the data set are drawn. In this way, both descriptive and inferential statistics are conducted on samples, but whereas descriptive statistics take the sample at face value, inferential statistics seek to leverage the sample to understand the world more broadly. The tools for doing so are new to us and will be the focus of this unit of the book. We will specifically focus on statistics for numeric variables, though there are techniques for working with categorical and character variables as well.

To illustrate the distinction between descriptive and inferential statistics, suppose we analyze the standardized test scores of a sample of students from a single school. Descriptive statistics communicate the average score, the highest and the lowest scores, and so on. Inferential statistics, however, would take these values and use them to come to conclusions about scores for the entire school.

There are two things that must always consider when conducting inferential statistics. The first is an estimate of uncertainty. For instance, when conducting descriptive statistics, we are confident of the average test score in our sample because we have those numbers. But how this information generalizes to the full school is imprecise. It is possible that this sample is a bit above or below average for the school as a whole. Thus, we need to quantify and account for this uncertainty when communicating our inferences. You have probably seen this when reading about a public poll for which the results have a "margin of error," which seeks to communicate the range that the "true" or population mean most likely falls within. A critical element to uncertainty is the size of our sample, which is represented by the symbol n. As n gets bigger, we are increasingly confident in the characteristics that we are observing, because the amount of error created by the presence or absence of a single case gets smaller.

Second, we must take into consideration how representative our sample is of the full population. Samples will naturally deviate from the characteristics of the population, but we want to draw samples that are as insulated from bias as possible. For example, if we took the sample exclusively from a single classroom, or only from students who attended a special afterschool program for advanced math, or only from students in a single neighborhood, we may have an unrepresentative sample that is systematically biased. The easiest way to avoid this is taking a random sample from all students—that is, each student has an equal likelihood of being included. If this is not possible, we need to evaluate how our sample might have been biased and how this should inform our interpretation of the results. This is very similar to our discussion of the data-generation process and our efforts to interpret the contents of naturally occurring data. As we will return to below, it is important to keep sample size and bias separate. A very large sample might on the surface appear to have very little uncertainty, but it could still suffer from bias that could lead to erroneous conclusions if not handled properly.

10.2.2 Distributions of Numeric Variables

To make inferences about a given numeric variable for a population, we need to first understand its *distribution*, or the range of values present and their relative frequency. We have seen distributions repeatedly throughout this book, and they are often represented elegantly through histograms.

10.2.2.1 The Normal Distribution

Distributions come in many forms. The most common, and the one with which you are probably most familiar, is the *normal distribution* (or "bell curve"), as illustrated in Figure 10.2. Many, many variables naturally take on a normal distribution.

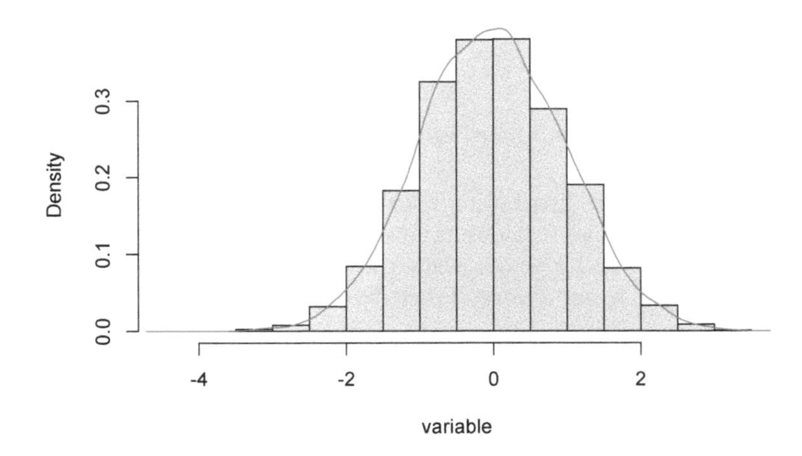

FIGURE 10.2
Histogram approximating the bell curve for mean = 1 and standard deviation = 0.

The normal distribution is an especially useful tool because its shape is defined by two statistics: it is centered at the mean; and its variation is defined by the standard deviation, such that 68% of cases sit within 1 standard deviation of the mean, 95% of cases sit within 2 standard deviations of the mean, and so on. This variation is symmetrical around the mean. In the example in Figure 10.2, the mean has been set to 0 and the standard deviation to 1.

10.2.2.2 Other Important Distributions

Not all variables are naturally normally distributed. The most common deviation from the normal distribution is referred to as *skew*, where the expected symmetry of the normal distribution is violated by outliers on the right side (i.e., high end) of the graph. This is demonstrated in Figure 10.3 and can be rather common. For example, income is well known to have positive skew, with certain individuals or communities having substantially higher incomes than would be expected in a normal distribution. There is also negative skew, in which the outliers are on the low end of the distribution.

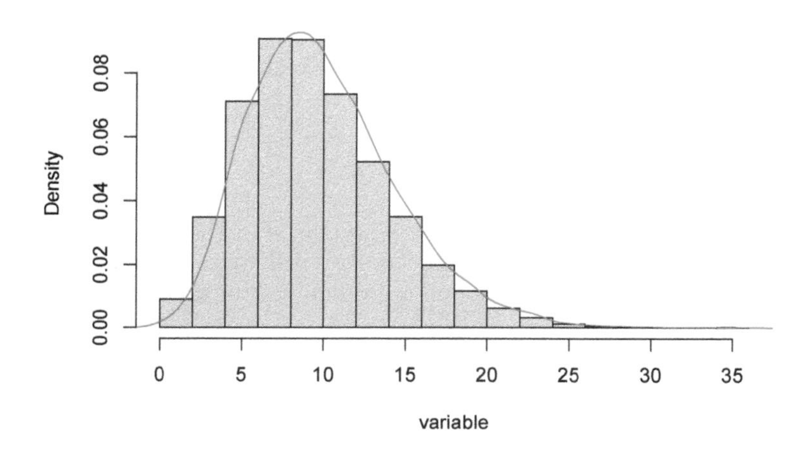

FIGURE 10.3
Histogram approximating a skewed distribution with outliers on the high side.

Last, we might consider a third distribution known as the Poisson distribution. The Poisson distribution is interesting because it is the natural distribution arising from a count variable. That is, if you have n events and x places where those events might occur and you distribute them at random, the result is not a normal distribution. Instead, it looks like what you see in Figure 10.4.

As you can see, a Poisson distribution has a large proportion of low values with a long tail of higher values. This is actually what would happen if you randomly distributed events across a set number of places (or people or things). This is a more complicated distribution whose analysis requires special techniques, but it is worth noting because it can come up rather frequently for variables that are based on counts, like the distribution of crimes across addresses.

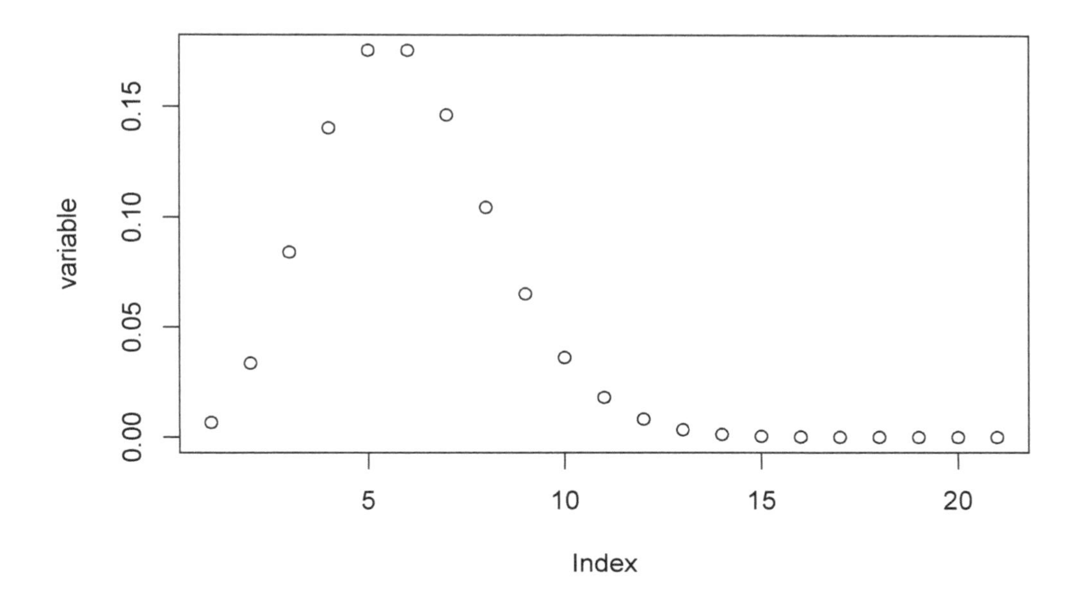

FIGURE 10.4
Plot of the values in a Poisson distribution with outliers on the high side.

10.2.3 Hypothesis Testing

"Testing hypotheses" is a phrase we often hear in common discourse, but what does it mean? It would be helpful to break down the concept into its two parts. A hypothesis is sometimes described as an "educated guess." This is accurate in spirit, though a bit simplistic. In inferential statistics, a *hypothesis* is a proposal that a certain characteristic of a single variable or relationship between two or more variables is true, with the proposal being based on some existing information or theory. The average standardized test score of this school is higher than 80% because of the large number of students with parents with high education levels. The crime level of a community is correlated with its median income, in part because of the social and economic challenges faced by lower income individuals that might lead to crime. Advertised rental prices might be higher during the winter because of lower supply. These are all hypotheses grounded in logical reasoning.

Once we have a hypothesis, we need to test it using inferential statistics. There are hundreds of different statistical tests for evaluating hypotheses and each is tailored to test a specific type of relationship. For example, in this chapter we are looking at correlations, which test whether two variables rise or fall together. Recall, however, that inferential statistics must always assess the uncertainty. It is not a matter of simply saying whether a hypothesis is "correct" or not. We must evaluate it relative to some benchmark. This benchmark is referred to as the *null hypothesis*. The null hypothesis, represented by H_0, assumes that there is no relationship between variables or that a single variable does not differ from expectations. The alternative hypothesis, represented by H_A, is then the proposal that such relationships or differences do exist. The alternative hypothesis does not typically predict a direction for the relationship, but simply assesses whether there is *any* relationship (e.g., positive or negative, above or below, etc.).

Returning to the three examples from the previous paragraph, H_0 would be: the average standardized test score of the school is 80%; the crime level of the community does not share a correlation with its median income; rental prices are the same in winter and summer. Meanwhile, H_A would be: the average standardized test score of the school is higher than 80%; the crime level of the community increases as its median income decreases; rental prices are higher in winter than summer. As you can see, the alternative hypothesis is often similar to how we would state our research question in simple English.

10.2.4 Effect Sizes and Significance

When we conduct a statistical test, the first thing we want to know is its *effect size*, or the magnitude of the relationship between two variables or characteristic of a single variable. The school has an average test score of 85%, which is 5% higher than 80%, for instance. For every increase in $10,000 in median income, how much does the crime rate of a community go down? How much more or less are rents between the summer and winter? All statistical tests communicate effect sizes in one or more ways. They are crucial to communicating the practical implications of our statistical test.

But how do we know if that difference is "real"? Again, uncertainty plays a critical role. We need to determine whether the patterns and relationships observed in the data are significant; that is, that they are strong enough that we should take them seriously and are not merely a product of chance. For instance, is an average score of 85% evidence that the school's performance is significantly different from 80%, or is it within the range of expected deviations? The significance of a statistical test depends on the effect size and the size of the sample. As the effect size becomes larger, significance is more likely because larger relationships are less likely to happen by chance. As the sample size becomes larger, significance is also more likely because we are less likely to have large deviations from expected outcomes. That is, if the average is 85% for 10 students, that is weaker evidence for difference from 80% than if the average were 85% for 100 students.

Significance is communicated with the *p-value* (for *probability*), which is the likelihood with which a certain outcome would occur if the null hypothesis were true. For example, $p = .15$ indicates that an effect of a given size or larger would occur 15% of the time by chance if the null hypothesis were true. The analyst needs to set a threshold for when the p-value is small enough for a result to merit classification as significant. We then tend to communicate significance relative to the established threshold (either being greater or less than the threshold) rather than in terms of their precise value. This is evaluated by calculating a *confidence interval* around the estimate of the effect size. The confidence interval is a range of possible relationships that might be likely given the effect size we saw.

The most common threshold for p-values is 5%, written as $p < .05$. The confidence interval in this case is called a 95% confidence interval (i.e., $1 - .05 = .95$, or 95%). If the confidence interval contains that point of comparison for the null hypothesis, then $p > .05$. If it does not contain that point of comparison, then $p < .05$. If $p < .05$, then we know that a given effect size would only occur 5% of the time (or 1 out of every 20 times) if the null hypothesis were true. To illustrate, given an average test score of 85%, if the confidence interval is 79–91%, which contains 80%, then the likelihood that the population mean is different from 80% has a p-value $> .05$. In such a case, we say that we "accept the null hypothesis." If the confidence interval were 83–87%, which does not contain 80%, the p-value would be $< .05$. We then "reject the null hypothesis" and say that "we have evidence for the alternative hypothesis."

There are situations when other thresholds for significance are more appropriate, especially when conducting many tests. Suppose you are testing 100 correlations. You would expect 5 of those correlations be significant *even if* there were no "true" correlation between the variables (i.e., the null hypothesis were true). In such situations we might lower the threshold to .01 or even further. There are multiple formal techniques for specifying an ideal p-value, though these are beyond the scope of this book.

In summary, when we conduct a statistical test, we always want to communicate both the effect size and the p-value. The effect size tells us the magnitude of the relationship, thereby quantifying its practical implications. The p-value communicates significance, which is the likelihood that an effect of that size would occur by chance, thereby evaluating whether we have evidence in support of our hypothesis or not.

10.2.5 Inferential Statistics in Practice: Analyzing "Big" Data

We have learned a lot in this section about inferential statistics and we are just about ready to start applying these concepts to our first statistical test: correlations. Before we do so, it is probably useful to summarize some of the main details while also relating them to the types of data that we have worked with throughout the book.

10.2.5.1 Samples with "Big" Data

A sample is a subset of a population. Often with naturally occurring data, however, we theoretically have access to the population, not a sample. Test scores for all students in a city, records of all crimes, tweets, and Craigslist postings, and demographics for all neighborhoods in a metropolitan region are technically populations, not samples. We still want to treat them as samples, though, and treat their features as having uncertainty. Why? Two reasons stand out. The first is philosophical in that even a population is just a sample of what is possible, influenced in a variety of unspecified ways. Thus, instead of assuming that everything we observe in the population is inherently true, we want to analyze it with the same rigor as any other sample. This is the case especially when the "population" is not all that large (e.g., a metropolitan region would probably be divided into a few hundred neighborhoods), which means it is vulnerable to chance relationships.

A second consideration is bias. Recall that when analysts source a sample they need to be sensitive to how the sample might be systematically biased by the sampling process. As noted repeatedly in this book, naturally occurring data may suffer from various forms of bias owing to the data-generation process. In this way the data-generation process and sampling are analogous processes, each of which forces us to consider how our data might differ from the ideal population that we are trying to understand. We want to be very careful about this with large data sets, which can suffer from the *big data paradox*. When we have such a large data set, we often have very low levels of uncertainty because the addition or subtraction of any given piece of information will have very little impact on the full data set. But what if the data set has fundamental biases arising from the sampling or data-generation process? This might then lead us to be overly confident about conclusions that are incomplete or inaccurate. As we have discussed in previous chapters, this is not to say that biased data is completely useless. However, it does place responsibility on the analyst to understand the potential sources of bias and the ways in which they should be considered when interpreting any analyses.

10.2.5.2 Distributions

Distributions will largely operate in the same way for naturally occurring data as for other forms of data. The one difference is that as data sets get larger, distributions tend to be cleaner. This means that there are fewer chance deviations from a normal distribution, and that we are more confident that skew or Poisson distributions are meaningful. It is important to note that nearly all statistical tests (and all tests learned in this book) assume that the data being analyzed are normally distributed. Arithmetically, this means they rely on the mean and standard deviation to reach inferences. Thus, they are not well-suited to evaluating other distributions. There are, however, alternatives that are not formally covered here.

There are workarounds for incorporating non-normal variables into statistical tests that assume a normal distribution. One is to transform a variable to be closer to a normal distribution. This is most often done by taking the natural logarithm of the variable to pull outliers closer to the rest of the distribution (i.e., *log(x)* or *log(x+1)* if any values are less than 1, which would otherwise produce hard to analyze negative outliers; if x has negative values, you need to do *log(x+1+min(x))*). You can also create categorical variables from a Poisson distribution based on a meaningful threshold. A classic technique for this is the "elbow test," in which the analyst identifies the point at which the curve flattens out into the tail, classifying everything beyond this point as scoring high on the variable.

10.2.5.3 Hypothesis Testing

Hypothesis testing itself works the same way regardless of the size or structure of the data set. There is one fundamental thing to always keep in mind, though: does the statistical test align with the question you are asking? Statistical tests are designed to evaluate very specific relationships. To communicate your results and their implications clearly, you need to be confident that you have selected the right statistical technique for testing your hypothesis appropriately.

10.2.5.4 Effect Size and Significance

Effect sizes are not typically sensitive to the size of a data set (though there are exceptions). As samples get larger, however, significance for the same effect size becomes smaller and smaller. This is because with more cases we can be increasingly confident that the effect sizes that we are observing are not by chance. Correspondingly, the confidence intervals are much smaller. For a very large data set with millions of cases, extremely small effect sizes are likely to have p-values of .001 or lower. This is not a problem in itself, but it further highlights the importance of communicating the effect size. If we only communicate significance when analyzing a large data set, then everything we observe is important. We have no way to determine which relationships have greater or lesser practical importance. Effect sizes allow us to make these determinations.

10.3 Correlations

We will now learn our first statistical test: the correlation. This and other statistical tests in this book will be presented in a consistent fashion. Each will start with conceptual information about the types of questions that the test can help answer and the effect size it reports. This will be followed by a worked example demonstrating how to conduct the test in R and interpret, report, and visualize the results. As described in Section 10.1, the worked example for this chapter will be to test correlations between various characteristics of the census tracts of Boston, including: demographic features associated with disadvantage and the social organization, such as median income and racial composition; and unwanted outcomes, such as gun violence and medical emergencies (as drawn from 911 dispatches). Each measure is drawn from the last time they were updated by BARI at the time of this writing (the 2014-2018 ACS estimates for demographics, 2020 for gun violence, and 2014 for medical emergencies). Despite slightly different time periods, all measures are largely stable over time so the inferences drawn from this illustration are generalizable to simultaneous measures.

10.3.1 Why Use a Correlation?

"Correlation" is a word we often see and hear in daily conversation and popular media. In statistics, a correlation is used to evaluate the relationship between two numerical variables, specifically whether they rise and fall together or whether one falls as the other rises and vice versa. The former is referred to as a positive correlation, the latter is a negative correlation. In the formal language of hypothesis testing, the null hypothesis is that two variables have no correlation. That is, the level of one variable is in no way predictive of the level of the other variable. The alternative hypothesis is that the two variables have *either* a positive or negative correlation. Note that the alternative hypothesis associated with a statistical test does not typically select a direction of a relationship, even if our own inspiration for the statistical test might be articulated this way.

You have probably heard the phrase, "correlation does not equal causation." It is a very true statement and should be taken seriously. A strong positive correlation tells us that two variables rise and fall together, but it does not tell us why that is true. We cannot definitively say that one variable causes the other, simply that they tend to coincide. We always need to be careful about this when interpreting correlations. We will attend to this more in Chapter 12.

10.3.2 Effect Size: The Correlation Coefficient (r)

The effect size measured by a correlation test is called, fittingly, a correlation coefficient, and it is represented by the symbol r. r is a representation of the strength of the relationship between two numeric variables and has a range of -1 to 1. When r is positive, the two variables rise and fall together. When r is negative, as one variable rises, the other falls, and vice versa. When r is further from 0, the correlation is stronger. When r is near 0, there is no meaningful correlation, which means that changes in one variable are not associated with changes in the other variable. It is also important to recognize that r and slope are not the same thing. Slope is a calculation of how much one variable increases (or decreases) on average every time the other variable increases by 1. r calculates how consistent this relationship is, but not the relationship itself.

To illustrate, let us take a moment and consider the well-known correlation between Celsius and Fahrenheit. This relationship follows the equation $F = C * 9/5 + 32$. Here the slope is 9/5 or 1.8. In other words, for every degree increase in Celsius, Fahrenheit increases by 1.8 degrees. The positive correlation between Celsius and Fahrenheit is perfect because they are defined by an equation. For this reason, the r between them is equal to 1.

To summarize, correlation tests communicate the consistency with which differences in one variable correspond to shifts in another variable. They have no way, however, of evaluating slope, and thus we cannot use them to communicate how much one variable is likely to differ based on changes in the other variable. We will return to slopes, though, when we learn about regressions in Chapter 12. We will now see how this works in practice with our worked example.

10.3.3 Running Correlations in R

There are a variety of ways to run correlations in R. We will learn three of them. `cor()` is the most basic way of running a correlation matrix, reporting the correlation coefficients between all pairs of variables in a set of two or more variables. It has its limitations, however. To address these, we also use `cor.test()`, which evaluates the size and significance of the correlation between a single pair of variables, and `rcorr()` from the `Hmisc` package (Harrell, 2021), which generates a correlation matrix for all pairs of variables in a set of two or more variables and reports their significance.

10.3.3.1 cor()

`cor()` is the simplest way to run a correlation in R. Let us start by running one on our `tracts` data frame.

```
cor(tracts)
```

```
##                   CT_ID_10     White      Black   Hispanic
## CT_ID_10        1.000000000 -0.2643588  0.3190802  0.1600927
## White          -0.264358837  1.0000000 -0.8507669 -0.6239106
## Black           0.319080196 -0.8507669  1.0000000  0.2545213
## Hispanic        0.160092734 -0.6239106  0.2545213  1.0000000
## MedHouseIncome -0.081310908  0.7626905 -0.5246184 -0.5141626
## Guns_2020       0.087688848 -0.6789560  0.7041435  0.2926918
## MajorMed_2014   0.003631076 -0.3210942  0.2839987  0.1006539
##                MedHouseIncome   Guns_2020 MajorMed_2014
## CT_ID_10          -0.08131091  0.08768885   0.003631076
## White              0.76269050 -0.67895596  -0.321094181
## Black             -0.52461841  0.70414355   0.283998689
## Hispanic          -0.51416260  0.29269185   0.100653893
## MedHouseIncome     1.00000000 -0.48100736  -0.253731258
## Guns_2020         -0.48100736  1.00000000   0.506307764
## MajorMed_2014     -0.25373126  0.50630776   1.000000000
```

This produces a matrix of correlation coefficients between all 7 variables in the data frame. This means 49 relationships (7*7), though note that it is a symmetrical matrix, with all values above the diagonal equal to their mirror image below the diagonal. This is because a correlation between two variables is the same regardless of whether one variable is the row or the column. Additionally, note that all correlations between a variable and itself are equal to 1 because it is the same variable. That means there are only 21 distinct, meaningful pieces of information in the matrix (7*6/2 = 21 unique pairs of variables).

There is an issue, though. `CT_ID_10` is included in the matrix even though it is a unique identifier. Because `CT_ID_10` is a numerical variable, `cor()` has treated it as such, even though this has no practical meaning. We can address this with a small tweak.

```
tracts %>%
  select(-'CT_ID_10') %>%
  cor()
```

```
##                    White      Black   Hispanic MedHouseIncome
## White          1.0000000 -0.8507669 -0.6239106      0.7626905
## Black         -0.8507669  1.0000000  0.2545213     -0.5246184
## Hispanic      -0.6239106  0.2545213  1.0000000     -0.5141626
## MedHouseIncome 0.7626905 -0.5246184 -0.5141626      1.0000000
## Guns_2020     -0.6789560  0.7041435  0.2926918     -0.4810074
## MajorMed_2014 -0.3210942  0.2839987  0.1006539     -0.2537313
##                Guns_2020 MajorMed_2014
## White         -0.6789560    -0.3210942
## Black          0.7041435     0.2839987
## Hispanic       0.2926918     0.1006539
## MedHouseIncome -0.4810074   -0.2537313
## Guns_2020      1.0000000     0.5063078
## MajorMed_2014  0.5063078     1.0000000
```

We have now removed `CT_ID_10` and can take a closer look at our correlation matrix. First, we see a strong positive correlation between the percentage of White residents in a neighborhood and the median income (r = .76) and a strong negative correlation between both the percentage of Black residents and Latinx residents and median income (r = -.52 and -.51, respectively).

Additionally, we see that gun violence and medical emergencies are more strongly concentrated in poorer neighborhoods, with correlations with median income of -.48 and -.25, respectively. Though smaller than some of the previous correlation coefficients, these are still substantial. Places with more Black and Latinx residents also experienced more of these challenges, with the former featuring correlations of .28 and .70, and the latter seeing correlations of .10 to .30. Last, hearkening to the clustering of "bad things" observed by Booth, DuBois, and the Chicago School, we see that gun violence and medical emergencies are correlated with each other, sharing a correlation of .51.

10.3.3.2 `cor.test()`

`cor()` is an effective tool for describing the relationships between our variables, but it has some deficiencies. The biggest is that it does not evaluate the significance of the correlation coefficients. In this sense, it might be characterized as a tool for descriptive statistics not inferential statistics. `cor.test()` is more thorough in conducting the hypothesis test associated with a correlation, but it can only do it for one pair of variables at a time.

Let's use `cor.test()` to further probe whether gun violence and medical emergencies are positively correlated with each other.

```
cor.test(tracts$MajorMed_2014, tracts$Guns_2020)
```

```
##
##  Pearson's product-moment correlation
##
## data:  tracts$MajorMed_2014 and tracts$Guns_2020
## t = 7.5646, df = 166, p-value = 0.000000000002531
## alternative hypothesis: true correlation is not equal to 0
## 95 percent confidence interval:
##  0.3843632 0.6108870
## sample estimates:
##       cor
## 0.5063078
```

Looking closely at the output, we see that the correlation at the bottom (labeled `cor`) is the same as in the `cor()` test we conducted above, indicating that the two functions are testing the same thing. Above that is a `95% confidence interval` with a range of .38–.61. This is the range of values that the population correlation might take with 95% likelihood given the correlation of .51. Note that this range does not include 0, which would suggest that the correlation is significant beyond a 5% chance. In other words, $p < .05$, and we have evidence that gun violence and medical emergencies are indeed correlated across census tracts. Going up to the top of the output, p indeed is less than .05—it is equal to $2.531 * 10^{-12}$, which is a really small number! In a formal report, we would communicate this result as: "Gun violence and medical emergencies shared a positive correlation across neighborhoods ($r = .51$, $p < .001$)." We could also say, "Neighborhoods with higher levels of gun violence also had higher rates of medical emergencies ($r = .51$, $p < .001$)."

10.3.3.3 `rcorr()`

`cor()` can report correlation coefficients between three or more variables in a matrix but does not evaluate them for significance. `cor.test()` can test the significance of the correlation coefficient, but only between a single pair of variables. `rcorr()`, a function in the package `Hmisc`, combines each of these capabilities, reporting a matrix of correlation coefficients and evaluating their significance.

`rcorr()` does have a few quirks we should anticipate. First, it can only process data as a matrix, meaning we will need to transform the format of our data before conducting the analysis. Second, on the positive side, it automatically does pairwise deletion, so we do not need to specify that (though we can specify other logics for deletion; see the function documentation of the function for more details). Third, it generates three separate matrices: one with the correlation coefficients, one with the pairwise n (sample) used to calculate each, and one with the p-value for each coefficient. If we run it directly, the output is all three matrices. For the sake of brevity, we will skip this (though you are welcome to try it on your own). We will instead save the result as an object. We can then call up any of the three matrices using bracket notation.

The matrices are stored in the order noted above (i.e., [1] contains the r values, [2] contains the pairwise sample (or n), and [3] contains the p-values). For instance, to see the correlation coefficients.

```
require(Hmisc)


rcorr_matrix<-tracts %>%
  select(-'CT_ID_10') %>%
  as.matrix() %>%
  rcorr()
rcorr_matrix[1]
```

```
## $r
##                    White      Black   Hispanic MedHouseIncome
## White          1.0000000 -0.8507669 -0.6239106      0.7626905
## Black         -0.8507669  1.0000000  0.2545213     -0.5246184
## Hispanic      -0.6239106  0.2545213  1.0000000     -0.5141626
## MedHouseIncome  0.7626905 -0.5246184 -0.5141626      1.0000000
## Guns_2020     -0.6789560  0.7041435  0.2926918     -0.4810074
## MajorMed_2014 -0.3210942  0.2839987  0.1006539     -0.2537313
##                Guns_2020 MajorMed_2014
## White         -0.6789560    -0.3210942
## Black          0.7041435     0.2839987
## Hispanic       0.2926918     0.1006539
## MedHouseIncome -0.4810074    -0.2537313
## Guns_2020      1.0000000     0.5063078
## MajorMed_2014  0.5063078     1.0000000
```

Starting with the matrix of correlation coefficients, we see that the numbers are the same as in the previous analyses, so there is not much to discuss here. Moving on to the second matrix of n for each pairwise calculation:

```
rcorr_matrix[2]
```

```
## $n
##                White Black Hispanic MedHouseIncome Guns_2020
## White            168   168      168            168       168
## Black            168   168      168            168       168
## Hispanic         168   168      168            168       168
## MedHouseIncome   168   168      168            168       168
## Guns_2020        168   168      168            168       168
## MajorMed_2014    168   168      168            168       168
##                MajorMed_2014
## White                    168
## Black                    168
## Hispanic                 168
## MedHouseIncome           168
## Guns_2020                168
## MajorMed_2014            168
```

all values are equal to 168, which is the number of census tracts analyzed.

In the third matrix:

```
rcorr_matrix[3]
```

```
## $P
##                      White                Black
## White                   NA 0.000000000000000000
## Black         0.0000000000                   NA
## Hispanic      0.0000000000 0.000870502127080286
## MedHouseIncome 0.0000000000 0.000000000002895462
## Guns_2020     0.0000000000 0.000000000000000000
## MajorMed_2014 0.0000219857 0.000190967927574670 1
##                          Hispanic       MedHouseIncome
## White         0.000000000000000000 0.000000000000000000
## Black         0.000870502127080286 0.000000000002895462
## Hispanic                        NA 0.000000000010145218
## MedHouseIncome 0.000000000001014522                   NA
## Guns_2020     0.000118090922862635 0.000000000412576640
## MajorMed_2014 0.194224604286196367 0.000904508610419130 2
##                         Guns_2020       MajorMed_2014
## White         0.000000000000000000 0.000021985702605587
## Black         0.000000000000000000 0.000190967927574670
## Hispanic      0.000118090922862635 0.194224604286196367
## MedHouseIncome 0.000000000041257664 0.000904508610419130
## Guns_2020                       NA 0.000000000002531308
## MajorMed_2014 0.000000000002531308                   NA
```

most of the p-values are quite small, often less than .001. There are some, however, that are non-significant. The proportion of Latinx residents in a community and medical emergencies, for instance, are not significantly correlated, with $p = .19$.

10.3.4 Visualizing Correlation Matrices

There are multiple ways to visualize correlations, some of which we have already seen. In Chapter 7 we learned how to create dot plots and density plots in **ggplot2**, which is the most traditional way to represent the correlation between two variables. In Chapter 9 we learned how to create correlograms, which are a visual version of a correlation matrix, using the **ggcorrplot** package. Correlograms are also useful in that they reorganize variables to be grouped by their strongest correlations. Here we will learn an additional tool for visualizing correlation matrices called gally plots, made possible by the `ggpairs()` function in the package **GGally** (Schloerke et al., 2021). Gally plots are similar to correlograms in that they communicate the full correlation matrix, but they do even more. They report r values with their significance in the upper half of the matrix and the dot plot relationships in the lower half. They also provide descriptive statistics for each of the individual variables along the diagonal (replacing the trivial reports of $r = 1$). As such, they communicate a lot of information in a single graphic—and from a single line of code.

```
require(ggplot2)
require(GGally)
ggpairs(data=tracts, columns=2:7)
```

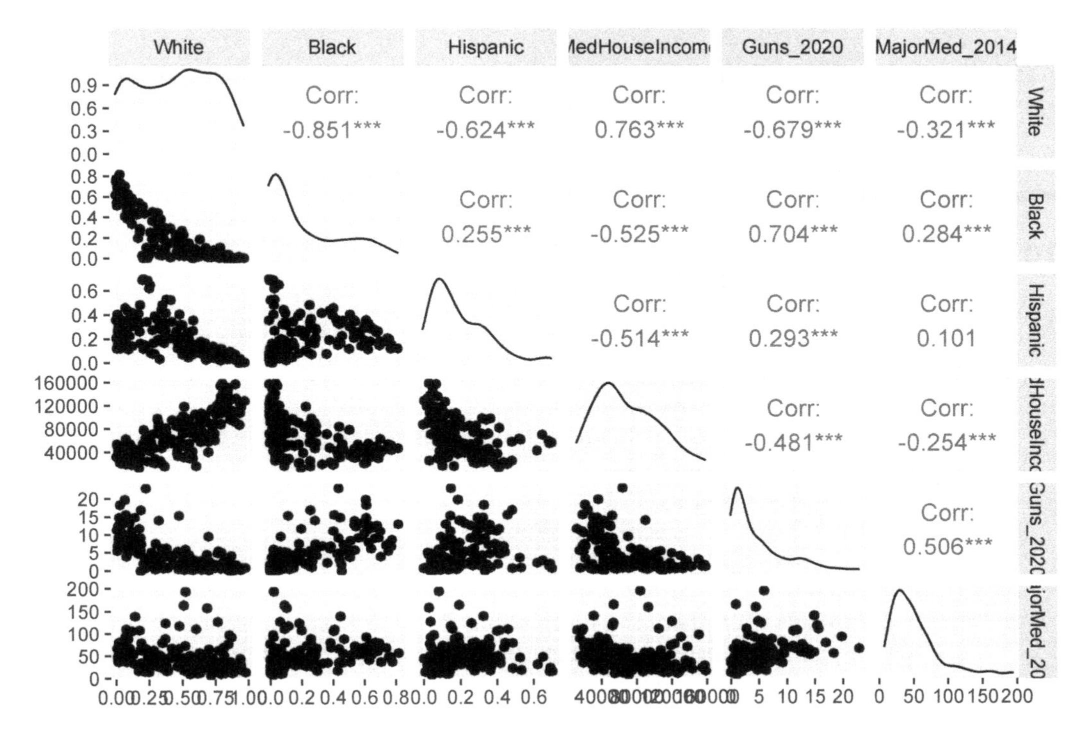

If you look closely at the output preceding the graphic (which I have suppressed here), you will note that it generates a series of `geom_point()` and `stat_density()` figures that it then combines. Turning to the graphic itself, it has all of the values we would expect above the diagonal. One star means $p < .05$, two stars means $p < .01$, and three stars means $p <$ .001, as is traditional in software packages and reports. We then see the distribution curve for each of our variables (many of which are non-normal, it turns out) along the diagonal. Below that are dot plots for all the pairs of variables. If you look closely, the shapes of these dot plots reflect the corresponding correlation coefficients, with strong correlations corresponding to more linear shapes on dot plots that point upward or downward, and weak correlations corresponding to noisier relationships that appear to be flat.

10.4 Summary

We have now learned the core concepts underlying inferential statistics and the practice of testing for relationships between variables. We have also applied them for the first time by evaluating the clustering of undesirable outcomes in those Boston neighborhoods that are populated by lower income residents and by historically marginalized populations that often suffer from higher rates of poverty, greater residential turnover, and more family disruption. In the process, we have learned to:

- *Consider the relationship between samples and populations*, including how the former do (and do not) help us reach conclusions about the latter;
- *Recognize a normal distribution* and deviations from it;
- *Interpret and communicate both the effect size and significance* of a statistical test;
- *Run and interpret correlations between pairs of variables* using `cor()`, `cor.test()`, and `rcorr()`;
- *Visualize a correlation matrix* with the `GGally` package.

10.5 Exercises

10.5.1 Problem Set

1. For each of the following pairs of terms, distinguish between them and describe how they complement each other during statistical analysis.
   a. Descriptive statistics vs. inferential statistics
   b. Sample vs. population
   c. Normal distribution vs. Poisson distribution
   d. Null hypothesis vs. alternative hypothesis
   e. Effect size vs. significance
   f. r vs. slope
2. You read a report that there is a correlation of $r = .5$ between income and health. The article uses this as evidence for saying that income has a causal effect on health. How do you interpret this statement? What critiques do you have?
3. Return to the data frame `tracts` analyzed in the worked example in this chapter. Select two correlations between variable pairs that you find interesting. For each:
   a. Report and interpret the r and p-value.
   b. Make a dot plot of the relationship.
   c. *Extra*: Relate the two correlations to each other, describing how together they tell a fuller story.
4. Return to the beginning of this chapter when the `tracts` data frame was created. Recall that the variables analyzed in the worked example were selected from a much larger collection of demographics and annual metrics of social disorder, violence, and medical emergencies through 911 dispatches.
   a. Create a new subset of variables that you think is interesting. Justify your selection.
   b. Run correlations between these variables. Interpret the results. Be sure to reference both effect size and significance. Note that if you have chosen a large number of variables, you will want to summarize the results without listing each individual correlation coefficient.
   c. Create a gally plot.
   d. Summarize the results with any overarching takeaways.

10.5.2 Exploratory Data Assignment

Complete the following working with a data set of your choice. If you are working through this book linearly, you may have developed a series of aggregate measures describing a single scale of analysis. If so, these are ideal for this assignment.

1. Choose a set of variables for a given scale of analysis. Justify why these variables are interesting together. It could be that there is good reason to analyze them collectively or that you have one variable of primary interest (e.g., that you developed for a previous exploratory data assignment) whose correlations with other variables you want to evaluate.
2. Run correlations among these variables and report and interpret the results. Be sure to reference both effect size and significance. Note that if you have chosen a large number of variables you will want to summarize the results without listing each individual correlation coefficient.
3. Create a gally plot of the relationships and at least one dot plot of the correlation between a particular pair of variables.
4. Describe the overarching lessons and implications derived from these analyses.

11

Identifying Inequities across Groups: ANOVA and t-Test

The *urban heat island effect* is the way that the structure and organization of cities—especially the extensive replacement of foliage with pavement—tend to trap heat. It has become increasingly apparent in recent years that this effect is responsible for more than just differences between cities and the surrounding suburbs and rural areas. It creates substantial temperature differences between the neighborhoods *within* a city as well. Especially concerning, a series of recent studies have demonstrated that these differences are highly correlated with race and income: Black and Latinx residents and those with lower incomes tend to live in warmer neighborhoods. This disparity can be life-threatening as such communities are then more exposed to extreme heat during the summer (Rosenthal et al., 2014).

The correlations between race, income, and heat were not an accident. They were part of the historical segregation of cities. In the 1930s, the federal government's Home Owners' Loan Corporation (HOLC) classified the neighborhoods of 239 cities according to their perceived investment risk. This practice has since been referred to as "redlining," as the neighborhoods classified as being the highest risk for investment were often colored red on the resultant maps. Majority Black neighborhoods were almost universally placed in this bottom category, meaning that residents were often unable to access mortgages and other loans related to real estate and development. Today, once-redlined neighborhoods tend to be hotter than the other parts of the city (Hoffman et al., 2020). Figure 11.1 contains an example of a redlining map from Portland, Oregon, and a comparison of current average temperatures by HOLC grades across cities.

Redlining was outlawed in 1968 as part of the Fair Housing Act. And yet its impacts persist. By preventing investment, it has condemned these neighborhoods to a future of concentrated poverty and all that comes with it: lower life expectancy, poorer health outcomes, and limited opportunity for local residents to advance their own opportunities through homeownership, entrepreneurship, or otherwise. These challenges have been inherited by each subsequent generation. And certain conditions, like hotter neighborhoods, make it particularly hard for communities to break away.

The legacy of redlining and the resultant correlation between temperature and race is a stark illustration of an *inequity*: when an individual or community is beset by challenges that make it difficult or impossible to achieve the same outcomes as others who do not face the same challenges. An important opportunity for urban informatics is to identify and reveal inequities statistically, shining a light on the ways in which some groups start at a deficit. Only then are we able to design policies, practices, and services to support them and even the playing field. In this chapter we will learn the basic statistical tools for identifying differences between groups (or, really, any categories we might define) and will apply them to the legacy of redlining and urban heat.

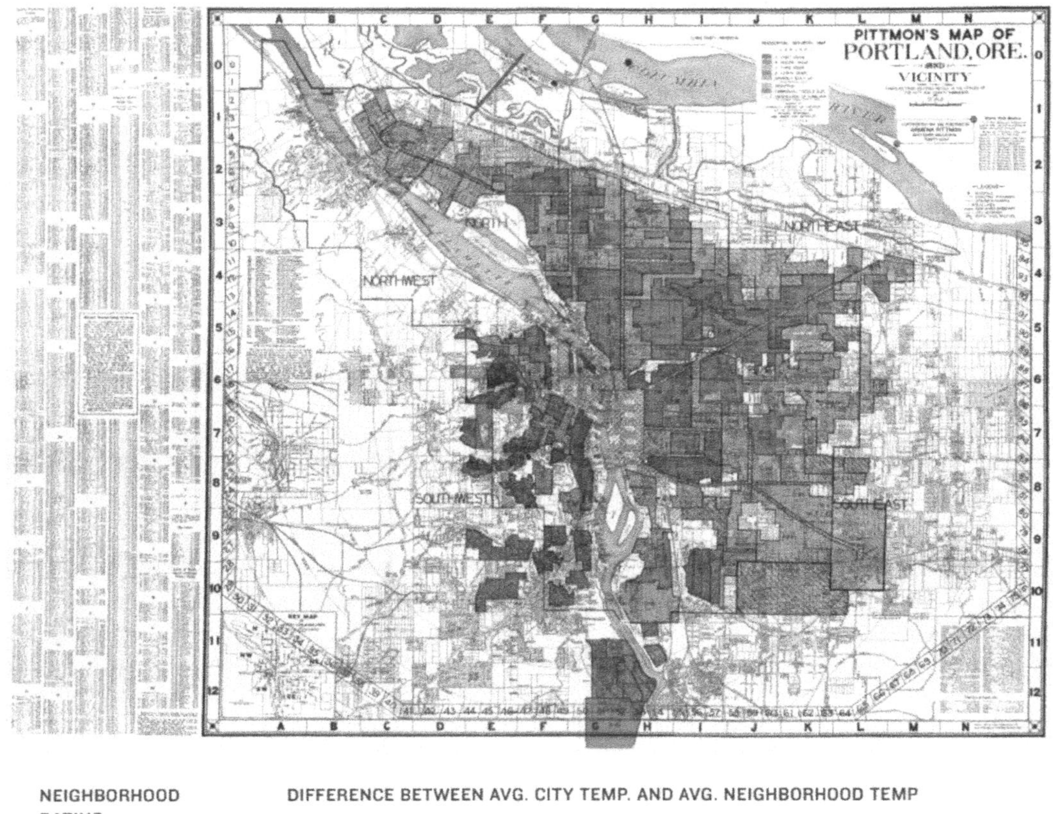

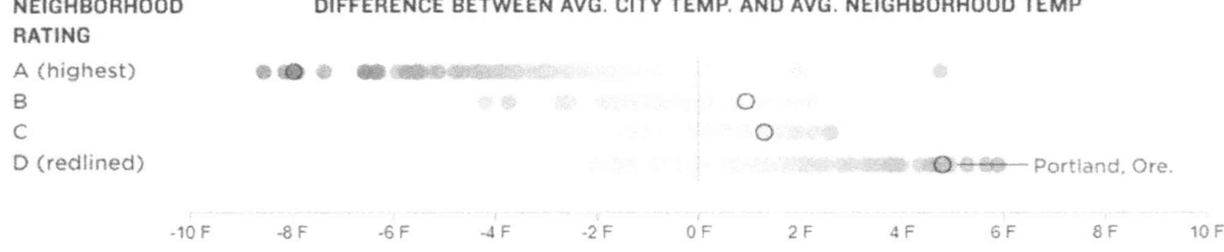

FIGURE 11.1
The federal government's Home Owners' Loan Corporation categorized the neighborhoods of 239 cities, including Portland, OR, (pictured, top) according to their investment quality (green being safest and red being riskiest) in the 1930s (top). "Redlined" neighborhoods tended to be those places occupied by communities of color, most notably Black Americans. Today, formerly redlined neighborhoods are still warmer than other neighborhoods in the same city, in Portland and elsewhere (bottom). (Credit: Source: Nelson, Winling, Marciano, Connolly, et al., Mapping Inequality; see image)

11.1 Worked Example and Learning Objectives

In this chapter we will reveal the relationship between redlining and urban heat in Boston . We will do so using a type of data we have not yet encountered in this book: remote sensing. Remote sensing data are gathered through imagery and other information on ground conditions that can be derived from a plane or satellite. These data are then processed to produce various measures, including the density of trees and pavement, estimates of population, and, of course, land surface temperature. In this chapter we will use the Urban Land Cover and Urban Heat Island Effect Database for greater Boston, curated by researchers at Boston University and available through the Boston Area Research Initiative's Boston Data Portal. The database includes land surface temperature as well as a collection of related metrics that we will return to in Chapter 12. We will combine these data with an HOLC redlining map of Boston from 1938, provided by the Mapping Inequality[1] project, a collaboration of faculty at the University of Richmond's Digital Scholarship Lab, the University of Maryland's Digital Curation Innovation Center, Virginia Tech, and Johns Hopkins University. The Mapping Inequality project has digitized and geo-referenced (the process of taking a historical map and matching its points to a modern projection) all 239 HOLC maps and made them publicly available. The Boston map has then been spatially joined to census tracts to determine how each was graded by the HOLC.

As we move forward, we will learn both conceptual and technical skills, including how to:

- Differentiate between inequalities and inequities;
- Conceptualize analyses that compare values across categories of people, places, and things, including identifying dependent and independent variables;
- Conduct *t*-tests that compare values on two groups;
- Conduct ANOVA tests that compare values across three or more groups;
- Visualize differences between groups in `ggplot2`.

Links:

Urban Heat Island Database: https://dataverse.harvard.edu/dataset.xhtml?persistentId=doi:10.7910/DVN/GLOJVA

Redlining in Boston: https://dataverse.harvard.edu/dataset.xhtml?persistentId=doi:10.7910/DVN/WXZ1XK

Data frame name: `tracts`

```
require(tidyverse)

uhi<-read.csv('Unit 3 - Discovery/Chapter 11 - Comparing Groups/Worked
Example/UHI_Tract_Level_Variables.csv')

tracts_redline<-read.csv('Unit 3 - Discovery/Chapter 11 - Comparing
Groups/Worked Example/Tracts w Redline.csv')
```

[1] https://dsl.richmond.edu/panorama/redlining/#loc=5/39.1/-94.58

```
tracts<-merge(uhi,tracts_redline,by='CT_ID_10',all.x=TRUE)
```

11.2 Identifying Inequities across Groups

Equity has been a buzzword of late, but what does it mean? And how is it different from equality? Turning to the dictionary, the literal definition of inequity is "lack of fairness or justice." When we pursue equality we intend for fairness and justice, but there are situations in which equality can still be unfair. For instance, in the oft-used meme in Figure 11.2, we see three people trying to watch a sports game over a fence. On the left, each has a box of the same size, but this is not sufficient for the shortest person, who is still unable to see the game. On the right, we see a more equitable situation, in which the shorter person is provided with a taller platform that allows him to see over the fence. From this perspective, an inequity is a hurdle or challenge that an individual or community faces or has experienced that makes them less capable than others of achieving a desired goal.

Starting with the Civil Rights Era of the 1960s, the stated objective in policy and practice has been to achieve equality. This is often defined as ensuring that everyone has equal inputs and "equal opportunity." This was an important transition from how society operated previously, as many populations, especially Black Americans, had been deliberately discriminated against and excluded from crucial resources, like education, transportation, and political representation. We have since discovered, however, that equality of inputs and opportunity without consideration of each individual or community's starting point is insufficient for realizing equal outcomes. To illustrate, redlining by HOLC and the real estate companies that used the maps was a clear violation of the principle of equality. The practice is now illegal, but multiple decades of non-investment has left long-standing disparities in the financial capital, infrastructure, and conditions in those same places and the Black communities that occupied them. These manifest in a multitude of ways, including more intense heat, which in turn has inequitable consequences for health.

The recognition of the limitations associated with an "equality-first" approach has inspired a paradigm shift toward equity. Instead of pursuing equal access to government resources and other opportunities, the goal has become to design programs and services so that all populations have the same ability to avail themselves of these opportunities. In some cases, this means offering more support to those who need it to succeed, like the taller platform for the shorter person trying to see over the fence.

Data analysis plays a crucial role in the realization of equity. First, correlations, like those we conducted in Chapter 10, can reveal inequities: do certain populations systematically experience a different set of conditions and challenges that lower their ability to succeed? Often, we are most concerned when these factors correlate with race and socioeconomic status, which are numerical variables. Sometimes, however, we are faced instead with categorical variables—for example, was a neighborhood redlined?—whose relationships require a different set of statistical tools. In either case, the job of the analyst is to identify the distinct challenges that different groups inherit that can hinder their ability to thrive, thereby inspiring further examination and action.

FIGURE 11.2
Providing equal inputs or opportunities for all does not guarantee that all individuals will have the same success. When three people of different heights receive supports of the same height, the shortest is unable to see over the fence (left). An equity approach would be to provide the resources to each that enables them to succeed, as in giving the shortest person a taller support to see over the fence (right). (Credit: Interaction Institute for Social Change (interactioninstitute.org) | Artist: Angus Maguire (madewithangus.com))

11.2.1 Making Statistical Comparisons across Groups

11.2.1.1 Assessing Averages and Errors

Statistical comparisons across groups often lead to sweeping statements. "Men are taller than women." "Low-income neighborhoods have more crime." "Redlined communities are hotter." These statements can be taken to imply that *all* men are taller than *all* women, that *all* low-income neighborhoods have higher crime than *all* high-income neighborhoods, that *all* redlined communities in a city are hotter than *all* non-redlined communities. Those statements are far too broad and demonstrably false. They also are inconsistent with what a statistical test is actually examining.

Let us take the example of comparing a particular measure in two groups, each of which we might treat as its own population. As we learned in Chapter 10, all populations have a range of values on any given numerical measure. This range often takes the form of a normal distribution, with values largely centered on the mean but spread out according to the standard deviation. Thus, each of our populations is its own normal distribution with its own characteristic mean and standard deviation, as we see here.

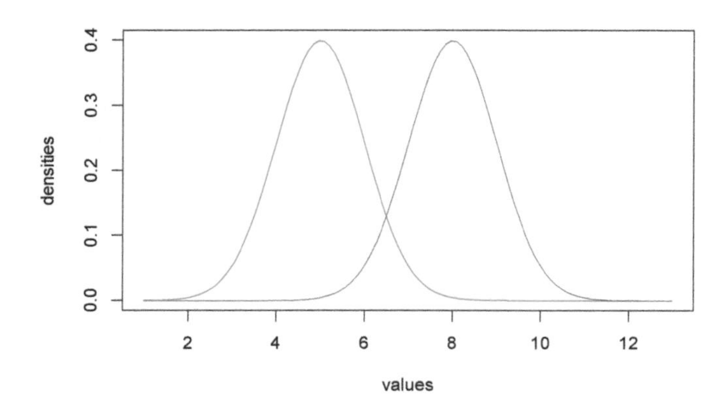

When we conduct a statistical comparison of values in two groups, the question we are actually asking is, "are the means of these two populations different?" Now, obviously, the means in the graph are different, but recall the distinction between samples and populations. When we are conducting statistical analyses, we assume that we are working with a sample of a population. As such, its mean is but an approximation of the population's "true" mean, shifted up or down by chance error. Statistical tests then take into account the standard deviation to estimate how much error is likely to be present. We are then able to assess the likelihood that two populations have different means given the differences we observe in samples drawn from each. As in Chapter 10, this likelihood is quantified as p-values and evaluates whether a difference is significant or not.

This reasoning has three implications for comparing groups:

1. Groups whose means are further apart are more likely to be significantly different. This becomes apparent visually as the distributions barely overlap, as in the following example.

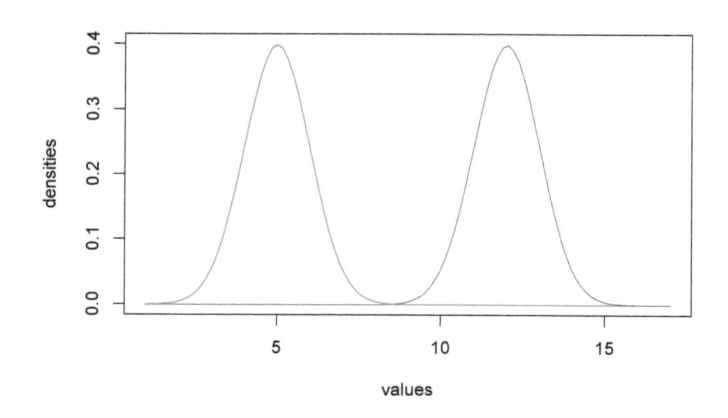

2. Groups with more variability are less likely to be significantly different from each other. This is because (a) individuals from each population are more likely to have similar values to individuals from the other, and (b) because of this greater variability, error can have greater impacts on our means, in which case we are less confident that the mean of the sample is representative of the mean of the population. Though the first of these considerations is more apparent in the visual example below, the latter has considerable impact on the arithmetic underlying statistical calculations.

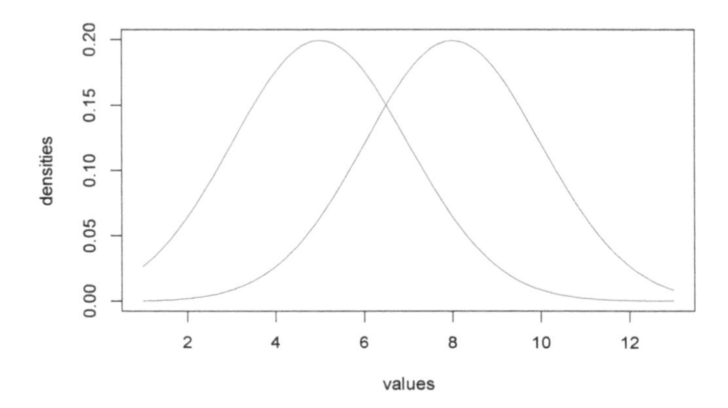

3. Larger sample sizes produce less error in the estimation of the population means. Less error in the estimation of the means makes us more confident in the observed differences between the samples. Arithmetically, this is because we do not use the standard deviation alone to compare samples. We use a new statistic called the *standard error of the mean*, which quantifies how far we expect the mean of our sample could differ from the population mean. The standard error of the mean (or standard error, for short) is calculated as the standard deviation divided by the square root of the sample size (represented by n, as we learned in Chapter 10). Consequently, returning to our first example graph, suppose we conduct two studies, one in which each sample has 10 cases and another in which each sample has 100 cases. Even if they have the same means and standard deviations, the latter will have a smaller standard error for the estimate of the mean, because as the sample size goes up, the standard deviation is divided by a larger value. Consequently, a statistical test will be more likely to identify a significant difference.

This arithmetic assessment of means and standard errors (as a function of standard deviations and sample sizes) is the heart of comparing groups. It is actually the heart of all statistical tests, including the correlations we conducted in Chapter 10. You might note that correlations examine the relationship between two numerical variables, not between a categorical variable and an outcome, and you would be right. That said, the same premise applies. A correlation test is essentially testing how the mean in one variable changes with the mean in another variable and evaluates the significance of these changes in light of the overall variability. The arithmetic of how this works is beyond the scope of this book, but it is important to keep in mind that whenever we are conducting statistical tests we are analyzing means and standard errors to determine effect size and significance.

11.2.1.2 Multivariate Analysis: Defining Independent and Dependent Variables

Comparing groups also raises a new consideration: distinguishing between *dependent variables* and *independent variables.* A dependent variable is our outcome of interest, and we want to use independent variables to explain why those outcomes are sometimes higher or lower. A statistical analysis models variation in our dependent variable as a function of the variation in one or more independent variables. Another, equally accurate way of describing this is saying that we are using our independent variables to predict variation in our dependent variables; semantically, "prediction" in this case does not necessarily mean anticipating future events but instead using a statistical model to "predict" what each value in the dependent variable might be given the corresponding values on the independent variables. For instance, to what extent does a history of redlining account for differences in neighborhood temperature? In this case, redlining is the independent variable and neighborhood temperature is the dependent variable.

One way to think about the difference between dependent and independent variables is in terms of an experiment. In an experiment, the researcher alters one or more variables and then tests the effect on one or more outcome variables. For instance, a clinical trial gives some individuals a medicine and others a placebo and analyzes their subsequent health outcomes. The variable that the experimenter altered, in this case the administration of medicine or a placebo, is the independent variable; and the outcome variables, in this case the health outcomes, are the dependent variables. Occasionally, we have the opportunity to conduct experiments in community-based research, such as placing a community garden in some neighborhoods and not in others and then testing the impacts. But often we have to work with the natural variation that is already present. This can create some interpretative complications that go beyond the scope of this book, but for the moment this framing can be useful to determining which are your dependent and independent variables. For the worked example in this chapter, redlining was not quite an experiment, but it was a way that the HOLC altered the treatment of certain neighborhoods, resulting in a variety of later outcomes, including higher temperatures. It would be harder to reason in the other direction that temperature explains why differences in redlining exist. In addition, thinking in terms of experiments might help us consider what effective programs and services might look like.

11.3 *t*-Test: Comparing Two Groups

11.3.1 Why Use a *t*-Test?

The simplest tool for comparing the distributions of multiple groups is the *t*-test, or Student's *t*-test, so called because it was published anonymously by Student in 1908. Student, whose real name was William Sealy Gosset, worked in quality control for the Guinness brewing company and had invented the *t*-test to rigorously compare batches of ingredients to each other. The company saw the work as its intellectual property and he was forced to publish it under the cover of a pseudonym.

The job of the *t*-test is to compare two values taking into consideration their variability, as measured by standard errors. It takes three primary forms:

- *Single-sample t-tests* compare the mean of a sample against a pre-established value of interest. These tests only take into consideration the standard error of the estimate of the mean for the sample as the pre-established value has no error. It is the precise value to which the analyst wants to compare. This test is not especially common because it requires a meaningful benchmark to compare against.
- *Two-sample t-tests* compare the means of two samples against each other, taking into consideration the standard errors for each. This is the most common form of t-test and what would be used in the examples of overlapping normal distributions featured in the previous section (Section 11.2.1).
- *Paired-sample t-tests* compare means on two different outcomes for the same set of items. The measures on each outcome are "paired," meaning we compare them only to their counterpart and then analyze the differences of all these pairings. This can be useful when two measures are highly correlated. For example, different types of air pollution tend to correlate from day to day, which makes it difficult to ask questions like "is there more carbon monoxide or sulfates in the air?" If they are going up and down together, their distributions may be hopelessly entangled But, if we ask if there is more of one or the other on each day, and then assess those within-day comparisons, we have a greater ability to observe differences. We will not illustrate a paired-sample t-test in this chapter, but it is a valuable tool worth knowing about.

For all t-tests, the null hypothesis is that there is no difference between the values evaluated, whether they be the mean of one group and a pre-established value or the means of two groups. The alternative hypothesis is that there is a difference between them. As we saw in Chapter 10 with correlations, the alternative hypothesis does not have a direction (i.e., if there is a difference, it does not stipulate which value is greater).

11.3.2 Effect Size: Magnitude of Difference in Means

The effect size for a t-test is highly intuitive. Most often, we want to tell our audience the "magnitude of difference in means." This is just a fancy way of saying, how much larger or smaller is one mean than the other? If two neighborhoods differ in their temperatures by 3, 5, or 10 degrees, that is the most easily interpreted and actionable piece of information that an analyst can provide.

A t-test assesses the magnitude of difference in means using the standard errors for each. This process generates the t-statistic, which is then used to evaluate significance. As noted in Section 11.2.1, we are more confident that a difference is significant when the samples are larger. For this reason, t-statistics are evaluated according to the sample size from a t-statistic table, like the one pictured in Figure 11.3. In the era of modern statistical analysis, programs like R have these tables programmed in and they conduct the evaluation themselves and report the difference in the means, the t-statistic, and the significance level.

Degrees of Freedom	p-value				
	.2	.1	.05	.01	.001
5	1.48	2.02	2.57	4.03	6.87
10	1.37	1.81	2.23	3.17	4.59
25	1.32	1.71	2.06	2.79	3.72
120	1.29	1.67	2.00	2.66	3.46
∞ (infinite)	1.28	1.64	1.96	2.58	3.29

FIGURE 11.3
Example t-value lookup table. Each cell contains the t-value that corresponds to the threshold for a given p-value for a given number of degrees of freedom. Any t-value higher than that value would be less likely than that p-value. For example, for 120 degrees of freedom, a t-value of 2.00 or greater would have a p-value of less than .05.

11.4 Conducting a t-Test in R

We will illustrate how to conduct single- and two-sample t-tests in R using the `t.test()` function to examine differences in land surface temperature between neighborhoods that were and were not redlined in Boston. (Note: Whereas `t.test()` works with a single outcome variable and a single categorical variable, a paired-sample t-test requires a different data structure with two outcome variables that are "paired" in that they each reference the same set of cases. For this reason, it also requires a different function, `pairwise.t.test()`.)

11.4.1 Single-Sample t-Test

A single-sample t-test is useful if we want to compare the mean of a variable to a pre-established value. We have not to this point made a statement about whether we would like to test the mean land surface temperature across neighborhoods against a certain benchmark, but doing so gives us an excuse to get to know our variable a bit better. (It would have been good practice to look closely at the mean, standard deviation, minimum, maximum, and other characteristics of the distribution of this variable before diving into statistical analyses. You might want to take a moment and do that.)

Let us ask the simple question, "Is the land surface temperature in the average Boston census tract different from 90°F during the summer?" This would be executed with the following code, in which we indicate our variable of interest (`tracts$LST_CT`) and our benchmark (`mu=`, as mu, or μ, is the Greek letter used to symbolize the mean).

```
t.test(tracts$LST_CT, mu=90)
```

```
##
##  One Sample t-test
##
## data:  tracts$LST_CT
## t = 31.751, df = 177, p-value < 0.00000000000000022
## alternative hypothesis: true mean is not equal to 90
## 95 percent confidence interval:
##  98.66290 99.81115
## sample estimates:
## mean of x
##  99.23702
```

The result offers us a few pieces of information. Starting from the bottom, the mean of our variable is 99.24; in other words, the average census tract in Boston has a land surface temperature of 99.24°F in the summer. The 95% confidence interval above it ranges from 98.66 to 99.81°F, which is approximately all values within 1.96 times the standard error of the mean (see Chapter 10 for a refresher on confidence intervals). This range does not include 90°F, so unsurprisingly we see that the p-value is less than .001 (it is quite a bit smaller than that at $< 2.2 * 10^{-16}$), offering support for our alternative hypothesis that the true (or population) mean is not equal to 90°F. This is based on the assessment of a rather large t-statistic of 31.75 (often values above 2 or 3 are significant, depending on sample size) and 177 degrees of freedom (which is calculated from the sample size).

One could take these results and write the following: "The average census tract in Boston, MA has a land surface temperature greater than 90°F during the summer ($mean = 99.24°F$, $t = 31.75$, $p < .001$)." Alternatively, one might write: "The average census tract in Boston, MA has a land surface temperature of 99.24°F during the summer, which was significantly greater than 90°F ($t = 31.75$, $p < .001$)." You are probably saying to yourself, "That's hot!" And it is. But it is important to keep in mind that land surface temperature is different from air temperature as we experience it because the ground absorbs a large amount of the heat generated by the sun. You should not interpret these numbers in the same way that you would a weather forecast.

11.4.2 Two-Sample t-Test

Now that we know the average land surface temperature across Boston's census tracts in the summer, we are ready to evaluate whether this mean differs between those that were and were not redlined. This can also be done with `t.test()`. In this case, we indicate two variables, `LST_CT` and `redline` with ~ between them. In this and other functions, ~ indicates that the variable on the left side is being analyzed as a function of the variable (or variables) on the right side; that is, the left side is the dependent variable and the right side contains all independent variables. We then indicate the data frame containing the variables.

```
t.test(LST_CT~redline,data=tracts)
```

```
## t = -7.698 df = 173.8 p-value = 9.999734e-13

## 95 percent confidence interval:

## -4.689 -2.775

## mean in group FALSE  mean in group TRUE
##              97.7299            101.4620
```

Again, let us work through these results from the bottom up (note that the output has been reformatted to fit the margins of the book and will look slightly different on your screen). We see that our `TRUE` group—that is, those for which `redline==1`, or those that were redlined—have an average summer land surface temperature of $101.46°F$, and our `FALSE` group-—i.e., `redline==0`, or never redlined—have an average land surface temperature of $97.73°F$. This is a magnitude of difference of 3.73°F. Though `t.test()` does not actually calculate this for us, it does report the 95% confidence interval for the difference as ranging from -4.69 to $-2.78°F$. This range does not contain 0, implying that the difference is significant at $p <$ .05. Indeed, the p-value is again very, very small at $1 * 10^{-12}$. This is based on a t-statistic of -7.70 and 173.84 degrees of freedom. Note that everything in this result is negative because R subtracts the mean from the `TRUE` group from the mean of the `FALSE` group.

In simple terms, we see strong evidence for our alternative hypothesis that there is a temperature difference between neighborhoods that were and were not redlined. We might write this as: "Redlined neighborhoods have summer land surface temperatures $3.73°F$ warmer than neighborhoods that were not redlined ($means = 101.46°F$ and $97.73°F$, $t =$ 7.70, $p < .001$)." Note that it is optional to maintain R's decision to communicate t-statistics as positive or negative. They can be treated as absolute measures of the effect and thus reported as positive.

11.5 ANOVA: Comparing Three or More Groups

11.5.1 Why Use an ANOVA?

An ANOVA, also known as an **An**alysis **o**f **Va**riance, picks up where t-tests leave off in the comparison of means across groups. Whereas t-tests are limited to comparing two groups to each other, ANOVAs can compare means across any number of groups. In practice, this includes two groups, though the t-test is often a more efficient way to examine such questions. This is because an ANOVA specifically analyzes the question, "Are there differences in the means between these groups?" It does not, in fact, evaluate any of the pairwise differences between those groups. This can be a little frustrating to the analyst because knowing that there is some overarching variation between 3, 5, 8, or 10 groups is only so informative, and certainly not very actionable. More often we want to know which of these groups differ from each other. This can be accomplished, as we will see, through post-hoc comparisons. For the special case of two groups, conducting a t-test is more straightforward and arithmetically identical.

It is worth noting that ANOVAs are quite versatile and take a variety of different forms. They can include multiple categorical variables, assess interactions between variables (i.e., does a second set of categories alter the differences between the original categories of interest?), and accommodate many different research designs. These are most commonly used in disciplines that do controlled experiments, especially psychology. In fact, graduate students in psychology programs often have to take one or more whole semesters dedicated exclusively to ANOVA! Here we will only conduct the most basic form of ANOVA, which is to examine differences in means across one set of categories, also known as the one-way ANOVA. In sum, the ANOVA as we will work with it in this chapter tests the null hypothesis that there are no differences in means between two or more groups. The alternative hypothesis is that there are differences between the means of these groups. If we find evidence for the alternative hypothesis and want to know which groups' means are different from each other, we can use post-hoc tests. The post-hoc tests include a series of pairwise comparisons, each with the null hypothesis that there is no difference between that pair of means and the alternative hypothesis that they do differ in their means.

11.5.2 Effect size: F-Statistic

An ANOVA reports an F-statistic, which evaluates the extent to which the means of groups differ from each other more than would be expected based on the natural variability in the data. This is also referred to as a comparison between *between-group variation* and *within-group variation*. The latter is also known as error variance or residuals, because it is the variation leftover once we account for groups. Much like the t-statistic, the F-statistic is a technical calculation that is combined with degrees of freedom (again based on sample sizes) to evaluate the p-value. It does not hold actionable meaning.

When conducting an ANOVA, there are two pieces of information we might include in the results to give our audience a more practical sense of the effect size. The first is again the magnitude of difference in the means themselves. Post-hoc tests allow us to evaluate their significance. Second, we can report the R^2, which is the proportion of variance explained by our groups. If groups have completely non-overlapping distributions—say, every value in Group A == 5, every value in Group B == 10, and every value in Group C == 15—then R^2 is equal to 100% or 1.0. This is because literally all variation is accounted for by groups. If, however, groups have identical distributions, R^2 is equal to 0, because there is no variation associated with groups. Most cases fall somewhere in between, and they are calculated with a set of values called *sums of squares* (SS). Sums of squares are calculations of variation, and the arithmetic of ANOVA relies on differentiating SS within and between groups. R^2 is then equal to SS between groups divided by Total SS; that is, it is the mathematical representation of the proportion of all variation in the sample that is accounted for by differences between groups.

11.6 Conducting an ANOVA in R

To extend our analysis of heat and redlining, we might recall the map in Figure 11.1. It is not a map of "red areas" and "everything else." It is a detailed grading system with four different levels of classification. The highest of these, or 'A' communities, were considered the most promising investments. The lowest, or 'D' communities, were considered high-risk investments. These were the communities we now refer to as "redlined" and that probably have suffered from the most intensive effects of this discrimination. But it is possible that these effects saw a gradient, with A communities at the top, D communities at the bottom, and B and C communities falling in between. We can test this with ANOVA.

11.6.1 ANOVA with aov()

The `aov()` function in R conducts ANOVAs. It actually creates objects of class "aov" "lm". The first part of the class is probably self-explanatory. The second part is a reference to linear models, which will make more sense in Chapter 12. That `aov()` creates objects of a special class will come in useful when we want to conduct post-hoc comparisons of means.

The arguments for `aov()`, at least one running a test of a single set of categories, is identical to that for `t.test()`. The only difference is that we will store the results in an object called `aov_heat` and use `summary()` to see our results. If you were to submit an `aov()` command directly, you would see that the results are limited and do not include everything we would like to know.

```
aov_heat<-aov(LST_CT~Grade,data=tracts)
summary(aov_heat)
```

```
##              Df Sum Sq Mean Sq F value           Pr(>F)
## Grade         3  700.9  233.62    20.6 0.0000000000184 ***
## Residuals   173 1961.9   11.34
## ---
## Signif. codes:  0 '***' 0.001 '**' 0.01 '*' 0.05 '.' 0.1 ' ' 1
## 1 observation deleted due to missingness
```

The results include an F-statistic of 20.6, which has a p-value of $1.84 * 10^{-11}$, which is quite small. This is based on the knowledge that 3 degrees of freedom are associated with the categories and 173 are associated with the rest of the variation. This provides substantial evidence for the alternative hypothesis that there are differences in land surface temperature across the four different grading levels created by the HOLC in the early 20th century. We can also calculate R^2 as $700.9/(700.9 + 1{,}961.9) = .26$.

We might report all of this as follows. "Census tracts with different investment grades created by the HOLC had significant differences in their land surface temperature ($F = 20.6$, $p < .001$). The grades explained 26% of the variation across census tracts." This is a nice start, but it does not articulate the meatier comparison of communities we would probably like, however. For that we need post-hoc tests.

11.6.2 Post-Hoc Tests with TukeyHSD()

ANOVA only evaluates whether there are differences between the means of multiple groups. Post-hoc tests allow us to dig into the actual cross-group differences underlying that evaluation, which are more likely to give us practical, actionable information. In a sense, post-hoc tests are just a series of t-tests comparing the means of pairs of categories. This, however, is an oversimplification because the significance is evaluated differently. Recall from Chapter 10 that if we use a p-value of .05 as our threshold for significance, we would actually expect a significant result 5% of the time (i.e., 1 of every 20 times) just by chance—meaning that we could erroneously identify a relationship that is not actually there. For this reason, we need to be careful about conducting too many comparisons at once. If, for example, we are comparing 5 groups, there are 10 pairwise comparisons. If we have 10 groups, there are 45 pairwise comparisons between groups! In the latter case we would expect at least two comparisons to be significant at $p < .05$ even if in reality there were no population differences.

Post-hoc tests have been designed to deal with the issue of multiple comparisons. There is no single way to address the issue, however, and statisticians have literally developed dozens of solutions that vary in their complexity, specificity, and their conservativeness (i.e., how worried they are about falsely identifying non-existent relationships). R and its packages have functions for running many of these. The most commonly used is the Tukey HSD (for Honestly Significant Difference) test, and it can be run by applying the `TukeyHSD()` function to our `aov` object.

```
TukeyHSD(aov_heat)
```

```
##   Tukey multiple comparisons of means
##     95% family-wise confidence level
##
## Fit: aov(formula = LST_CT ~ Grade, data = tracts)
##
## $Grade
##          diff        lwr       upr     p adj
## B-A  6.588242 -2.5362265 15.712710 0.2434591
## C-A  8.534387 -0.2484688 17.317242 0.0602325
## D-A 11.981306  3.1848355 20.777777 0.0029332
## C-B  1.946145 -0.8392843  4.731574 0.2709477
## D-B  5.393065  2.5649974  8.221132 0.0000105
## D-C  3.446920  2.0755722  4.818267 0.0000000
```

Now we have a full table comparing the means across all four categories of HOLC investment grades, represented as differences. These each have a 95% confidence interval (represented as `lwr` and `upr`) and associated p-value. Note that this is labeled as an adjusted p-value, reflecting that it is not a pure t-test, but one that accounts for multiple comparisons. For instance, communities graded B had a land surface temperature $6.59°F$ greater than communities graded A, though this difference had a confidence interval ranging from -2.54 to $15.71°F$. Because the confidence interval contained 0°F, it is unsurprising to see that the p-value = .24, meaning the difference is non-significant and we accept the null hypothesis of no difference in temperature between these groups.

Digging deeper into the table, we see that the only significant differences were between communities graded as D and communities with all other grades. Communities graded C were nearly significantly warmer than communities graded A, but not quite. Though the magnitude of difference would seem large at 8.53°F, the p-value is .06. Interestingly, the difference is substantially greater than that between D and B neighborhoods at 5.39°F, which is highly significant at $p = .00001$. Why? Let us take a quick look at the number of census tracts in each of these categories.

```
table(tracts$Grade)
```

```
##
##  A  B  C  D
##  1 11 93 72
```

There is only one census tract graded as A! Historically speaking, this tells us just how unfavorably the HOLC saw investing anywhere within the City of Boston, probably preferring to invest in the surrounding suburbs. For our purposes here, though, it shows just how small the sample for A neighborhoods is. For this reason, even a substantial difference might be non-significant because we cannot be fully confident that it is not a result of random error.

11.6.3 Communicating ANOVA Results

Let us expand on our brief summary of the initial ANOVA test from above to incorporate our post-hoc tests and tell a full story. A reader typically does not want to read through a number-by-number reiteration of six pairwise comparisons, so we need to distill the results. Often, a table might accompany the write-up below so that readers can inspect the results themselves. Or we could provide a graph, which we will learn to do in Section 11.7.

"Census tracts with different investment grades created by the HOLC had significant differences in their land surface temperature ($F = 20.6$, $p < .001$). The grades explained 26% of the variation across census tracts. Post-hoc tests found that only communities graded D were significantly warmer than the others (differences in means $= 3.45 - 11.98°F$, all p-values $<$ .001). Communities graded A, B, and C did not have significantly different temperature (all p-values $>$ .05). It is worth noting, though, that only one census tract in Boston was graded A, making it hard to make a meaningful comparison between this and the other grades."

11.7 Visualizing Differences between Groups

Sometimes, in addition to a statistical test, the easiest way to communicate differences between groups is with a bar chart. This is especially true for ANOVAs, where the numerous pairwise comparisons of post-hoc tests can be hard to absorb from a table or text. We will do this in a series of steps, moving up to increasingly sophisticated ways of communicating our data.

11.7.1 Representing Means

Creating a bar chart of means in `ggplot2` is a touch more complicated than you might expect. Because our data consist of the individual records that contribute to the means, we need to first aggregate the data ourselves, taking means for each group. We can then visualize those means.

```
means<-aggregate(LST_CT~Grade,data=tracts,mean)
```

11.7.2 Adding Variability

As we have discussed, means are not the whole story. Variability is key to evaluating whether differences in those means are meaningful. To add this consideration to our bar chart, we want to use standard errors to represent our confidence intervals, which are approximately equal to the standard error * 1.96 in each direction of the mean. This is calculated through the first line below, and then added to the the graph with the `geom_errorbar()` command at the end of the block of code.

```
ses<-aggregate(LST_CT~Grade,data=tracts,
      function(x) sd(x, na.rm=TRUE)/sqrt(length(!is.na(x))))
names(ses)[2]<-'se_LST_CT'
means<-merge(means,ses,by='Grade')
means <- transform(means, lower=LST_CT-1.96*se_LST_CT,
                   upper=LST_CT+1.96*se_LST_CT)

bar<-ggplot(data=means, aes(x=Grade, y=LST_CT)) +
  geom_bar(stat="identity",position="dodge", fill="blue") +
  ylab('Land Surface Temp.')
bar + geom_errorbar(aes(ymax=upper, ymin=lower),
             position=position_dodge(.9))
```

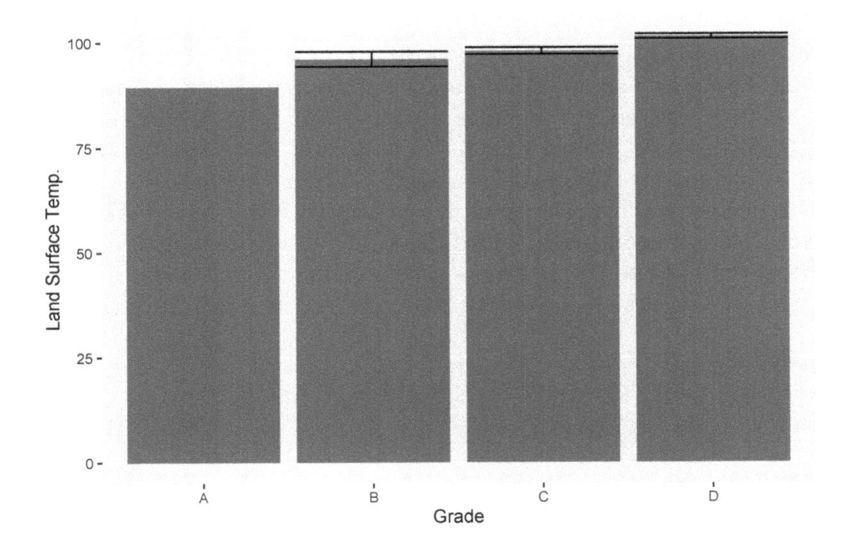

Here we can see that the variability would suggest that B and C are not all that difference from each other but that D is consistently warmer than the others. Meanwhile, A has no standard error because it cannot be calculated on only one case. Having a y-axis that ranges from 0 to 100 makes it challenging to interpret this graph, however, as all of the variation is constrained in the top part of the graph. We can fix this with `coord_cartesian()`. (You might be inclined to use `ylim` to do this, but in a bar graph, it considers the entire bar as part of the data point, so if you limit the y-axis above zero, it deletes the entire bar.)

```
bar + geom_errorbar(aes(ymax=upper, ymin=lower),
                    position=position_dodge(.9)) +
    coord_cartesian(ylim=c(80,105))
```

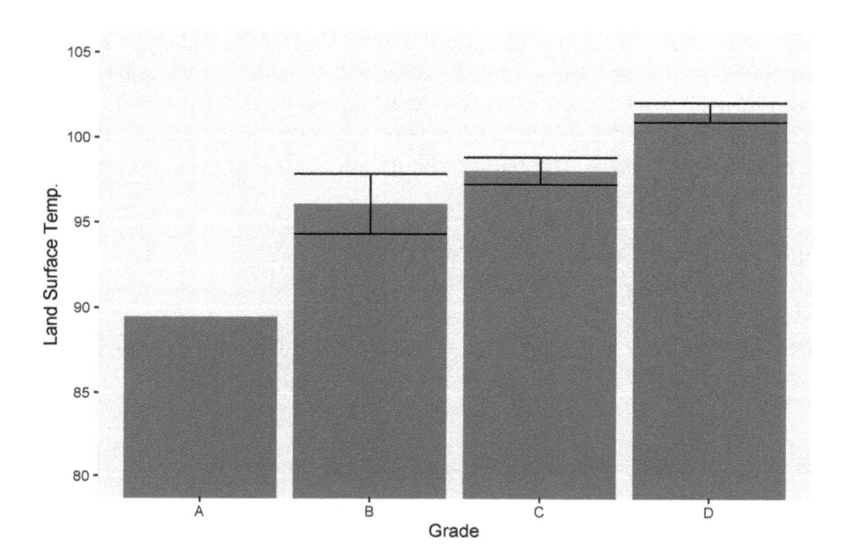

11.7.3 Comparing Multiple Variables across Groups

So far we have worked entirely with land surface temperature, but recall that the urban heat island database has a series of measures that are potentially related to temperature. Two of these, canopy cover and impervious surface coverage, are also measured on the same scale of 0–100. As a teaser to next chapter's analysis of these variables and temperature, we might want to compare them simultaneously across neighborhoods. This is a bit more complicated as we need to create separate means for each variable across each neighborhood grade. We will need a new function, called `melt()` from the `reshape2` package (Wickham, 2020) to convert our data so that it treats each combination of measure and HOLC grade into its own row. We will then use `aggregate()` on this new data frame.

```
require(reshape2)
melted<-melt(tracts[c("Grade",'CAN_CT','ISA_CT')],
          id.vars=c("Grade"))
means2<-aggregate(value~Grade+variable,data=melted,mean)
names(means2)[3]<-"mean"
ses2<-aggregate(value~Grade+variable,data=melted,
        function(x) sd(x, na.rm=TRUE)/sqrt(length(!is.na(x))))
```

```
names(ses2)[3]<-'se'
means2<-merge(means2,ses2,by=c('Grade','variable'))
means2<-transform(means2, lower=mean-se, upper=mean+se)

ggplot(data=means2, aes(x=Grade, y=mean, fill=variable)) +
  geom_bar(stat="identity",position="dodge") +
  geom_errorbar(aes(ymax=upper, ymin=lower),
                position=position_dodge(.9)) +
  ylab("Land Surface Temp.")
```

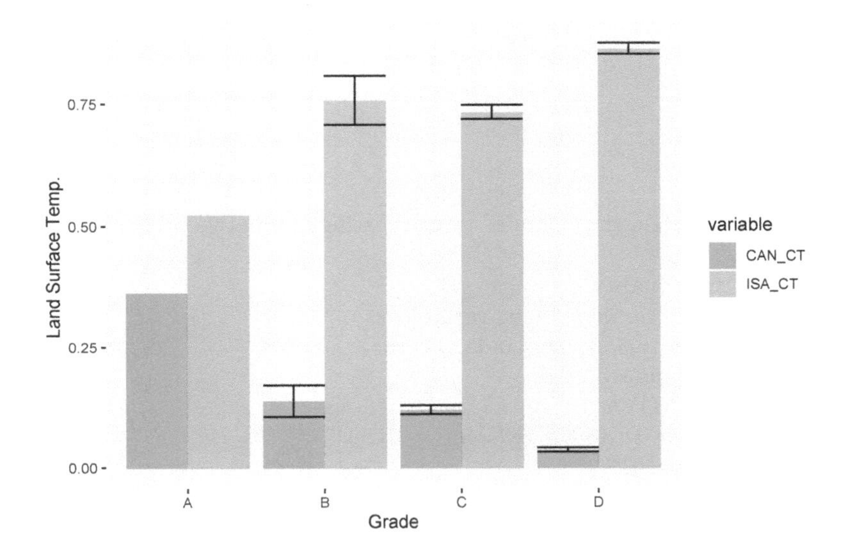

As you can see, differences in tree canopy and impervious surface correspond to differences in temperature. They also negatively correlate with each other. As there is more pavement, there is less tree canopy. And that tradeoff is most apparent in redlined neighborhoods, where there is very little canopy and lots of pavement. We will use additional statistical tests in the next chapter to evaluate the extent to which these structural differences explain temperature.

11.8 Summary

In this chapter we used the legacy of redlining and the consequent inequities as an excuse to learn the statistical tools that evaluate differences between groups. Specifically, we used t-tests and ANOVAs to assess how the grades that the HOLC gave neighborhoods in the first half of the 20th century continue to be associated with the urban heat island effect, with "redlined" neighborhoods having the hottest summers. In the process, we have learned to:

- *Distinguish between equity and equality*;
- *Conceptualize the comparison of groups* as an assessment of means and standard errors;

- *Identify independent and dependent variables* when conducting an analysis;
- *Conduct a t-test* to compare the means of two groups, or one group against a pre-established benchmark;
- *Conduct an ANOVA* to evaluate differences between three or more groups;
- *Conduct post-hoc tests following an ANOVA* to evaluate whether pairs of groups have different means;
- *Visualize differences across groups* with bar charts with standard errors, including comparing multiple variables simultaneously.

11.9 Exercises

11.9.1 Problem Set

1. For each of the following pairs of terms, distinguish between them and their roles in analysis.
    a. Equity vs. equality
    b. Independent vs. dependent variables
    c. Standard deviation vs. standard error
    d. Mean vs. standard error
    e. t-test vs. ANOVA
    f. Single-sample t-test vs. two-sample t-test
    g. F-statistic vs. R^2
2. For each of the following scenarios, indicate whether the proposed analysis is correct or whether you would do something different and why.
    a. Conducting an ANOVA to compare the mean canopy between main streets and non-main streets.
    b. After running an ANOVA, running a series of t-tests to compare mean between pairs of groups.
    c. Using `t.test()` to compare whether neighborhoods with above-average income have more supermarkets than neighborhoods with below-average income.
3. Return to the beginning of this chapter when the tracts data frame was created. Recall that I have worked previously with multiple data sets with many more measures describing tracts, including demographic characteristics from the American Community Survey, metrics of physical disorder, engagement, and custodianship from 311 records, and social disorder, violence, and medical emergencies from 911 dispatches. For the following you can merge any of these variables into the `tracts` data frame.
    a. Select an outcome variable of interest. Explain why you are interested in this variable.
    b. Select at least one categorical variable (or create one using thresholds or other logics). Explain why this variable is interesting and might be related to your outcome variable of interest.
    c. Run either a t-test or ANOVA for each of the categorical variables you selected or created to see if there are any differences across groups. Be certain to use the appropriate statistical tool and to conduct post-hoc tests if necessary.
    d. Visualize the relationship between your dependent variable and at least one

of the independent variables, presumably focusing on relationships that were the most interesting.

e. Summarize the results with any overarching takeaways, including whether the differences should be considered inequities.

11.9.2 Exploratory Data Assignment

Complete the following working with a data set of your choice. If you are working through this book linearly you may have developed a series of aggregate measures describing a single scale of analysis. If so, these are ideal for this assignment.

1. Select at least one outcome variable of interest. Explain why you are interested in this variable.
2. Select at least one categorical variable (or create one using thresholds or other logics). Explain why this variable is interesting and might be related to your outcome variable of interest.
3. Run either a t-test or ANOVA for each of the categorical variables you selected or created to see if there are any differences across groups. Be certain to use the appropriate statistical tool and to conduct post-hoc tests if necessary. Try to find a way to conduct both a t-test and an ANOVA as part of this assignment, even if it takes creating a new variable.
4. Visualize the relationship between your dependent variable and at least one of the independent variables, presumably focusing on relationships that were the most interesting.
5. Summarize the results with any overarching takeaways, including whether the differences should be considered inequities.

12

Unpacking Mechanisms Driving Inequities: Multivariate Regression

In Chapter 11, we observed how a legacy of redlining has left some communities warmer than others. This is an inequity that makes it harder for local residents to stay cool and safe during summer heat, leading to more medical emergencies. This is just one of many inequities stemming from the history of redlining. But once we know about an inequity, what can we do to address it? Clearly, drawing lines on a map did not make certain neighborhoods hotter, nor can we go back in time and undraw those lines. Instead, we need to understand how the patterns of investment that followed instigated forms of community development that created hotter places. In understanding that history, we might then craft strategies that counteract its consequences.

A thorough "equity analysis" aims to uncover the specific mechanisms that are creating or perpetuating the inequity. This entails leveraging additional information to reveal which mechanisms are driving and sustaining said correlation. In the current case, we would be able to better understand *why* redlined neighborhoods are warmer, thereby guiding us toward effective interventions that might fix the issue. There are three infrastructural features of neighborhoods that have been found to be particularly important for heat. The first is the amount of canopy cover from trees. Canopy can deflect sunlight and create more air flow, each of which have cooling effects. The second is the amount of impervious surfaces, which is largely synonymous with pavement, as they trap and emanate heat. The third is albedo, or the tendency of surfaces to reflect light energy, which has its own additional cooling effect. For example, white roofs have become popular because they help cool off communities by reflecting large amounts of heat. Figure 12.1 maps the first two of these variables across Boston.

An equity analysis requires unpacking the relationships between multiple variables to understand which are most responsible for the phenomenon at hand. The most common statistical tool for this task is a *multivariate regression*, which assesses how multiple variables explain variation in a single variable of interest. Because the levels of canopy, impervious surfaces, and albedo are all interrelated, we will use regression to disentangle their relationships and better understand if and how they explain the differences in temperature across neighborhoods, including whether they account for the effects of redlining.

12.1 Worked Example and Learning Objectives

Building on Chapter 11, we will examine three infrastructural features and their role in driving the urban heat island effect: tree canopy, pavement (i.e., impervious surfaces), and albedo. We will then evaluate the extent to which they explain the correlation between heat

FIGURE 12.1
Maps of canopy cover (left; darker green indicates more canopy) and land surface temperature (right; darker red indicates higher temperature) across the neighborhoods of Boston, MA. (Credit: Urban Land Cover and Urban Heat Island Database, the data set used in this chapter)

and redlining. In doing so, we will learn both conceptual and technical skills, including how to:

- Propose and test potential mechanisms that might account for inequitable conditions, generating insights that might guide action;
- Conduct and interpret multivariate regressions that unpack the relationships between multiple variables, including:
  - Identifying the dependent variable of interest and potentially informative independent variables;
  - Determining when a regression is preferable to a correlation and the distinct interpretations;
  - Building a regression model that will be most informative;
  - Visualizing a regression's results.

We will continue to work with the same data set we constructed for the analysis in Chapter 11, consisting of two components. First, the Home Owners Loan Corporation's 1938 map grading the investment quality of the neighborhoods of Boston , provided by the Mapping Inequality[1] project, a collaboration of faculty at the University of Richmond's Digital Scholarship Lab, the University of Maryland's Digital Curation Innovation Center, Virginia Tech, and Johns Hopkins University. This map has been spatially joined to census tracts so that we know how each tract was classified. The Urban Land Cover and Urban Heat Island Effect Database is a product of researchers at Boston University available through the Boston Area Research Initiative's Boston Data Portal that combines remote sensing data from numerous sources

[1] https://dsl.richmond.edu/panorama/redlining/#loc=5/39.1/-94.58

that capture land surface temperature and a variety of related conditions, including the three we are concerned with here.

Links:

Urban Heat Island Database: https://dataverse.harvard.edu/dataset.xhtml?persistentId=doi:10.7910/DVN/GLOJVA

Redlining in Boston: https://dataverse.harvard.edu/dataset.xhtml?persistentId=doi:10.7910/DVN/WXZ1XK

Data frame name: `tracts`

```
require(tidyverse)


uhi<-read.csv('Unit 3 - Discovery/Chapter 12 - Regressions and
Equity Analysis/Worked Example/UHI_Tract_Level_Variables.csv')
tracts_redline<-read.csv('Unit 3 - Discovery/Chapter 12 - Regressions and
Equity Analysis/Worked Example/Tracts w Redline.csv')

tracts<-merge(uhi,tracts_redline,by='CT_ID_10',all.x=TRUE)
```

12.2 Conducting an Equity Analysis

Chapter 11 described the difference between equity and equality. To briefly summarize, equality often focuses on equal inputs, access to programs and services, or opportunity. However, such an approach ignores how individuals and communities often differ in their ability to take advantage of these resources and opportunities. The hurdles or challenges that cause some to be less able to succeed despite supposed equality are known as *inequities*. (For more, feel free to flip back to Chapter 11.) From this perspective, redlining instigated multiple decades of non-investment that has left long-standing inequities on the financial capital, infrastructure, and conditions in certain neighborhoods.

Analysts play a crucial role in contributing to policies, programs, and services that pursue equity. Obviously, data analysis is necessary in revealing and confirming inequities, as we saw with correlations in Chapter 10 and *t*-tests and ANOVAs in Chapter 11. We can go further, however, by helping to understand how those conditions and challenges have arisen and how they persist. As in the worked example for this chapter, which aspects of the infrastructure of redlined neighborhoods make them hotter? Such insights make it possible to design the policies, programs, services, and other forms of action that will counteract them. From this description, it is probably apparent how equity analyses are relevant to the work of public agencies and community-serving non-profits, but it is worth emphasizing that they can also be central to the work of academic institutions and private corporations. In any of these contexts, it can be worthwhile to ask if all of the populations that an institution serves have the same ability to succeed in leveraging its resources and services, and, if not, how resources and services should be designed to ameliorate those differences.

There are a few methodological considerations required when conducting an equity analysis. The first is that an equity analysis moves from a two-variable, or *bivariate*, perspective to one that unpacks the relationship between many variables. This is called a *multivariate analysis*. For example, the observation that redlining predicts neighborhood temperature is a bivariate inspiration for an equity analysis, but we will need to incorporate aspects of neighborhood design and infrastructure, including tree canopy, density of pavement, and albedo to further unpack that relationship through a multivariate analysis. In doing so, we need to be mindful of how we determine our dependent and independent variables and model their relationships. Second, we need to interpret the relationships arising from the multivariate analysis with an eye toward the distinction between correlation and causation. We will explore each of these considerations in turn.

12.2.1 Multivariate Analysis: Modeling Dependent and Independent Variables

When we analyze multiple variables at once, we almost always need to decide which variables are our outcomes of interest and which variables we want to use to explain variation in those outcomes. As we learned in Chapter 11, these are respectively referred to as our *dependent variables* and our *independent variables*. At that time, it was rather apparent which was which. The independent variable was our categorical variable, in this case redlining; and our dependent variable was our numerical outcome variable, temperature. When we have numerous numerical variables, though, it takes critical thinking to determine which variable or variables make more sense as "predictors" and which we are trying to predict.

In Chapter 11 we discussed how independent and dependent variables can be understood in terms of experiments. The independent variable is something that an experimenter altered in order to observe changes in one or more dependent variables. Applying this logic to our current question, are we asking ourselves if changing the temperature of the neighborhood will alter the amount of tree canopy and pavement, or whether changing the amount of tree canopy and pavement would alter the temperature? The latter is how we have conceptualized the problem. In addition, thinking in terms of experiments might help us to consider what effective programs and services might look like. This also hints at the larger question of causation—more than just correlating with temperature, is one or more of these features causing differences between neighborhoods?

12.2.2 Multivariate Relationships: Correlation and Causation

You have almost certainly heard the phrase, "correlation is not causation." This is one of the most important adages for interpreting statistics. Take the inspiration for the worked example in this chapter. Temperature is higher in redlined neighborhoods that have historically been majority Black communities. This is not because Black people live there. Sure, Black people live in these neighborhoods, which led banks to withhold investment as a form of discrimination, which led to community design and development decisions that made the neighborhoods hotter. In that sense, the presence of Black people indirectly led to higher temperatures, but think like an experimenter for a moment: if more Black people moved into the neighborhood, or people of other racial backgrounds moved in, would that change the temperature of the neighborhood? The answer is no.

For analysts the guiding objective is always to reveal the independent variable or variables that are *causing* our dependent variable. This is difficult, in large part because almost all statistical tools can only reveal correlation. Without an experiment and the ability to control everything else that might somehow impact the dependent variable, we cannot be 100% certain that we are observing causation. The careful use of multivariate tools, however, can bring us closer to causal interpretations and at least allow us to dismiss certain independent variables as being "merely correlated" with the dependent variable. For instance, in our task at hand, we want to understand which aspects of community design might account for the correlation between temperature and race. In other words, we are confident that the latter correlation is not a reflection of causation, and we want to test other variables that might in fact play a causal role. As we will see, multivariate techniques can help us tease these relationships apart, revealing which variables are, if not necessarily "causal," more directly relevant to the outcome variable. We call these *proximal predictors* because they are closer to the variable in the sense of causal relationships. The variables whose role in causality is then "explained away," as it were, are called *distal predictors* because they are further away in terms of causality.

The debate between correlation and causation often confuses technical and conceptual issues. In short, correlation is a statistical association between two variables. Nearly all statistical tools, in the absence of controlled experimentation, are only capable of communicating correlation (though, to be fair, there are techniques that can get us pretty close given the right data structure and research design, though they are beyond the scope of this book). Moving toward causation is a matter of interpretation, though that terminology can distract from what we want to accomplish. The goal of an analyst whose intent is to support action is to reveal mechanisms. How does something occur? The better we are at identifying proximal predictors of an outcome, the more information we have about which mechanisms might be in play. If we can articulate those potential mechanisms and how to reinforce them, mitigate them, or otherwise intervene, then we have a set of options for enacting change (or maintaining the status quo, if that is the goal). This is key to an equity analysis because, as we said at the beginning, it is one thing to know that redlined neighborhoods tend to be warmer, but the next step is to be able to do something about it. You may also recognize here echoes of a discussion in Chapter 6: whereas some might argue that the data can lead us to the desired answers, we are best able to inform action if we understand why the world works the way it does. Understanding "why" entails revealing the underlying mechanisms.

12.3 Regression

12.3.1 Why Use a Regression?

Regression is one of the most popular statistical tools and likely the most commonly known after a correlation. Regression and correlation are closely related conceptually and technically. The underlying arithmetic, in fact, is almost identical as both are designed for the analysis of numeric variables. There are two major distinctions, though. First, the purpose of a correlation is to describe whether two variables share variation or not. It does not differentiate between a dependent variable and independent variable. In contrast, regression uses the variation in an independent variable to explain (or "predict") the values in a dependent variable. The determination of dependent and independent variable must be made before the analysis.

Second, a correlation can only occur between two variables whereas a regression must have one dependent variable but as few or as many independent variables as the analyst might desire. When a regression has only one independent variable, it is referred to as a bivariate regression and is nearly identical to a correlation, though with some distinctions in how the effect size is reported. Any regression with two or more independent variables is referred to as a multivariate regression.

The design of a regression, including all of the independent variables included and any other customization to the analysis, is also referred to as a *model*. For example, we might say that we modeled the dependent variable as a function of three independent variables. There are other ways to customize a regression, but for our purposes we will primarily consider how a model is determined by a set of independent variables. Importantly, a regression cannot accommodate multiple dependent variables. An analysis of multiple dependent variables requires conducting a separate regression for each.

12.3.2 Interpreting Multiple Independent Variables

The interpretation of a correlation is relatively straightforward. If it is positive and significant, we can say that, as one variable increases or decreases, the other tends to do the same. Interpreting a bivariate regression is very similar. A multivariate regression is a little more complicated because we now have a model with multiple independent variables each explaining our dependent variable. How do we interpret each of these relationships? The key is keeping in mind that these relationships were evaluated simultaneously and not independently. Thus, if an independent variable's relationship with the dependent variable is positive and significant in the regression model, we can say that the dependent variable tends to rise and fall with the independent variable *holding all other variables in the model constant*. Though this lacks the brevity of describing a correlation, it is critical to interpreting the data accurately. For instance, it might be that one case is above average on the independent variable but below average on the dependent variable, which would seem to be at odds with a positive relationship; however, upon closer scrutiny, the values on all of the other independent variables are more likely to be responsible for the lower value.

The nuance of interpreting each variable in the context of the others—also known as their *independent effects*—is also part of regression's strength as a multivariate tool. Recall from above that part of the value of regression is to try to reveal which variables are more proximal predictors, meaning they better explain variation in our outcome variable than other variables that are also known to be correlated with the outcome variable. By estimating the independent effects of all of the independent variables holding all other variables constant, the regression can give us insights on which variables are most meaningful. For example, suppose we find that redlining is no longer a positive, significant predictor of heat after we account for tree canopy. That means that, holding tree canopy constant, redlined communities are no longer warmer, meaning that tree canopy is more likely to reveal the mechanism responsible for that inequitable relationship.

12.3.3 Effect Size: Evaluating Each Independent Variable and the Model

A regression model tests effect size and significance at two levels. First, it tests the relationship of each independent variable with the dependent variable. Second, it evaluates how effective the full model—that is, the combination of all of the independent variables—is at explaining variation in the dependent variable. There are different ways of communicating the effect sizes at each of these levels.

12.3.3.1 Regression Coefficients (B and β)

A regression tests a linear model of the following form:

$$y = B_1 * x_1 + B_2 * x_2 + B_3 * x_3 + c$$

This may look familiar from algebra class. The dependent variable (y) is modeled as a function of a series of independent variables (the xs) and an intercept (c). As noted, there can be as many xs as the analyst wants, and this example has three. Each x is multiplied by a B, or beta. Beta estimates how much y is likely to change if the corresponding x increases by 1. This is also known as the slope of the relationship between a given x and y. A regression estimates beta for all independent variables and estimates the intercept, which is the expected value of y when all xs are equal to 0. It also evaluates whether each of these estimates are significantly different from 0.

The betas included in the linear equation, which quantify slopes, are also referred to as unstandardized betas. These can be arithmetically transformed into standardized betas that take into account the distribution of both the independent and dependent variables. These are represented with the Greek letter β. β is a direct analog of the correlation coefficient, r, that we encountered in Chapter 10, and is interpreted the same way. Just like r, it ranges from -1 to 1 and estimates how many standard deviations of change we should expect in our dependent variable for each standard deviation of change in the independent variable. Note that β for the sole independent variable in a bivariate regression will be the same value as r in a correlation between the same two variables.

12.3.3.2 Model Fit and R^2

Sometimes we want to know not only about the relationship between each independent variable and the dependent variable but also how well the full model—that is, all of the independent variables together–explains variation in the dependent variable. Regressions evaluate this with the R^2 value. In arithmetic terms, R^2 is a calculation of the percentage of the variation the model explains by considering the values of the independent variables rather than assuming that all cases have the same value on the dependent variable. This is also known as a test of model fit.

You may note that the statistic quantifying model fit uses the same letter as the correlation coefficient. This is because in the bivariate case R^2 is the square of r. If $r = .5$, then $R^2 = .25$, meaning that 25% of the variation in our variable of interest is explained by this relationship. The same would be true of β in our bivariate regression. In multivariate regressions, the relationship between R^2 and the βs are a bit more complicated, but the interpretation of the value itself is the same: it captures the percentage of variation in the dependent variable explained by the model. Also, recall that in Chapter 11 we used R^2 to evaluate the strength of an ANOVA model. We will not go deep into it here, but R^2 is calculated the same way in both techniques, with total variation quantified as the sums of squares and the model explaining a certain portion of this variation. The only difference is that, in an ANOVA, variation is explained by the categories of the independent variable; in a multivariate regression, variation is explained by the numerical values of the multiple independent variables. For the same reason, the overall fit of the model is evaluated for significance using an F-statistic, which you may recall is the primary statistic for the evaluation of ANOVAs.

12.4 Conducting Regressions in R: `lm()`

Base R includes a function for running regressions called `lm()`. The name references another term for a standard regression: *linear model*. We will illustrate the use of lm and associated functions through our worked example of redlining, community design and infrastructure, and neighborhood temperature in Boston's census tracts. The key variables will be land surface temperature (`LST_CT`), tree canopy (as a percentage of space covered, from 0-1; `CAN_CT`), impervious surface (a near-synonym for pavement, as a percentage of space covered, from 0-1; `ISA_CT`), and albedo (or the proportion of light energy reflected and not absorbed by surfaces, from 0-1; `ALB_CT`). Later we will re-incorporate whether a census tract was redlined or not.

12.4.1 Bivariate Regression

The simplest regression is one in which a single independent variable is used to explain variation in the dependent variable. This is arithmetically identical to a correlation between two variables. The differences are: (1) the specification of an independent and dependent variable; and (2) the results communicate not only the strength of association between the two variables but also a slope describing their linear relationship. To get things started, let us examine the relationship between the tree canopy and neighborhood temperature as a bivariate regression.

```
lm(LST_CT~CAN_CT, data=tracts)
```

```
##
## Call:
## lm(formula = LST_CT ~ CAN_CT, data = tracts)
##
## Coefficients:
## (Intercept)       CAN_CT
##      102.86       -40.72
```

We see here the structure of the arguments in the lm command. It begins with a formula, with the dependent variable (`LST_CT`) on the left, followed by `~`, and then the independent variable (`CAN_CT`). We then specify the data frame of interest, in this case tracts.

The output includes an estimate of the intercept and the unstandardized beta for `CAN_CT`. The intercept is estimated as $102.86°F$ degrees. Given the logic of a regression, this is the estimated average when the percentage of canopy cover is 0%. This value is estimated to drop by 40.72 for every change of 1 in canopy. Because this latter variable is a $0-1$ measure of the proportion of canopy coverage, the change of $40.72°F$ is the rather dramatic (and unreasonable) expected shift in temperature between a neighborhood with no canopy and one with 100% canopy cover.

The output, however, is a little simplistic, telling us only the estimates for the intercept and the unstandardized beta for `CAN_CT`. There is no significance communicated for these parameters nor a mention of model fit. Like many other functions in R that conduct statistical tests, lm is designed to generate objects whose contents are best accessed through the `summary()` command. Let us try again.

```
reg1<-lm(LST_CT~CAN_CT, data=tracts)
summary(reg1)
```

```
##
## Call:
## lm(formula = LST_CT ~ CAN_CT, data = tracts)
##
## Residuals:
##     Min      1Q  Median      3Q     Max
## -5.2698 -0.9127  0.0203  0.8920  5.1705
##
## Coefficients:
##             Estimate Std. Error t value             Pr(>|t|)
## (Intercept) 102.8613     0.1861  552.61 <0.0000000000000002 ***
## CAN_CT      -40.7176     1.5099  -26.97 <0.0000000000000002 ***
## ---
## Signif. codes:  0 '***' 0.001 '**' 0.01 '*' 0.05 '.' 0.1 ' ' 1
##
## Residual standard error: 1.718 on 176 degrees of freedom
## Multiple R-squared:  0.8051, Adjusted R-squared:  0.804
## F-statistic: 727.2 on 1 and 176 DF,  p-value: < 0.00000000000000022
```

`reg1` is an object of class `lm` and it can be analyzed with `summary()`, giving us a lot more to work with. We see the same parameter estimates, but that they are both significant at $p < .001$. This is trivial for the intercept, as it is simply evaluating whether the average neighborhood with no tree canopy has a temperature different from 0 (unsurprisingly, it does). The B for `CAN_CT` is significant as well, indicating that the shift of -40.72 is greater than would be expected by chance.

We also now have the opportunity to assess model fit. The R^2 is 0.81, indicating that tree canopy accounts for 81% of the variation in temperature across neighborhoods. The adjusted R^2 is similar. This is actually very high for social science research. Rarely are we able to explain anywhere near that much variation. This latter statistic is helpful when there are lots of variables in the model, as variables are likely to explain a certain amount of variation just by chance. The F-statistic indicates that the R^2 is significant at $p < .001$. Note that this p-value is identical to the one for the beta for `CAN_CT`. This is because in a bivariate regression, the sole beta is the only piece of information determining model fit.

12.4.2 Calculating Standardized Betas with `lm.beta()`

Interestingly, `lm()` does not report a standardized beta. There are two ways of doing this. The simplest is the `lm.beta()` command from the `QuantPsyc` package (Fletcher, 2012). It takes your `lm` object as input.

```
require(QuantPsyc)
lm.beta(reg1)
```

```
##     CAN_CT
## -0.8972984
```

This tells us that the standardized Beta for CAN_CT is -.90. To confirm that this is the same as the *r* correlation coefficient, we can run `cor.test()`.

```
cor.test(tracts$CAN_CT,tracts$LST_CT)
```

```
##
##  Pearson's product-moment correlation
##
## data:  tracts$CAN_CT and tracts$LST_CT
## t = -26.967, df = 176, p-value < 0.00000000000000022
## alternative hypothesis: true correlation is not equal to 0
## 95 percent confidence interval:
##  -0.9226171 -0.8642804
## sample estimates:
##        cor
## -0.8972984
```

As you can see, the results are identical. The *r* is equal to our standardized beta and the *p*-value is the same.

If you would rather not dabble with another package or `lm.beta()` runs into issues (which can happen being that it was developed under earlier versions of R), you can extract standardized betas yourself by scaling your variables by their standard deviations. Remember that a standardized beta is the change in standard deviations in the dependent variable for every change in standard deviations in the independent variable. This can be accomplished by applying the `scale()` function to every variable in the `lm()` function. However, it is important to remove all NAs for this comparison to work. `lm()` removes NAs pairwise, only including complete cases, because otherwise it cannot estimate the model. The standardization of beta then depends on the values in this subset of complete cases. If we did not remove these before applying `scale()` to our variables, the standardization of one or more variables might include values not included in the linear model.

```
tracts_temp<-tracts %>%
    filter(!is.na(LST_CT) & !is.na(CAN_CT))
reg_stand<-lm(scale(LST_CT)~scale(CAN_CT), data=tracts_temp)
summary(reg_stand)
```

```
##
## Call:
## lm(formula = scale(LST_CT) ~ scale(CAN_CT), data = tracts_temp)
##
## Residuals:
##      Min       1Q   Median       3Q      Max
```

```
## -1.35770 -0.23516  0.00523  0.22981  1.33211
##
## Coefficients:
##                               Estimate              Std. Error
## (Intercept)    -0.000000000000001485  0.033180032841127596
## scale(CAN_CT) -0.897298355795607794  0.033273629734652463
##                t value              Pr(>|t|)
## (Intercept)       0.00                     1
## scale(CAN_CT)   -26.97 <0.0000000000000002 ***
## ---
## Signif. codes:  0 '***' 0.001 '**' 0.01 '*' 0.05 '.' 0.1 ' ' 1
##
## Residual standard error: 0.4427 on 176 degrees of freedom
## Multiple R-squared:  0.8051, Adjusted R-squared:  0.804
## F-statistic: 727.2 on 1 and 176 DF,  p-value: < 0.00000000000000022
```

Again, we see an identical result, with β = -.90.

12.4.3 Multivariate Regression

We have observed that neighborhoods in Boston with more tree canopy are cooler than other neighborhoods. Canopy, pavement, and albedo are intertangled, however. More pavement means less canopy. Less canopy also means less albedo. Which of these infrastructural variables are most important for temperature? To address this, we must move from a bivariate regression to a multivariate one. To do so, we simply expand our list of independent variables in `lm()`.

```
reg_multi<-lm(LST_CT~CAN_CT + ISA_CT + ALB_CT, data=tracts)
```

Before we move to the results, note that adding these variables is a lot like an equation. Remember that a linear model is simply the estimation of an algebraic formula. In fact, ~ indicates a formula, with the left side as the dependent variable and the right side as the independent variables.

```
summary(reg_multi)
```

```
##
## Call:
## lm(formula = LST_CT ~ CAN_CT + ISA_CT + ALB_CT, data = tracts)
##
## Residuals:
##     Min      1Q  Median      3Q     Max
## -5.4115 -0.8090  0.1198  0.8822  4.0028
##
## Coefficients:
##              Estimate Std. Error t value             Pr(>|t|)
## (Intercept)    96.883      3.432  28.232 < 0.0000000000000002 ***
## CAN_CT        -28.731      2.559 -11.226 < 0.0000000000000002 ***
```

```
## ISA_CT          7.991       1.910   4.185            0.0000453 ***
## ALB_CT        -11.096      18.271  -0.607                0.544
## ---
## Signif. codes:  0 '***' 0.001 '**' 0.01 '*' 0.05 '.' 0.1 ' ' 1
##
## Residual standard error: 1.591 on 174 degrees of freedom
## Multiple R-squared:  0.8349, Adjusted R-squared:  0.8321
## F-statistic: 293.4 on 3 and 174 DF,  p-value: < 0.00000000000000022
```

The new regression has estimated three betas, one for each of our independent variables. These indicate that the density of the tree canopy and impervious surfaces are the primary predictors of temperature in a neighborhood. Both are significant at $p < .001$. Temperature now is estimated to drop by 28.73 degrees as canopy goes from 0 to 1 (i.e., no trees to complete coverage), meaning some of the effect we saw in the bivariate regression is accounted for by the other two variables. Meanwhile, the effect of impervious surfaces is a little less dramatic, with temperature rising by 7.99 degrees as they go from 0 to 1 (i.e., no pavement to 100% pavement and other impervious surfaces). Albedo is a non-significant predictor with a p-value of .54.

The fit of our model has grown as well, with R^2 increasing to .83. That means that canopy and pavement account for over 83% of the variation in temperature across neighborhoods, and that the addition of pavement accounts for 2% more variation than canopy alone. Unsurprisingly, the F-statistic finds this to be significant at $p < .001$.

To finish, we can look at the standardized betas of our independent variables.

```
lm.beta(reg_multi)
```

```
##      CAN_CT      ISA_CT      ALB_CT
## -0.63314619  0.28956203 -0.03155982
```

As we can see, the standardized beta for canopy is still quite large, albeit smaller than before, with β = -.63. Impervious surface's standardized beta is substantially smaller at $\beta = .29$, although clearly still meaningful. The standardized beta for albedo is very close to 0, which is consistent with its lack of significance.

12.4.4 Reporting Regressions

12.4.4.1 Describing Regression Parameters

There is a lot of information in a multivariate regression, especially when there are many variables. There are some tricks, though, to communicating this information in an efficient way. First, as we did in Chapter 10 when describing the results of correlations, we noted that the numbers do not always need to be part of the sentence but can be tucked in parentheses. For instance, we might say, "Neighborhoods in Boston with more canopy cover had lower temperatures (B = -28.73, $p < .001$)." Alternatively, we could state the standardized beta in the parentheses. We do not typically report both in the text.

In larger multivariate models the interpretation becomes more complicated, as in "Neighborhoods with greater tree canopy coverage had lower temperatures, holding pavement and albedo constant (β = -.64, $p < .001$)." That said, we often assume that the reader understands that a regression holds the other independent variables constant and omit that phrase. There may be times, though, when it is prudent to include this reminder.

12.4.4.2 Reporting Unstandardized and Standardized Parameters

Unstandardized betas can be especially helpful for communicating the practical implications of your findings. For instance, we might say that, "For every 10% increase in tree canopy, neighborhoods saw a drop in temperature of nearly 3 degrees (B = -28.73, $p < .001$)." There are other times, however, when the point you are trying to make or the expectations of the audience make standardized betas the preferable metric for communication. This is especially true when you want to communicate the strength of relationship between the independent and dependent variables in a way that can be compared to other analyses using different measures. Indeed, one of the great advantages of standardized betas is the ability to compare separate statistical tests to each other regardless of the distributions of the original variables.

12.4.4.3 Summarizing Results

As noted, multivariate regressions produce a lot of information: unstandardized betas, standardized betas, and significance levels for each independent variable and R^2 for model fit. Your audience might also want to know the number of cases (rows) in the analysis and other details. There are two techniques that may come in handy for distilling this complexity into an accessible form.

First, it is useful to focus the text on the findings that most matter. For instance, on might describe the independent variables that had significant relationships with temperature in detail but then summarize the others in brief (e.g., "Other predictors were non-significant."). You can also find other efficiencies that concentrate on meaning rather than statistics, such as, "Environmental variables like less tree canopy (B = -30.15, $p < .001$) and more impervious surface (B = 7.66, $p < .001$) predicted higher temperatures."

Second, Figure 12.2 illustrates an efficient way to communicate all of the information in a regression in an organized fashion. This can then be referenced in the text so that a reader can dig into the details that you do not address directly. This enables you to use efficient writing techniques without worrying that you have made your results too opaque.

12.4.5 Incorporating Categorical Independent Variables

Thus far, we have described regressions as a tool for using numerical independent variables to predict the variation in numerical dependent variables. Regressions can also incorporate categorical independent variables. The categorical variable simply needs to be converted into a series of dichotomous variables reflecting the separate categories, each of which will then have its own beta in the model. We create a dichotomous variable for all categories in the independent variable except for one because we need something to compare other categories to. We call this the *reference category*. Each of the betas is then the difference of that category's mean from the reference category's mean. Best of all, if we enter a categorical variable into `lm()`, it can automatically create these dichotomous variables for us.

Let us start by entering the HOLC grade variable (`Grade`) into a model predicting land surface temperature. In one sense, this is a bivariate regression because we have only entered

	Unstand. Beta (std. error)	Stand. Beta
Prop. Tree Canopy	-28.73*** (2.56)	-.63
Prop. Impervious Surfaces	7.99*** (1.91)	.29
Albedo	-11.10 (18.27)	-.03
R^2	**.83**	

FIGURE 12.2
Example for organizing and reporting multivariate regression results, including unstandardized betas, standard errors, standardized betas, and significance for all independent variables and R^2 for the full model fit. Significance is traditionally represented as: * - $p < .05$, ** - $p < .01$, *** - $p < .001$).

one variable. But because Grade has four categories (A, B, C, and D), our regression has three variables, with communities graded A as our reference category.

```
reg_grade<-lm(LST_CT~Grade, data=tracts)
summary(reg_grade)
```

```
##
## Call:
## lm(formula = LST_CT ~ Grade, data = tracts)
##
## Residuals:
##      Min       1Q   Median       3Q      Max
## -10.5562  -1.8257   0.0014   2.2248   8.7436
##
## Coefficients:
##             Estimate Std. Error t value              Pr(>|t|)
## (Intercept)   89.481      3.368  26.571 < 0.0000000000000002 ***
## GradeB         6.588      3.517   1.873             0.062741 .
## GradeC         8.534      3.386   2.521             0.012614 *
## GradeD        11.981      3.391   3.533             0.000526 ***
## ---
## Signif. codes:  0 '***' 0.001 '**' 0.01 '*' 0.05 '.' 0.1 ' ' 1
##
## Residual standard error: 3.368 on 173 degrees of freedom
##   (1 observation deleted due to missingness)
## Multiple R-squared:  0.2632, Adjusted R-squared:  0.2504
## F-statistic:  20.6 on 3 and 173 DF,  p-value: 0.0000000000184
```

We see here that census tracts graded D or C were significantly hotter than Grade A neighborhoods by 11.98°F ($p < .001$) and 8.53°F ($p < .05$), respectively. Neighborhoods graded B were close behind at 6.59°F warmer than A neighborhoods, but not quite significant. If you look closely and compare to the results of our ANOVA in Chapter 11, the results are very familiar. The unstandardized betas are equal to the differences in means reported in the post-hoc tests. R^2, the F-statistic, and the p-value for the full model are the same as those from the ANOVA as well. This is because the ANOVA and the categorical regression are arithmetically identical. The one difference is the significance of the regression parameters. This is because the post-hoc tests adjusted for multiple comparisons, but the regression does not. The adjustments were especially meaningful given how small the sample of grade A neighborhoods was.

As a last step, we might examine the extent to which canopy, pavement, and albedo account for the differences in temperature between neighborhoods with different HOLC grades. As we have already seen, a regression limited to the former three factors explained far more variance than the HOLC grades did (83% vs. 26%). This suggests that the environmental factors are probably more proximal predictors than the HOLC grades.

```
reg_grade2<-lm(LST_CT~Grade + CAN_CT + ISA_CT + ALB_CT,
               data=tracts)
summary(reg_grade2)
```

```
##
## Call:
## lm(formula = LST_CT ~ Grade + CAN_CT + ISA_CT + ALB_CT, data = tracts)
##
## Residuals:
##     Min      1Q  Median      3Q     Max
## -5.5021 -0.8924  0.0524  0.9329  3.8710
##
## Coefficients:
##             Estimate Std. Error t value             Pr(>|t|)
## (Intercept) 98.29662    3.79896  25.875 < 0.0000000000000002
## GradeB      -1.74841    1.66440  -1.050                0.295
## GradeC       0.01725    1.62067   0.011                0.992
## GradeD       0.06251    1.65999   0.038                0.970
## CAN_CT     -26.98596    2.71056  -9.956 < 0.0000000000000002
## ISA_CT       7.89329    1.89695   4.161            0.0000502
## ALB_CT     -22.51662   18.29857  -1.231                0.220
##
## (Intercept) ***
## GradeB
## GradeC
## GradeD
## CAN_CT      ***
## ISA_CT      ***
## ALB_CT
## ---
## Signif. codes:  0 '***' 0.001 '**' 0.01 '*' 0.05 '.' 0.1 ' ' 1
##
## Residual standard error: 1.55 on 170 degrees of freedom
##   (1 observation deleted due to missingness)
## Multiple R-squared:  0.8466, Adjusted R-squared:  0.8411
## F-statistic: 156.3 on 6 and 170 DF,  p-value: < 0.00000000000000022
```

In this model we see that HOLC grades are no longer predictive of land surface temperature, but canopy and pavement have very similar betas as before and explain an overwhelming amount of the variation. As such, we are able to say with some confidence why redlined neighborhoods are warmer. It is because they have less canopy and more pavement (see Section 11.7 for a visual depiction of this latter relationship). In turn, we now have multiple mechanisms by which we might address the issue.

12.5 Some Extensions to Regression Analysis

12.5.1 Dealing with Non-Normal Dependent Variables

As noted in Chapter 10, not all variables are normally distributed. This can create issues for traditional statistical tests, linear regression included. In fact, a linear equation may not be an appropriate model for fitting a variable. If, for example, it has a Poisson distribution, we need to assume an entirely different arithmetic relationship between the independent and dependent variables. There are also times when we want to predict a dichotomous (i.e., 0/1) variable. Each of these special cases is possible using the `glm()` function, which stands for generalized linear model. Generalized linear models extend the logic of the linear model to estimate models that predict dependent variables with non-normal distributions. The specific model needed can be specified with the `family =` argument (e.g., `family = 'Poisson'`). We will not go further into `glm()` in this book, but it is a tool you are welcome to explore if relevant to the variables you are looking to analyze.

12.5.2 Working with Residuals

The point of a regression is to fit a linear model that can explain variation in a variable of interest. Sometimes we might be as interested in the places where the model *failed* to explain values as we are in the model itself. Returning to our multivariate model above, which neighborhoods are warmer or cooler than expected taking into consideration their canopy and pavement coverage? These differences from expectation are known as *residuals*, and they can be extracted from an `lm` object using `$residuals`. They can be positive (above expectations) or negative (below expectations). We can also visualize them (say, with a map) or analyze them in conjunction with other variables by merging them back into the original data set with the following example code.

```
tracts<-merge(tracts,data.frame(reg_multi$residuals),
              by='row.names',all.x=TRUE)
```

12.5.3 Building a Good Model

Conducting a solid multivariate analysis requires thinking critically about how you have selected your independent variables. There are three suggestions I might make on this front, though there are certainly other considerations one might make:

1. *Be careful that the implied independent-dependent variable relationship is appropriate.* Does it make sense for your selected independent variable to be predicting your dependent variable? Describe the relationship in words to check that it is logical and is the relationship you want to be testing.
2. *Try not to include multiple independent variables that are overly similar.* For example, median income and poverty rate of a neighborhood are nearly the same measure. Statistically, this can create problems with running the model. Conceptually, it makes interpretation difficult, especially if the similarity of the variables causes one to be positive and the other to be negative, which would seem to be logically impossible.
3. *Consider confounding variables.* Are there other features related to one or more of your independent variables that might help tell the story even better? Further, are there confounds that are muddying the interpretation? For example, many analyses of crime separate out predominantly residential from predominantly commercial neighborhoods because those two contexts have distinct social patterns. You can include such variables in the model as a control. However, it is sometimes necessary to consider subsetting your data and running the models by limiting each regression to a single group to confirm that such issues are not obscuring all results.
4. *Do not overload your model.* Having access to lots of variables is a major advantage for telling a full story. However, too many variables in a model can become confusing, sometimes obscuring meaningful predictors or overstating the importance of others. A telltale sign of this situation is having no significant predictors but a significant R^2. This means that some set of variables is explaining your dependent variable, but they are dividing their impact. A classic example is including median income and percentage of residents under the poverty line in a neighborhood in the same model. These two variables have nearly identical variations, creating issues for interpretation. It is important then to think critically about what should be in the model and which variables are complicating matters and therefore should be trimmed.

12.6 Visualizing Regressions

Visualizing regressions can be tricky. How does one visualize the effect of multiple independent variables on a single dependent variable? The answer is, short of a multi-dimensional plot, you cannot. Instead, we have to find ways to communicate our multivariate results with bivariate (and sometimes trivariate) visualizations.

In Chapter 7, we learned the most traditional way to show the relationship between two numeric variables: the dot plot. We are going to replicate that here and add a new wrinkle that captures the linear relationship between the variables more explicitly. To do so, we will return to `ggplot2`.

```
base<-ggplot(data=tracts, aes(x=CAN_CT, y=LST_CT)) +
    geom_point() + xlab("Prop. Canopy Coverage") +
    ylab("Land Surface Temperature")
base
```

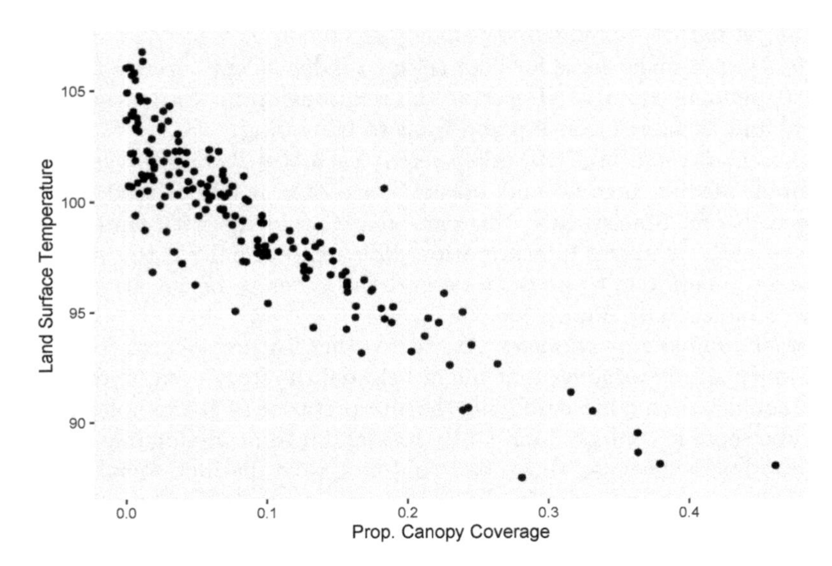

As we might have expected given the large, negative standardized beta, we see a sharp drop in land surface temperature in neighborhoods with more canopy coverage. If we want to visualize this relationship in the linear sense embodied by a regression, though, we can add the `geom_smooth()` command, with `method=lm`. `geom_smooth()` summarizes a bivariate relationship with a single line, and the `lm` method, as you may have guessed, uses the slope and intercept from a bivariate regression to do so.

```
base + geom_smooth(method=lm)
```

```
## `geom_smooth()` using formula 'y ~ x'
```

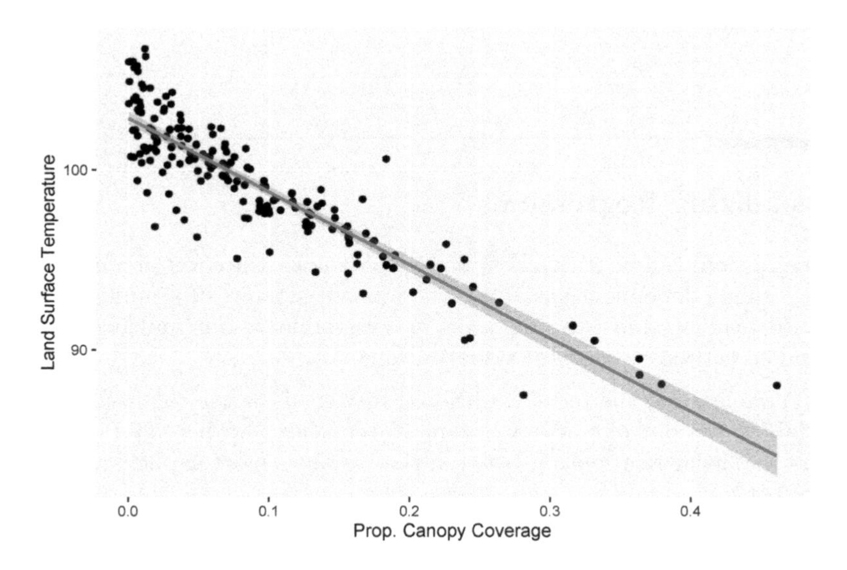

As we can see, the line summarizes the graph with a single line and confirms the strong negative relationship between the two variables. Also, returning to Section 12.5.2, we can observe residuals visually. Dots that fall above the line are those that are warmer than expected given their canopy cover, and dots that fall below the line are those that are cooler than expected given their canopy cover. Again, we might choose to extract these residuals from a regression and map them to see how they are distributed across the city or further analyze them to see what additional factors might be driving land surface temperature.

12.7 Summary

The goal of this chapter was twofold. Conceptually, we wanted to conduct an equity analysis that took a bivariate relationship between a single characteristic (i.e., redlining) and an outcome (e.g., temperature) and revealed the mechanisms that might be perpetuating that relationship. In a technical sense, this was a vehicle for conducting our first multivariate statistical analyses. We learned that the correlation between redlining and temperature could largely be explained by the design and infrastructure of the community, specifically the presence (or absence) of tree canopy and pavement and other impervious surfaces. In the process, we have learned to:

- *Distinguish equity and equality*;
- *Identify independent and dependent variables* when conducting a multivariate analysis;
- *Propose and test mechanisms* that might be responsible for perpetuating inequities;
- *Conduct and interpret a regression* using the `lm()` command, including:
  - *Design a model* including the appropriate independent variables for explaining a given dependent variable;
  - *Generate, interpret, and report unstandardized and standardized betas*;
  - *Generate, interpret, and report model fit statistics*;
  - *Visualize a bivariate regression relationship.*

As a final thought, it is worth noting that the correlation between race and temperature in Boston is not as concerning today as this analysis might suggest. Black communities have largely been displaced from their historical neighborhoods by gentrification, meaning that the hottest parts of the city are now occupied by relatively affluent, predominantly White populations that often have a disproportionate number of young professionals. This is not necessarily true across the United States, especially in cities in the southern United States where summer heat is quite dangerous and hot neighborhoods are unlikely to attract gentrification. Also, cities that have not seen the economic boom that Boston has over the last 30 years may not have seen the same demographic turnover. Further, even in Boston, the issue of the urban heat island effect and health goes beyond exposure to heat. It is also a matter of vulnerability to medical emergencies when exposed to heat, which tends to be higher in communities of color, and the ability to mitigate that exposure (e.g., escaping to air conditioning), which tends to be lower in communities of color. Thus, although communities of color are not systematically warmer in Boston, the relationship between redlining and infrastructure still plays a role in the impacts of the urban heat island effect and other racial inequities in the United States.

12.8 Exercises

12.8.1 Problem Set

1. For each of the following pairs of terms, distinguish between them and describe when each is the most appropriate tool.
   a. Correlation vs. causation
   b. Correlation vs. regression
   c. Beta vs. R^2
   d. Unstandardized beta vs. standardized beta
   e. ANOVA vs. regression
2. Which of the following statements about r and β are true?
   a. r controls for other variables.
   b. r and β are on the same scales.
   c. β is generated by correlation tests.
   d. None of the above
   e. All of the above
3. You read a report that how income predicts health, with the beta being -2.5. The article does not state if it is a standardized or unstandardized beta. Which is it? How do you know?
4. For each of the following pairs of variables, if you were to conduct an analysis with one predicting the other, which do you think should be the independent variable and which should be the dependent variable? If you think it could go in both directions, say so. Justify your answer by proposing at least one mechanism that could run in that direction.
   a. Distance from city center and population density
   b. Race and locally-owned businesses in a neighborhood
   c. Crime on a subway line and the number of people who ride it
   d. Redlining and medical emergencies per 1,000 residents
   e. Academic achievement in high school and career success in adulthood
5. Return to the beginning of this chapter when the `tracts` data frame was created. Recall that we have worked previously with multiple data sets with many more measures describing tracts, including demographic characteristics from the American Community Survey, metrics of physical disorder, engagement and custodianship from 311 records, and social disorder, violence, and medical emergencies from 911 dispatches. For the following you can merge any of these variables into the tracts data frame.
   a. Select an outcome variable of interest. Explain why you are interested in this variable.
   b. Run correlations between this variable and demographic characteristics to see if there are any disparities that might qualify as inequities.
   c. Select a set of independent variables that might partially explain this correlation. Justify this selection and test it in a regression model. Be sure to describe the findings for the individual variables and for model fit.
   d. Visualize the relationship between your dependent variable and at least one of the independent variables, presumably focusing on relationships that were the most interesting.
   e. Summarize the results with any overarching takeaways.

f. *Extra*: Extract the residuals from your regression model and map them across the tracts of Boston.

12.8.2 Exploratory Data Assignment

Complete the following with a data set of your choice. If you are working through this book linearly you may have developed a series of aggregate measures describing a single scale of analysis. If so, these are ideal for this assignment.

1. Select an outcome variable of interest. Explain why you are interested in this variable.
2. Run correlations between this variable and demographic characteristics to see if there are any disparities that might qualify as inequities.
3. Select a set of independent variables that might partially explain this correlation. Justify this selection and test it in a regression model. Be sure to describe the findings for the individual variables and for model fit.
4. Visualize the relationship between your dependent variable and at least one of the independent variables, presumably focusing on relationships that were the most interesting.
5. Describe the overarching lessons and implications derived from these analyses.

Discovery: *Unit III Summary and Major Assignments*

Summary and Learning Objectives

Unit III has focused on the technical and interpretive skills associated with inferential statistics, including the ability to generate and evaluate evidence for hypotheses regarding the characteristics of individual variables and relationships between multiple variables. We focused especially on the application of these tools to questions of equity.

Technical Skills

- Conduct and interpret multiple statistical tests, including:
  - Correlations, which evaluate the strength of association between two numerical variables;
  - *t*-tests, which compare values on two groups;
  - ANOVA tests, which compare values across three or more groups;
  - Regressions, which use one or more independent variables (numerical or categorical) to predict the values of a single, numerical dependent variable;
- Create visualizations illustrating the results of these statistical tests, including:
  - Matrices representing the distributions of a set of variables and the correlations between them;
  - Bar charts that represent the differences between groups;
  - Dot plots with linear relationships represented.

The packages we have learned include:

- `Hmisc`, which includes a sophisticated way to test multiple correlations simultaneously;
- `Ggally` for generating a matrix that visualizes the distributions of a set of variables and the correlations between them;
- `reshape2` for reorganizing data to realize a distinct structure required for a specific analysis or visualization;
- `QuantPsyc`, which includes a function for calculating the standardized betas from linear models.

Interpretive Skills

- Conduct and interpret inferential statistical tests, including:
  - The logic of using samples to reach generalizable conclusions;
  - The reporting and interpretation of both effect size and significance;

- Differentiate between inequalities and inequities and design and interpret analyses intended to explain the mechanisms driving the latter.
- Conceptualize independent and dependent variables and building the appropriate model with the appropriate statistical tool when generating and testing hypotheses.

Unit-Level Assignments

Community Experience Assignment

The community exploration assignments in this book are designed to align skills you have been learning with real-world contexts. They are most useful in conjunction with the Exploratory Data Assignments at the end of each chapter, especially when you have been working through them with a single data set. They provide an opportunity to "ground truth," or really evaluate the assumptions and objectives that have guided your analysis thus far. There will be one in each unit. These can also be combined with a service-learning or capstone oriented course.

This third community exploration assignment will return to the direct experience of a neighborhood. Please:

1. Select a neighborhood based on something notable in your analyses regarding the direction you anticipate for your final project, with an eye towards providing "ground truth" relative to something that has intrigued or challenged you. You might also consider how this particular "ground truthing" exercise will enable you to evaluate the potential impact or public value of your analyses.
2. Visit and explore this neighborhood either in person or through whatever virtual tools you find useful (including Google StreetView, BostonMap, public media, etc.), seeking to observe how the characteristics of the data and your analyses manifest themselves in the real world. (Note: A visit that lasts less than a half-hour would be unlikely to generate enough observations to support a high-quality set of insights).
3. Write a 3-5 page memo describing the logic for why you visited this place, what you discovered, and what this tells you about the interpretation of your data. This last part should include or be followed by a broader discussion of how your perspective on the relevance of your work for communities has evolved as you have progressed through this book or when analyzing these data in general. The memo should include images from your exploration and maps with data describing the region.

Post-Unit Assignment: A Full Research Study

This unit of the book has focused on using inferential statistics to "discover" relationships between variables, especially focusing on how we identify questions and subsequent analyses that are meaningful and can have public impact. This builds upon our previous efforts to reveal basic information and to create custom measures of interest by enabling us to formally test the hypotheses that naturally follow. This paper will bring this work together in the format of a public report, consisting of:

- A brief *Executive Summary* that details the main points of your analysis and findings. Think of the audience for this being someone who would benefit from the insights but might not have the time to read the whole paper or the desire to wade through methods.
- An *Introduction* that describes the conceptual inspiration for your analysis and why it might be interesting, both conceptually and practically. You might also include hypotheses if appropriate. The Introduction should include a few citations to fully justify and motivate your analyses.
- A brief *Data & Methods* section that describes the content of your data set, how you calculated any new measures, and other data sources you used. **This is not your complete Methods section, but just enough for a reader to be able to understand the content that follows. It should reference the Appendix (see below).**
- *Results & Discussion* section, broken up into one or more sub-sections, where you describe your analyses. This will include (1) descriptive statistics (e.g., what is the distribution of a critical measure across the city, what is the average, maximum, etc.), (2) inferential statistical tests, and (3) illustrative visualizations. This part of the paper should strive to tell one or more interesting stories.
  - It does not need to be titled Results & Discussion. The title of the section and sub-sections should capture what you discovered.
- A *Conclusion* that briefly interprets what you've found and suggests implications for research and policy.
- A *Methodology Appendix* that describes the content of your data set and then summarizes how you calculated any new measures and from where you accessed other measures. Keep in mind that this section should not include detail for detail's sake but should provide the information necessary for an expert to (a) fully understand what you did and (b) replicate the work if they so desired.

Suggested Rubric (Total 12 pts.)

Executive Summary: 1.5 pts.

Introduction: 1.5 pts.

Data & Methods: .5 pts.

Results & Discussion: 3 pts.

Conclusion: .5 pts.

Methodology Appendix: 2 pts.

Visuals: 1.5 pts.

Details: 1.5 pts.

Unit IV

The Other Tools

13

Advanced Analytic Techniques

Assuming you have moved through this book in a linear fashion, you have learned a lot about how to access, manipulate, analyze, visualize, and interpret data, especially of the large, messy, under-documented sort that are naturally occurring through administrative processes, internet platforms, and otherwise. But as often happens with learning, you have probably realized that there is so much more to learn, and that you now have the foundational skills to pursue that additional knowledge in an informed, strategic way that will allow you to build yourself as a practitioner of urban informatics. Obviously, a single book cannot impart every skill one might need, and this is intended as an introductory text. But I want to conclude with an initial roadmap for some directions that you might go next.

This last unit of the book, *The Other Tools*, consists of a series of primers on some of the most buzzworthy and influential trends in the field of urban informatics and the popular push for "smart cities." Artificial intelligence. Predictive analytics. Sensor networks. 5G. You have probably heard of all of them, but you may not fully understand what they are, how they work, their strengths and weaknesses, and their applications. Presenting that information is the goal of this unit, which is divided into two chapters: Chapter 13 presents analytic techniques; Chapter 14 presents emergent technologies. In some ways this distinction is a bit arbitrary, as data and their analysis are often central to the function of novel technologies. Here we make the distinction by including in Chapter 13 things that could be fully contained within R and other software packages and keeping for Chapter 14 technologies that require engagement with additional external hardware. Though these chapters will not make you a practitioner of these tools quite yet, they provide the initial insights that will enable you to reason about them and their uses in informed ways and to easily learn more if you so choose.

13.1 Structure and Learning Objectives

The chapters in this unit will break away from the structure of those in the previous three units. Instead of working through a specific example to learn a set of related technical and interpretive skills, each of these chapters will walk through multiple tools that are shaping the practice of urban informatics and the popular pursuit of "smart cities." Nonetheless, in the spirit of those previous chapters, the description of each tool will be replete with real-world examples and illustrations. This chapter will focus on three advanced analytic tools that have opened up new opportunities for data analysis and its application:

- *Network science*, or the analysis of the ways in which people, places, and things are linked to each other;

- *Machine learning and artificial intelligence*, or techniques that allow the computer to determine the analytic model and its extensions;
- *Predictive analytics*, or the use of analytic models to forecast future events and conditions.

The chapter will offer a conceptual overview on each of these three analytic techniques that will be broken up into six parts: (1) What it is; (2) How it works; (3) When to use it; (4) Ethical considerations; (5) Major applications; and (6) Additional reading. These sections will not present the technical skills needed to practice each of these skills—each would require an entire textbook of its own! They will, however, provide enough base knowledge for you to discuss them, consider how they are relevant to your own community or work, and even to find additional resources by which you could learn how to use them yourself.

13.2 Network Science

13.2.1 What It Is

It is almost trite at this point to say that we "live in the social network." Social media has made this readily apparent as we "friend" each other, "like" and comment on each other's posts, and watch information disseminate across millions of people in mere moments. But we have always lived in the social network. Our friends, family, and acquaintances, and the institutions that coordinate services, like schools and local government, constitute the relationships that undergird society. There are also critical infrastructural networks, like the power grid or the interstate highway system. The analysis of these networks is distinct from any analytic technique we have learned so far in this book because we are no longer studying the people, places, and things themselves—we are studying the linkages between them. This requires a whole new data structure and set of analytic tools for examining it. The practice of these tools is called network science.

13.2.2 How It Works

Network science is the analytical study of the connections between a set of elements. These elements are often referred to as *nodes*. The connections between them are referred to as *links*. Network science was originally developed to study the interactions of children in classrooms. In these early studies, the nodes were the children and the links were whether they were friends with each other. If two students were deemed friends, either through a survey or observational protocol, there was a link between them in the network. If not, there was no link. Since then, network science has been applied to all sorts of topics, from the management of transportation systems to behavior on Facebook to the physical interactions of microscopic particles.

Network science requires a different data structure than those we have worked with in this book. Our analyses have utilized a traditional structure wherein a series of cases each have values on one or more variables. Cases are rows and variables are columns. Network science, however, is focused not so much on the attributes of the individual cases but on the linkages between them. This requires a matrix in which each cell is the relationship between two nodes. This is often represented as

$$A = \begin{bmatrix} a_{11} & \cdots & a_{1n} \\ \vdots & \ddots & \vdots \\ a_{n1} & \cdots & a_{nn} \end{bmatrix}$$

where a_{ij} describes the link (or lack thereof) between nodes i and j.

Note that A is a square matrix wherein all of the nodes are represented in both the rows and columns. In the simplest cases, all a_{ij} are equal to either 1 or 0 and a_{ij} and a_{ji} are equal to each other—either there is a linkage between the two nodes or there is not. For instance, saying that child i is friends with child j is the same as saying that child j is friends with child i, and each statement indicates the same linkage (i.e., $a_{ij} = a_{ji} = 1$). This same basic logic might be applied to a variety of topics: Are stops in a public transit system adjacent to each other?; Which individuals in a sample of Twitter users have responded to each other's tweets?; Which members of a city council have co-sponsored bills together? The list goes on.

There are more complex cases, however. First, it is possible to have a linkage that is measured on a gradient. Second, it is possible that the linkage is directional or asymmetric meaning the linkage from i to j can be different from the one from j to i. To illustrate, let us take the popular example of using cellphone mobility data to analyze flows between neighborhoods. In this case, a_{ij} would be set equal to the proportion of residents of neighborhood i that visited neighborhood j. Thus, a_{ij} and a_{ji} would take on distinct values based on the movements of the residents of each neighborhood. You can compare the representation of this measurement strategy with the simpler one seen in the previous paragraph in Figure 13.1.

The matrix structure of a network data set complicates the process of analysis. No longer do we have rows and columns. In fact, our matrix only contains *one* variable. That is, it contains the values for a specific type of linkage. What if we have multiple types of linkage we want to understand? For instance, a lot of work in recent years has used cellphone-generated mobility records to examine racial segregation. Understanding this phenomenon fully would require not only a matrix of movement between each pair of neighborhoods i and j, but also the physical distance between each pair of neighborhoods, whether they are connected by public transit, and so on. Each of these would be their own matrix.

13.2.3 When to Use It

When we analyze networks, we are generally trying to analyze one of three things.

1. *Characteristics of the linkages themselves.* This can include trying to explain why some linkages are present and others are absent, or, for linkages that have a gradient, why some are stronger and others are weaker. For instance, to what extent does the distance between neighborhoods explain the tendency of people to move between them? How does race or income explain these movements? When we have longitudinal data, we can study how these linkages emerge or strengthen and how they dissolve or weaken.
2. *Attributes of nodes IN LIGHT OF the structure of the network.* This might include how outcomes and impacts can travel along interconnections. For instance, recent work on segregation has found that mobility between high-poverty neighborhoods can reinforce local vulnerability to crime. A related area of inquiry is to discover how the attributes of elements cluster or become clustered within the network, or how characteristics spread throughout the network. For example, mobility data

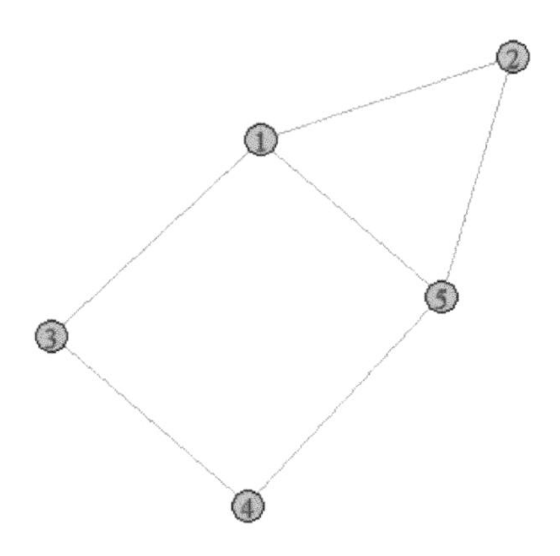

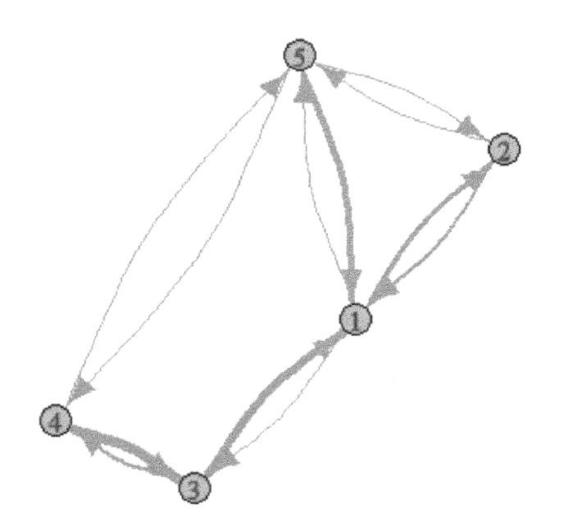

FIGURE 13.1
Networks can have varying levels of complexity, as illustrated by these two dummy networks of 5 Facebook users. In the simplest form, we can represent relationships as 0/1 and being shared by two nodes, for instance, whether two users are friends (top). This is a symmetrical (or undirected) and unweighted network because the relationship goes both ways and either exists or does not. A more sophisticated representation might be the number of times each user has liked a post made by the other user (bottom). This is an asymmetrical (or directed), weighted network because the linkage can be different depending on which user the activity is going "to" or "from" and it can take a variety of values.

was used intensively during the onset of the COVID-19 pandemic to track how infections moved between communities. Note that I have written "in light of" the structure of the network in caps. This is because thinking in terms of a network can be tricky. It is easy to slip back into traditional hypotheses that use attributes of a node to predict an outcome. The key is to make certain that the independent variables describe the node's placement in the network, for instance, the number or strength of linkages it has, the nodes to which it is linked, whether it is in the center of periphery of the network, or otherwise.

3. *Structure of the network.* To network scientists, the network itself is an important subject of study. For instance, the interconnections in a racially segregated city will look very different from those in a more integrated city. Examining the structure as a whole can tell us much about the underlying dynamics of the system. Further, it can provide insights on emergent properties—that is, systemwide outcomes. For instance, does segregation tell us something about how a city would evacuate when faced with a major storm?

As you can see, the opportunities for analyzing network data are many. Luckily, statisticians have developed a variety of software packages, including multiple for R (including one named `network`; see Section 13.2.6), that can analyze networks in a multitude of ways. These include tools that are analogous to the statistical tests we have learned in this book (e.g., regressions that use one type of linkage to predict another) and a large variety that are more specific to the special opportunities created by networks.

13.2.4 Ethical Considerations

The ethical considerations associated with network science are less about its use as a methodology and more about the types of data sets that tend to be popular in network science, such as social media posts and cellphone generated mobility data. Many of these data raise questions around privacy. There have also been concerns about how Facebook and other social media companies capitalize on these data to design and promote services that prioritize profit over the well-being of their users. To do so, they mine our personal relationships to generate insights, but they then implement these insights in ways that further manipulate the operation of the social network. These are not critiques of network science itself, of course, but of the data sources and use cases to which it can be applied, which have become increasingly intimate and powerful.

13.2.5 Major Applications

As you have probably grasped at this point, network science can be applied to many, many different topics relevant to cities and communities. At the risk of potentially leaving out some opportunities, I will provide a short list of four areas of inquiry that I see as the most promising at this time. Some of these are illustrated in Figure 13.2.

1. *Within-community social dynamics.* As we saw in Chapter 10, scholars have long been interested in how the relationships between neighbors emerge and operate, also known as the social organization. Network science enables new directions on this subject. For instance, a recent series of studies have identified

"ecological networks," demonstrating how the tendency of residents to visit the same institutions (e.g., schools, stores) can lead to more relationships between them, indicating the important role that interactions at these institutions play (Browning et al., 2017).

2. *Between-neighborhood flows.* As we have already discussed, scholars have been studying how individuals move between neighborhoods. This has been used to better understand the origins and consequences of segregation (Wang et al., 2018), the transmission of infection across communities (Badr et al., 2020), economic development (Eagle et al., 2010), and other critical aspects of community well-being.
3. *Institutions as conduits for causal dynamics.* Institutions create networks in a variety of ways. Transportation systems connect neighborhoods. Many cities have school assignment systems creating geographic and demographic mixtures at schools, thereby linking communities to each other. Are there outcomes that "travel" along these linkages? For instance, do low-income students who attend mixed-neighborhood schools have more access to opportunity and learning? A common refrain against public transit is that crime will be more likely to migrate from high- to low-crime communities. Network science can probe whether this is actually true.
4. *Emergence of community-level properties.* Is a community "resilient" during a disaster? If so, or if not, why? Is it a product of the local social network? How well is the community connected to institutions? These and related questions that use the characteristics of a local network to predict major outcomes for the community as a whole are crucial and rooted in network science.

Each of these four areas of inquiry capitalizes on the unique advantages of network science. As you can probably see, they are not only interesting questions analytically, they also promise insights that are actionable, potentially informing new or refined policies, programs, and services for communities. As always, these advances could come from the public, private, or non-profit sectors.

13.2.6 Additional Reading

The New Science of Cities by Michael Batty presents a network approach to studying cities (2013, The MIT Press).

Big Data of Complex Networks by Matthias Dehmer, Frank Emmert-Streib, Stefan Pickl, and Andreas Holzinger presents the ways that big data might be leveraged to study complex networks (2020, Chapman Hall / CRC Press).

`network` package in R, by Carter T. Butts, includes many functions for constructing and analyzing data structures for network science.

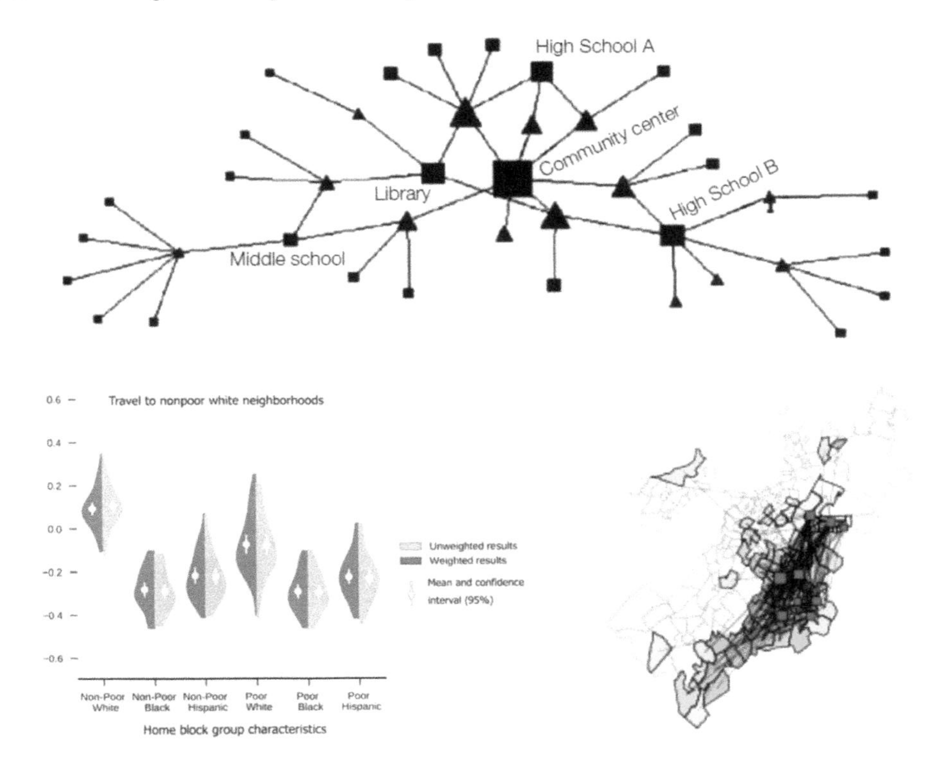

FIGURE 13.2
Network science can be applied to many different questions. We can analyze the social organization in communities (top) through the tendency of residents (triangles) to visit the same places (squares). We can study segregation between communities with different racial and socioeconomic characteristics, including how often they visit each other (bottom left). We can study how much the neighborhoods of a city are connected to each other through attendance at different public schools (bottom left). (Credit: Browning et al., 2017; Wang et al., 2018; BARI)

13.3 Machine Learning and Artificial Intelligence

13.3.1 What It Is

A lot has been made of machine learning in popular discourse. Even more is made of the ways that artificial intelligence (AI) is going to transform society. Alexa and Siri hear our voices, interpret our words, and serve up the desired information. Autonomous vehicles will change transportation as we know it by navigating streets and communicating with each other via WiFi. Investment strategies are now based on "bots" that predict good bets. In fact, many of those bots were designed by other bots. In response to this hype, some circles mock the vagary of machine learning and AI, as in the cartoon in Figure 13.3, which pokes fun at the idea that these tools are just fancy window dressing for traditional statistics. Who is right?

We can square the futuristic and critical perspectives on machine learning and AI by understanding two things. First, these techniques are based on traditional statistical tools. Second, they advance the power of these statistical tools by allowing the vast computational power of modern technology to determine and test the models. These two facts, respectively, explain both the hesitance and excitement. The computers can build models that are far more sophisticated than what a human would develop, a sophistication that can then build on itself to generate fundamentally distinct insights and products. But the statistical tests are still subject to the same weaknesses that we have seen throughout the book, including bias and overinterpretation, and the power of machine learning and AI will only amplify these issues and can accidentally encode them into the new technologies that have been hailed as so transformative.

FIGURE 13.3
A meme that pokes fun at the way machine learning and artificial intelligence use traditional statistics, highlighting how the issues with statistics are still present, they are just being covered up by fancy framing. As with most cartoons, this one simplifies the situation, but the lesson is instructive. (Credit: https://www.instagram.com/sandserifcomics/).

13.3.2 How It Works

Machine learning and AI both begin by asking a computer to make sense of a large data set. This process is called *training* and often involves a surprisingly limited amount of human intervention. Of course, there are dozens if not hundreds of different machine learning techniques, each with its own assumptions and underlying mathematics guiding the process by which the computer learns the contours of the data set. Some of the more well-known examples include random forests, which represent the organization of data as a series of decisions to place data in either of two categories; and neural networks, which try to mimic the structure of the human brain. For our purposes, training requires an analyst to specify: (1) the outcome or dependent variable that the analyst wants the computer to be able to recognize; (2) a large set of cases, each described by a large number of variables, or *features* as they are called in machine learning terminology; and (3) a technique (e.g., random forest) by which the computer will learn how features and their combinations predict the desired outcome.

The second step of machine learning is *testing*. At this stage, the researcher provides the computer with an additional data set. Most often, they split the data set into separate training and testing data sets before the analysis. Whereas during the training process the computer uses information on the outcome variable to identify the best-fitting model, during testing the computer uses that model to predict the outcome for each case. These predictions are compared to the actual data and the accuracy of these predictions is used to assess the power of the model. The more accurate the model is when making these predictions, the better.

We might illustrate machine learning with the example of facial recognition, which has been a subject of much attention in recent years. The first step of developing a model for facial recognition is to provide a computer with a vast number of pictures, some of faces, some of other things (e.g., cats, chairs, musical instruments, construction tools, etc.). In the training process, the computer develops a model of the features in a picture that help it determine whether something is a human face or not. We would then test the power of this model against a new set of pictures. Obviously, facial recognition has gone far beyond this initial step. Subsequent advances have involved machine learning of facial expressions, identification of individuals (like Face ID on iPhones), and so on. Accomplishing these additional levels of sophistication has been through the same basic process of testing and training.

It is important to distinguish between machine learning, traditional statistics, and AI. In traditional statistics, the analyst pre-specifies the model—that is, the independent variables whose relationships with the dependent variable are being tested. In machine learning, the computer builds this model itself. In some cases, the final model may look a lot like the types of regressions we learned about in Chapter 12, it just happens to have been built by the computer based on the strongest relationships in the data set rather than the insights of the analyst. More often, it entails a variety of complex combinations (or interactions) between variables that analysts would rarely attempt to specify themselves. Thus, machine learning makes the fundamental tools of statistical modeling available to the vast computational capacity of computers.

AI is when machine learning processes are incorporated into a larger decision-making system. That is, the model developed through machine learning on past information is instigating actions based on new events and information. Typically, AI places machine learning in a feedback loop that uses new experiences to refine the model. That is, each time the system is asked to make a decision, it saves the data around the event and its outcomes, adding it to an ever-growing "training" data set. In this way it is able to continue evaluating and adjusting its own model for future events.

13.3.3 When to Use It

Machine learning is very powerful, but it has some limitations. First, it requires lots of data. The definition of "lots" varies as more sophisticated approaches often require more data, but in most situations an analyst will want thousands of cases to conduct a training and testing process. Also, for it to be a meaningful improvement over traditional statistics, machine learning needs lots of features (i.e., variables) that may share complex relationships with each other and the outcome of interest. Machine learning is then able to make sense of this complexity in service of building a model for predicting a desired outcome. For instance, pictures have many, many ways their pixels can be organized, including shapes, color contrasts, etc. As such, defining the rules for "what is a face?" manually would be painstaking if not impossible. And yet, with enough well-designed data sets and training processes, a computer can build this rule set itself and distinguish a face from other objects with near 100% efficacy.

A second limitation that an analyst must always keep in mind stems from the complexity that machine learning can achieve. Often, it entails combinations of variables that are difficult to meaningfully communicate. And many machine learning techniques do not report what these combinations are. For this reason, machine learning can predict things, like if something is a face or not, but it might not be able to tell us *why* something is a face. We have to derive this information post-hoc, but sometimes even this is tricky. Consequently, machine learning is not always the idea tool for advancing our own understanding of a phenomenon, which, as discussed in Chapter 6, complicates its potential as the basis for designing new, more effective strategies for action. It also creates some anxiety regarding a lack of transparency, especially when developed into AI, as we will explore further in Section 13.3.4.

13.3.4 Ethical Considerations

There has been a lot of anxiety about machine learning and AI, and for good reason. I want to group these concerns under two main umbrellas. The first might be summarized as the "garbage in, garbage out" critique. A colleague and leading scholar in machine learning once said to me, "When I conduct an analysis, I don't care if your data are right or wrong. I only care that they can predict something." This is how statistical tests work in general, but it highlights a deep vulnerability to machine learning. The development of science and science-based practice has depended on colleagues looking at each other's data sets and models and pointing out weaknesses in interpretation and assumptions. But what if we are allowing the computers to build the models themselves? Who is policing the misinterpretations of the computer?

Suppose, for instance, that we give a computer thousands of images of faces and other objects in order to develop an algorithm for facial recognition but all of the faces are from White people. The computer will then build a model for facial recognition with faces defined as White faces. By providing the computer with a biased training data set, we produce a biased model. This actually happened in the early stages of facial recognition software, some of which were unable to recognize non-White faces as faces. Similar examples continue to occur, though there is increased awareness around this issue of data equity (Klare et al., 2012) and other forms of *algorithmic bias.* This awareness is also becoming more sophisticated, matching the complexity of the process itself. Extending our example, what if certain demographic groups have more prominent or distinctive facial features? If so, the model will be inherently more capable of identifying those individuals, which means we need training data sets that overrepresent groups whose faces are harder for the computer to learn. Whatever the application of machine learning or AI, one must consider how a biased training data set will construct a data-driven model and associated systems that will embody that same bias.

The second concern is how we design the interactions between humans and AI systems, a debate that reveals multiple tensions. On the one hand, there is the danger of entrusting critical societal processes to computers, especially when we are incapable of fully understanding how the computers are coming to these conclusions. Science fiction authors have already provided us with all sorts of dystopic scenarios that could come of that. On the other hand, we have found that the computers are better than humans at lots of tasks that we assume require specialized expertise. When built properly, machine learning models are more accurate and less biased at predicting places that are at-risk for crime, criminal defendants who are likely to reoffend, children who are likely to experience child abuse, and even driving cars in standard situations. But how comfortable are we replacing crime analysts, judges, social workers, and drivers with computers? And what are the economic consequences? These are the transformative questions that AI forces us to consider.

FIGURE 13.4
Facial recognition depends on a computer identifying the geometry of a face through a variety of visual "features". These precise features may differ, however, by sex, race, age, and culture, requiring a diverse training data set for any such technology to be developed and implemented equitably. (Credit: Microsoft).

13.3.5 Major Applications

Machine learning and AI can be applied to a variety of subjects in cities and communities. Some are straightforward and benign, like the use of AI to interpret open text 311 reports and classify them by the type of services needed. This is a rather simple, Alexa-like improvement that allows an existing system to be more flexible in how it receives and responds to communities and their needs. Others are more complex, like autonomous vehicles. Despite some persistent limitations to the technology—it turns out that some of the many events that a driver might experience are difficult for a computer to learn, such as the effects of snowfall—but we are not far off from a world in which autonomous vehicles are *safer* drivers than most humans. What that will mean for transportation infrastructure is yet to be seen.

The applications that raise the greatest concerns are the ones that touch sensitive parts of our lives. By accessing increased complexity, AI unlocks the potential to do a lot of things for which we do not currently have rules because we never imagined them possible, and some of them will inevitably invade our privacy by using data to reveal our individuality. Facial recognition is now at the point that unique individuals can be identified. This can be highly invasive and can give law enforcement, for example, an unprecedented amount of power. Numerous communities around the United States have consequently decided to ban law enforcement from using facial recognition software. Additionally, as we will explore in Section 13.4 on predictive analytics, there are some hesitations for AI systems that direct human resources, especially in the criminal justice system.

13.3.6 Additional Reading

Public Policy Analytics: Code and Context for Data Science is a nice complement to this book in that it applies machine learning to public policy-oriented questions (2021, Chapman Hall / CRC Press).

R has a multitude of packages for machine learning, including a few we have already seen in this book, such as `tm` for text mining. Some of the other recommended ones are `CARAT`, which is especially made for training and testing processes, `randomForest` for random forest analyses, and `neuralnet` for running neural networks (though you may want to try some simpler techniques before making this leap).

13.4 Predictive Analytics

13.4.1 What It Is

In the 2002 movie *Minority Report*, Tom Cruise played a police officer whose "partners" were clairvoyants who could see the future and foretell the next crime. He and other members of the police department would use these prophecies to detain would-be perpetrators before the crime occurred. This aspiration of precise prevention has long been a dream of practitioners. If only we knew what was going to happen, we could be proactive instead of reactive. We have become increasingly adept at this for weather, for example, where we can anticipate events days in advance. For a lot of social phenomena, it has been more difficult, but many institutions are still working hard to develop predictive analytic systems that can be incorporated into daily and long-term processes.

13.4.2 How It Works

Predictive analytics is not really a new analytic technique in the way that network science or machine learning is. Instead, it is more of a shift in how all of the other statistical techniques are used. Whereas Unit 3 in this book used "prediction" to mean "explaining variation in a dependent variable," here the word takes on its more forward-looking meaning. It does so by taking statistical models built on the past to forecast future events and conditions. This is very similar to the idea of testing machine learning models after training or incorporating machine learning models into AI. It is also perfectly possible with simpler techniques, such as regression. We can take the parameters of the original model, enter values for new cases, and see what the model suggests.

Let us use the real-world version of the *Minority Report* example. Predictive policing has become very popular over the last decade or so. It leverages highly detailed models to forecast where crime is likely to occur. Most often, these models are based largely on land use patterns (e.g., a bar is more likely to generate a violent assault than a church or a single-family house) and previous crimes (i.e., events tend to recur at the same or nearby places), as well as weather and other circumstantial factors. A predictive policing algorithm then takes all the variables of interest for today, enters them into the model, and reports out the places most likely to experience a crime.

13.4.3 When to Use It

The answer to the question of when to use predictive analytics is pretty simple: when you want to anticipate something and, presumably, proactively respond to or prepare for it. Similar to machine learning, building a predictive system requires a large historical data set to ensure that the model is strong enough to make meaningful predictions. Often, the construction of the model leverages machine learning techniques. That said, even large amounts of data are sometimes insufficient for a predictive model to be as accurate or precise as we might prefer, as we will explore in Section 13.4.4.

13.4.4 Ethical Considerations

Predictive analytics has generated lots of debate. One of the major issues is a specific extension of a concern described above for machine learning and AI: "garbage in, garbage out." In this case, the output is a decision-making system. Predictive policing has been subject to this critique. It is well established that, in many parts of the United States, if the same potentially criminal event were to occur in a predominantly Black neighborhood and a predominantly White neighborhood, the former is more likely to result in an arrest. As a result, crime records are a racially biased representation of where crimes occur, overemphasizing Black communities. (Similar biases exist for other racial minorities and for low socioeconomic status.) If we were to then use crime records to predict where the next crime was likely to occur and deploy police accordingly, we would be perpetuating that same racial bias. This is a major problem, and no one has solved it in the case of predictive policing. Similarly, a major algorithm used to predict individual offenders was removed from use because, after further scrutiny, it was predicting ethnicity as much as it was predicting offending.

There are three additional concerns that are specific to predictive analytics and their applications:

1. *Precision.* In *Minority Report*, the date, time, place, and perpetrator were spelled out in detail. We often like to hope that predictive analytics can reach this level of specificity, but in reality we fall far short. Rarely can we realize the same precision in space and time. Can we predict which neighborhoods will have the most crime next year? Definitely. Can we predict which household will experience a crime tomorrow? No. Of course, we can build risk models that tell us which places are *more likely* to experience crime in the near future, but we have to respect the limitations of our models and operate accordingly.
2. *Implementation.* Once we know the strengths and limitations of our predictive model, we need to design the action that it instigates. This follow-up is often ignored in discussions of predictive analytics. This has been a critique in predictive policing as well, with many asking, once we have the model, what do we do with the information it generates? Some preliminary studies suggest that deploying police officers to pass through at-risk areas can in fact diminish crime, potentially by deterring delinquent individuals or interactions (Ratcliffe et al., 2021). But we also need to design these implementations with a respect for how precise of predictions we are actually able to make, per the previous concern.
3. *What if the prediction is wrong?* This is a tricky philosophical issue for predictive analytics. It is great if we can anticipate events and respond to them appropriately. But if the prediction is wrong, which it will be at times, what are the consequences? I tend to think of this in terms of whether the action instigated by the prediction could help or harm. I was part of a project that helped Boston Public Schools redistribute funds according to predicted academic achievement. In this case, errors in prediction would result in more funds supporting certain students. Those students were not harmed in the process. If a predictive policing algorithm instigates a wrongful arrest, that is a much more troubling error. This means that we need to be exceptionally careful with predictive models that would inform actions that could harm someone and consider how to design implementation that is sensitive to these risks.

13.4.5 Major Applications

There are lots of applications for predictive analytics, some of which are illustrated in Figure 13.5. In a sense, any situation in which an institution is using past data to anticipate future events or conditions is predictive analytics. We might group them into two categories, though there are likely applications that do not fit neatly into either of these groupings. First, lots of predictive analytics algorithms seek to anticipate events. Predictive policing is one of the most prominent examples. There are related models for predicting offending in individuals or reoffending in parolees. There are also similar models for house fires and other destructive events.

A second class of algorithms seeks to anticipate the need for human services. For example, Allegheny County in Pennsylvania has used the vast contents of their Integrated Data System (IDS; see Chapter 7) to predict which calls to child protective services are most likely to portend actual child abuse. Similar models based on IDSes attempt to foresee homelessness and other conditions that will require government support or intervention. As noted in Section 13.4.4, in all of these cases there is the need to tailor action following the results of the model to the overall accuracy of the model and the potential benefit or harm of the eventual intervention.

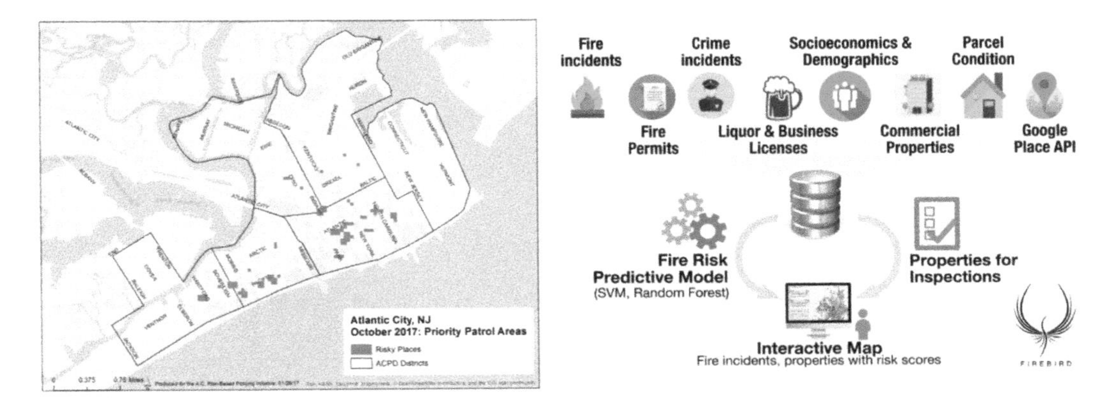

FIGURE 13.5
Predictive analytics can be applied to multiple use cases. Some of the most familiar include predicting places that are at risk for crimes, or "risky places," in this example that uses past crime events and land use to model the "risk terrain" (left); and coordinating dozens of data sources describing events and context to predict where fires will occur (right). (Credit: https://www.riskterrainmodeling.com/blog; https://firebird.gatech.edu/)

13.4.6 Additional Reading

Foundations of Predictive Analytics by James Wu and Stephen Coggeshall walks through many of the techniques described in this chapter and their application to prediction (2012, Chapman Hall / CRC Press).

Applied Predictive Analytics by Dean Abbott describes how to use analytics to predict business processes. Admittedly it is focused on business-based applications. (2014, Wiley Publishing).

13.5 Summary

In this chapter we have learned about three advanced analytic techniques that are increasingly used in urban informatics. We have learned how they work, when to use them, as well as ethical considerations and major applications. These include:

- *Network science*, or the analysis of the ways in which people, places, and things are linked to each other;
- *Machine learning and artificial intelligence*, or techniques that allow the computer to determine the analytic model and its extensions;
- *Predictive analytics*, or the use of analytic models to forecast future events and conditions.

The goal here was not to make you an expert just yet, but to expose you to the basic knowledge needed to meaningfully discuss these tools and their applications, as well as enable you to go out and learn more about them.

13.6 Exercises

13.6.1 Problem Set

1. For each of the following pairs of terms, distinguish between them and their roles in analysis.
   a. Machine learning vs. artificial intelligence
   b. Inferential statistics vs. machine learning
   c. Inferential statistics vs. predictive analytics
   d. Traditional statistics vs. network science
2. For each of the three techniques learned in this chapter, summarize what you think is the most important ethical concern.
3. For each of the three techniques learned in this chapter, research and describe one additional application.

13.6.2 Exploratory Data Assignment

Research a particular application of one of the three analytic techniques presented in this chapter (or that combines two or more of them). Write a short memo that describes:

1. The purpose of the policy, program, service, or product.
2. How it uses the technique (or techniques) in question.
3. How it does or does not address potential ethical considerations.
4. How well the technique has been applied to this situation.

14

Emergent Technologies

This is a book about accessing, manipulating, analyzing, visualizing, and interpreting data, but urban informatics is more than just data and analysis. The definition I offered in Chapter 1 was "the use of data and technology to better understand and serve communities." It is important not to lose sight of the word "technology" in that definition, as there are numerous technologies that have been incorporated into policy, practice, and services in recent years and promise to offer important advances. These include things like sensor networks, 5G cellular service, and blockchain. Of course, these technologies are not independent from data. In fact, they help us to generate, organize, and use data. Thus, a full understanding of the potential of urban informatics requires understanding the opportunities for these technologies. Introducing these technologies is the focus of this chapter.

14.1 Structure and Learning Objectives

This chapter will focus on three emergent technologies that have opened up new opportunities for policies, practices, services, and other products related to communities:

- *Sensor networks* that track environmental conditions across a city;
- *5G*, the newest generation of the cellular network;
- *Blockchain*, which is an innovative way to create security in tracking the history of items.

The chapter will offer a conceptual overview on each of these three technologies with a five-part structure very similar to the previous chapter: (1) What it is; (2) How it works; (3) Ethical considerations; (4) Major applications; and (5) Additional reading. Note that this chapter does not have the "When to use it" sub-section. Whereas statistical tools have specific data structures and research questions for which they are best suited, the appropriateness of technologies is more easily summarized in terms of their applications. Again, these overviews will not make you an expert just yet but will provide enough base knowledge for you to discuss them, consider how they are relevant to your own community or work, and even to find additional resources by which you could learn how to use them yourself.

14.2 Sensor Networks

14.2.1 What It Is

Imagine opening a weather app in the morning to find out the current conditions, but instead of entering your town or city, you can enter your street block, or the address where you work. The web site would then tell you the temperature, humidity, precipitation, and air quality at that precise location, plus the current rate of pedestrian, car, and bike traffic. It might even forecast what these metrics will look like throughout the day. This is the vision for sensor networks in cities. With sensors placed on every street corner, we could be aware of conditions throughout the city with unprecedented precision at all times. This could be used by practitioners to deploy services, drivers to plan their commute, community organizations to prepare for storms, and so on. An example from Chicago called the Array of Things is depicted in Figure 14.1, including the design of a single sensor node and the placement of nodes around the city.

14.2.2 How It Works

In some ways, sensor networks are deceptively simple. They consist of multiple individual sensor boxes, or nodes, each one containing at least one sensor. Each sensor collects a single type of information. There are sensors that measure light, sound, temperature, the rate of precipitation, one of numerous forms of air pollution, and more. There are sensors that note the presence of smart devices as people pass with phones in their pockets. Some sensor boxes include video cameras. Thus, a single sensor box might include any mixture of these. Typically, every node in a sensor network contains the same collection of sensors, though there might be reasons to have some nodes contain more or fewer sensors than others depending on their placement or cost considerations.

Sensors generate measures at a predetermined frequency (e.g., every minute, every hour, etc.). These measures are then combined by the node and transmitted by WiFi and cellular networks to a centralized server. The centralized server receives these from all nodes in the network and organizes them into a giant database that includes measures for all conditions, from all locations in the network, for all timepoints at which data were generated. As such, it is a uniquely detailed archive of conditions across space and time. The scope of these observations is determined by the number and diversity of nodes, the number of conditions measured, and the frequency of reporting.

The original measures often need to be processed in some way, which can happen before transmission to the server by computer chips inside the sensor box, or after transmission once the data are on the server. The nature of this processing varies by the measurement type. Pollution needs to be adjusted for wind speed, for example. Meanwhile, machine learning algorithms can be used to tag video content as containing people, cars, or bikes, thereby tracking different forms of traffic.

14.2.3 Ethical Considerations

The ethical considerations of sensor networks vary with their ambitions. Tracking temperature and air quality does not raise many eyebrows beyond the anticipation that the data will

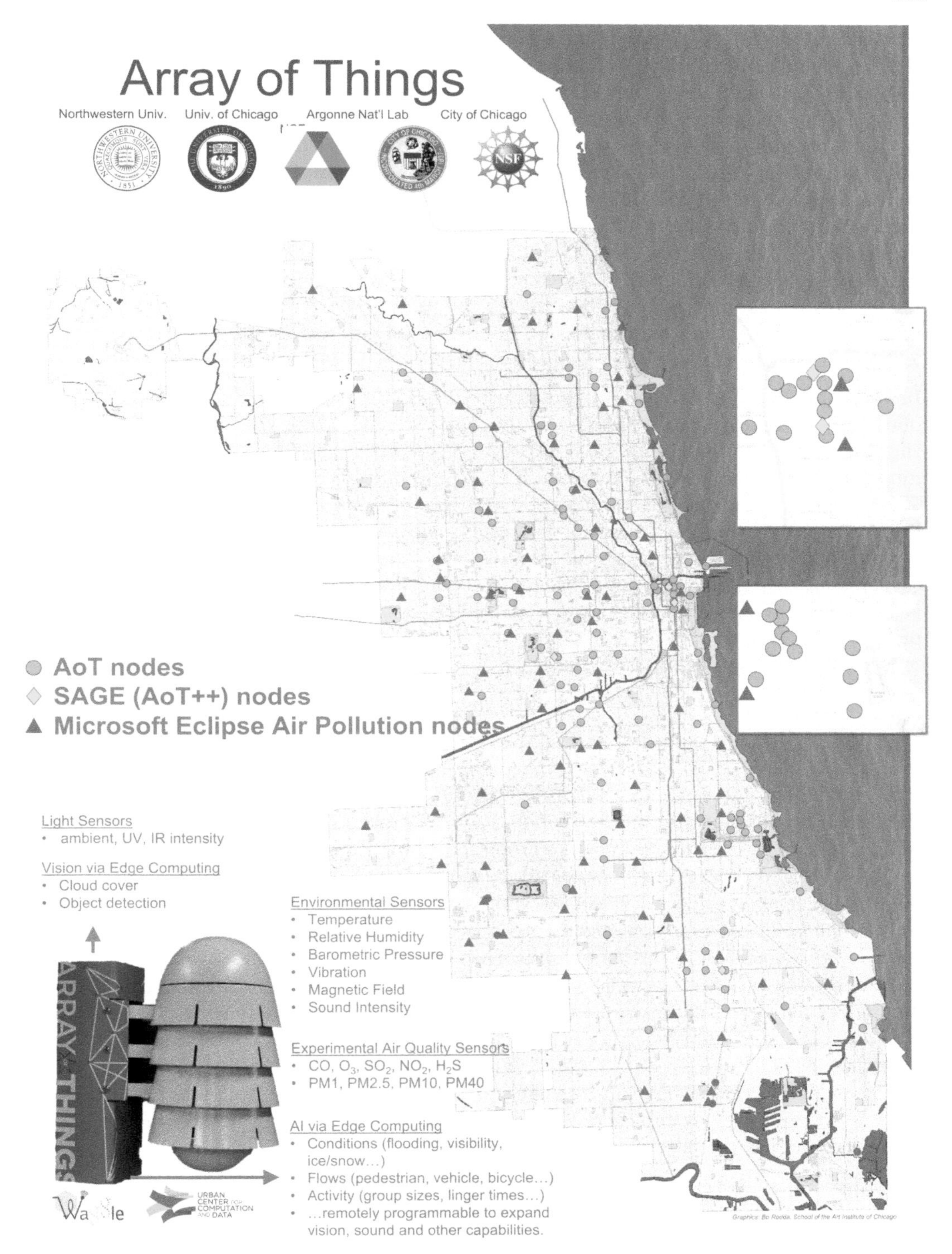

FIGURE 14.1
The Array of Things project has installed sensor "nodes" (see inset for what an individual node looks like) across Chicago, IL, to capture various conditions locally to gain a composite view of temperature, humidity, sound, light, and air quality, among other things, throughout the city. (Credit: http://arrayofthings.github.io)

be used appropriately to maximize public welfare. Tracking radio frequency identifications (RFIDs) from smartphones or taking videos of pedestrians is much more sensitive and potentially alarming. Many communities have actively rejected efforts to put such technologies in public spaces owing to fears that they can violate privacy. A sensor network designed by the University of Chicago and the City of Chicago called Array of Things (the same one depicted in Figure 14.1) had to remove RFID trackers from their nodes because of public backlash. They did this even though the trackers in the proposed nodes had been designed to guarantee anonymity, highlighting just how uncomfortable such technologies can make people. The same project, though, offered a potential solution to the tradeoff between videos offering both uniquely valuable and uniquely invasive information. They developed algorithms for pre-processing video within the node to extract information like "number of cars" and "number of pedestrians." The node then transmits these non-invasive metrics while deleting the original video footage.

In the end, the biggest question pertaining to sensor networks in public spaces is: What does the community want? This question has two sides to it. The first is comfort with the technology. If the community is comfortable with video cameras and RFIDs, then implementing them could be just fine. If the community is not, then that is not an option. The second is how the sensor network and the information it generates might benefit the community or address local challenges. Ideally, these solutions might be co-designed by community members, technical experts who would build them, and the agencies or corporations that would implement them. A difficulty around sensor network implementation is that they have been imagined and deployed by corporations in a top-down fashion. This process often fails to incorporate community voices to co-design the sensor networks to best reflect community interests and concerns.

14.2.4 Major Applications

Array of Things was billed as a "fitness tracker for the city" in Chicago. Efforts in other cities from private corporations have offered similar promises. And the promise is substantial. Again, imagine being able to know and forecast a broad range of conditions for every street block in the city in real time. The question, though, is whether the ambitions (and costs) of placing sensor nodes on every street block match the need and promised benefits. The answer to this is mixed and relies on specific use cases. For instance, recent work (by the Boston Area Research Initiative , in fact) found that summer temperature varies not just across neighborhoods but from street to street *within* neighborhoods, and that these localized differences drive disparities in medical emergencies during heat advisories (O'Brien et al., 2020). Preliminary analyses suggest that there are similar "microspatial inequities" in air pollution as well (Gately et al., 2017). These findings provide a justification for putting sensor nodes on every street, but to what end? We do not yet have programs or services that are able to capitalize on information that is this precise. More often policies are designed to be uniform across the city or, at most, neighborhood specific.

For the reasons described in this section and the previous (Section 14.2.3), sensor networks have yet to realize their potential. They have not been developed and designed with community needs in mind. They assume microspatial inequities but are needed to prove that those same inequities are present; otherwise the sensor networks are not needed. There is then the question of what types of action they could inform and instigate through the information they generate. But without evidence for the spatial and temporal variation in conditions that only sensor networks can track, no one has developed programs and services that are responsive to them. This is not to say that solving these issues is impossible, but it does capture how sensor networks, once seen as the infrastructure of the future, are trapped in a chicken-or-egg dilemma. Do we need sensor networks to address issues that only they can observe to enable activities that we have not developed because we do not know if we need them? The jury remains out.

14.2.5 Additional Readings

Distributed Sensor Networks by S. Sitharama Iyengar and Richard R. Brooks is a series of textbooks that explores the technical and applied aspects of sensor networks (2016, Chapman Hall / CRC Press).

14.3 5G Cellular Networks

14.3.1 What It Is

5G is the newest generation of the cellular network. At the time of this writing, most of the world is still operating at 3G, 4G, or earlier cellular systems. 5G is estimated to be 100 times faster than 4G, to have the capacity to accommodate thousands more devices over the same geographic scale, and to have 1/200th the latency (that is, the time it takes the network to respond to a request). These three metrics, apart from sounding like a sales pitch, reflect rather remarkable advances to the three parts of a cellphone network's function that matter: speed, capacity, and latency.

14.3.2 How It Works

Cellular networks work on radio waves, which can have low, medium, and high frequency. Low frequency is the traditional tool for covering large areas and have been the basis of TV, radio, and all generations of cellphone network through 3G. Medium frequencies are specifically used in WiFi. They were incorporated into 4G and have proven very useful in adding capacity locally. High frequency radio waves are used especially for sensors and satellites. 5G is the first generation of the cellular network that uses high frequency radio waves.

Using high frequency radio waves has multiple advantages. The simplest is that high frequency waves can support higher speeds. They also target each device with a greater precision that is not available from medium or low frequency waves. This allows 5G to support more devices more efficiently, reaching as many as 1,000,000 devices per square kilometer. Further, this precision and its efficiency will work better for devices that have small batteries, like smart watches or glasses, as the energy required to stay connected to the network is drastically lower. Last, high frequency waves provide a quicker back-and-forth between devices and the network, which reduces latency.

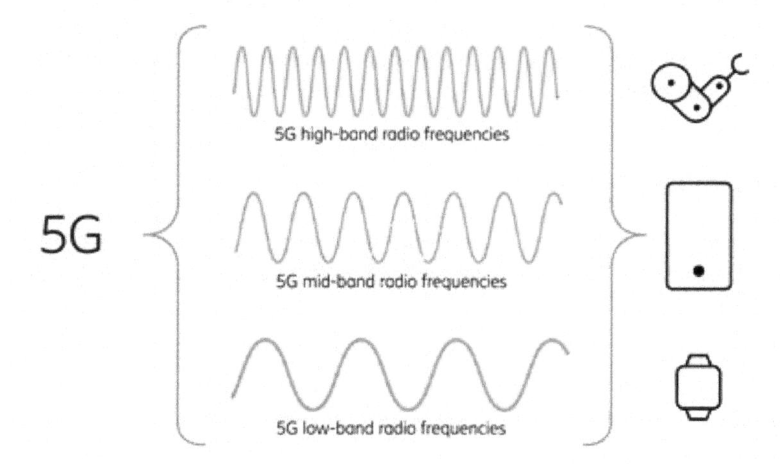

FIGURE 14.2
5G cellular networks incorporate low-, medium-, and high-band radio frequencies in order to enhance speed and capacity and lower latency. This will enable to greater usage of "smart" technologies that require efficient, rapid connections. (Credit: Ericsson).

14.3.3 Ethical Considerations

5G technology itself does not raise a lot of ethical considerations. It simply improves speed, capacity, and latency in the cellular network. There are, of course, potential concerns over the technologies that it facilitates, like augmented reality, but those should be applied to critiques of those specific technologies rather than 5G. That said, the rollout of 5G should be watched closely. Right now, 5G is only available on a small proportion of the worldwide cellular network. As that changes, though, it is always possible (and even likely) that it will be available in more affluent communities in cities long before it reaches other places. This exacerbates digital divides between rich and poor and urban and rural. This is something for which we will need to be vigilant if the advances of 5G prove to be sufficiently important to give communities with access to it an unfair advantage relative to those without such access.

14.3.4 Major Applications

Some of the benefits of 5G are incremental. Greater speed for our devices is always welcome. Who would say no to being able to download a whole HD movie in a matter of minutes? Capacity is nice, too, though not necessarily game-changing. Cellphone companies advertise how everyone in a stadium can engage with the cellphone network simultaneously. That sounds nice but does not transform society. (A better case might be made around cellular networks not bogging down when everyone is trying to call others during an emergency.)

There are some more transformative opportunities, though, created by 5G. The lower latency is critically important for unlocking the power of AI-based devices that communicate with each other or otherwise respond to information received through the network. For example, one of the important parts of a transportation system based on autonomous vehicles is the ability for them to communicate with each other. Whereas humans use blinkers to indicate that they will turn or shift lanes, errors can be made that lead to collisions. In contrast, cars would tell each other when they were going to turn, change lanes, or otherwise shift their trajectory with perfect fidelity, leading to smooth coordination and the virtual elimination of crashes. These split-second communications and responses, however, require the low latency of 5G. Similarly, robots and other AI-based technologies also require low latency to gather, respond to, and transmit information, especially when they are intended to communicate with each other.

Additionally, the efficiency of connection with 5G expands opportunities for devices with small batteries. Innovations like smart glasses currently have power limitations, as they cannot accommodate batteries large enough to last more than a short period of time. But if they were able to connect to the network for a fraction of the energy expenditure, a battery could last a full day or even longer. This lays the groundwork for these and other small smart devices to proliferate in number and type.

14.3.5 Additional Readings

5G and Beyond by Parag Chatterjee, Robin Singh Bhadoria, and Yadunath Pathak examines how 5G will influence technology, especially "smart" devices (2022, Chapman Hall / CRC Press).

14.4 Blockchain

14.4.1 What It Is

Bitcoin and other forms of cryptocurrency are a remarkable feat in that they have established markets and the basis for property in a completely virtual space. How is this possible in a world with increasing concerns around cybersecurity and reports of ransomware attacks? Is the system vulnerable to toppling with the clever invasion of a single hacker? The answer is a new technology called *blockchain*. Blockchain is a way of storing data in a secure way that is protected from manipulation by a single bad actor *without* being reliant on a single governing body or third party.

14.4.2 How It Works

Blockchain is a distributed database shared among the nodes of a computer network. We are going to break down what this means, but first it is useful to understand that it operates at two levels, each of which contributes to its unique system of security. The first is at the database level, the second is through the network that manages this database. How a blockchain works is also depicted in Figure 14.3, specifically for cryptocurrency.

The structure of a blockchain database is designed to ensure historical consistency, which is important for security in a variety of applications. Data in a blockchain are organized in "blocks" of a pre-determined size. Once a block is filled, it is closed and linked to previous blocks (hence, the chain). This creates an irreversible timeline that can be read but not be edited. Also, to alter one block in the chain would require changing all subsequent blocks to be consistent with the earlier change, which would be computationally difficult if not impossible given the way blocks are encoded based on each other. In these ways, the blockchain is protected against alteration.

Suppose a bad actor succeeds in breaking the read-only aspect of the blockchain database structure. This would undermine the promised security. The distribution of the database across multiple computers in a network offers a second level of security that guards against this. Every computer on the network has its own independent copy of the historical record of the blockchain. These copies are supposed to be identical, and the computers periodically confirm this with each other. If any computer disagrees with the others on the historical record—say, because someone attempting to modify the record for their own personal gain—the consensus overrides it and all copies revert to the agreed upon record. Someone looking to defeat this safeguard and alter the historical record would have to control more than 50% of the computers on the network. In a large network, like a cryptocurrency, this is practically impossible.

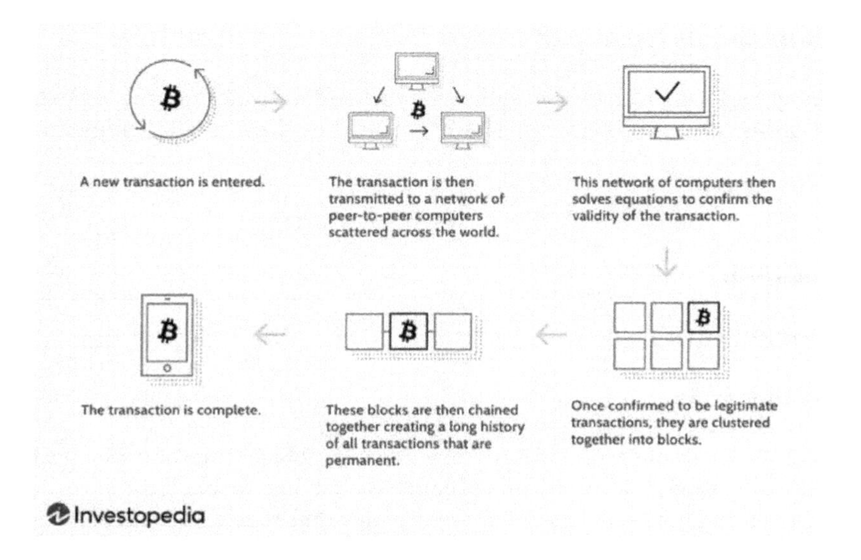

FIGURE 14.3
A depiction of how blockchain operates as a database of transactions, consisting of blocks, distributed across many computers (or nodes) on a network. (Credit: Investopedia)

14.4.3 Ethical Considerations

Much like 5G, there are not a lot of ethical considerations surrounding blockchain itself. It is a technology that facilitates certain types of activities and products. Some have raised ethical concerns about these specific products. Most notably, there are questions about the moral considerations of cryptocurrencies in undermining existing national currencies. If this were to accelerate in meaningful ways, it could have catastrophic impacts on the global economy, with the greatest consequences for those populations not currently holding any bitcoin, which are most likely to be those who are already poor. There are also questions about the implications of replacing societal functions with blockchain processes. For instance, smart contracts (see Section 14.4.4) are a technology-driven alternative to existing legal processes in commerce. Though this might seem innocuous on its own, one must keep in mind that contracts are governed by existing legal regulations. The regulation of a smart contract is dependent on the computer code embedded in the design of the blockchain. This code governs the contract in ways that may or may not be consistent with established law. Also, because these aspects of the contract are embedded in code, it may be difficult for the average user to investigate whether all of the features of the contract are as expected. This is similar to concerns about the "user agreements" to which most people blindly consent when using a new web site.

14.4.4 Major Applications

Cryptocurrency is the most prominent application of blockchain but far from the only one. As noted in the previous section (Section 14.4.3), smart contracts are another popular use of blockchain for managing the stages of signature and ownership. The latter is especially important for the transfer of digital property, like "non-fungible tokens" (NFTs). NFTs enable people to own digital media, like viral YouTube videos, in the same way that they might want to own a painting or sculpture. In fact, art dealers and collectors are often faced with the difficulty of ensuring the chain of ownership of an item when trying to guarantee its authenticity. Blockchain offers a solution not only for the ownership of digital media but also for this long-standing challenge for non-digital media.

The specialized security of blockchain has inspired some to consider how it might be applied to voting. Though this concept is still in its infancy, it appears to be promising. If every individual or voting machine were its own node in the system, then once a vote is tallied it cannot be altered. This would protect elections from being compromised by hacking.

A final application capitalizes primarily on the data structure of blockchain to track the history of objects in a longitudinal database while maintaining information about their status at all points along the way. This has proven extremely valuable for tracking food through the supply chain. In this case a blockchain tracks the location of a batch of food from source to final delivery. If, for instance, there is an *E. coli* outbreak, authorities can quickly identify the source and any other potential points of contamination, as well as other batches that passed through the same locations and might be contaminated, too. A similar logic has been applied to tracking diamonds to ensure their ethical sourcing and to immunity in COVID-19 patients owing to their vaccination or previous infection.

14.4.5 Additional Readings

Blockchain Applied by Stephen Ashurst and Stefano Tempesta explains the technical underpinnings of blockchain across multiple applications (2022, Chapman Hall / CRC Press).

14.5 Summary

In this chapter we have learned about three advanced analytic techniques that are increasingly used in urban informatics. We have learned how they work, when to use them, as well as ethical considerations and major applications. These include:

- *Sensor networks* that track environmental conditions across a city;
- *5G*, the newest generation of the cellular network;
- *Blockchain*, which is an innovative way to create security in tracking the history of items.

The goal here was not to make you an expert just yet, but to expose you to the basic knowledge needed to meaningfully discuss these tools and their applications, as well as enable you to go out and learn more about them.

14.6 Exercises

14.6.1 Problem Set

1. For each of the following pairs of analytic techniques and emergent technologies, describe how the first might be applied to or enabled by the latter (you may need to revisit previous chapters, especially Chapter 13).
   a. Network science and sensor networks
   b. Predictive analytics and sensor networks
   c. Artificial intelligence and 5G networks
   d. Inferential statistics and blockchain
2. For each of the three technologies learned in this chapter, summarize what you think is the most important ethical concern.
3. For each of the three technologies learned in this chapter, find and describe one additional application.

14.6.2 Exploratory Data Assignment

Research a particular application of one of the three analytic techniques presented in this chapter (or that combines two or more of them). Write a short memo that describes:

1. The purpose of the policy, program, service, or product.
2. How it uses the technique (or techniques) in question.
3. How it does or does not address potential ethical considerations.
4. How well the technique has been applied to this situation.

The Other Tools: *Unit IV Summary and Major Assignments*

Summary and Learning Objectives

Unit IV has presented a series of advanced analytic techniques and emergent technologies that have been rapidly influencing policy, practice, services, and products in communities. Though we did not learn how to implement these tools, we did learn their purpose, how they work, and when they are most valuable.

Technical Skills

Although we did not learn how to implement any new tools in this unit, we did learn about the technical basis and potential applications of the following:

- *Network science*, the analysis of the ways in which people, places, and things are linked to each other;
- *Machine learning and artificial intelligence*, techniques that allow the computer to determine the analytic model and its extensions;
- *Predictive analytics*, the use of analytic models to forecast future events and conditions.
- *Sensor networks*, which track environmental conditions across a city;
- *5G*, which is the newest generation of the cellular network;
- *Blockchain*, which is an innovative way to create security in tracking the history of items.

Interpretive Skills

- Determine when these tools are most useful;
- Identify and overcome weaknesses of the tools, including ethical challenges they create.

Unit-Level Assignments

Community Experience Assignment

The community exploration assignments in this book are designed to align skills you have been learning with real-world contexts. They are most useful in conjunction with the Exploratory Data Assignments at the end of each chapter, especially when you have been working through them with a single data set. They provide an opportunity to "ground truth," or really evaluate the assumptions and objectives that have guided your analysis thus far. There will be one in each unit. These can also be combined with a service-learning or capstone-oriented course.

This fourth community exploration assignment will explore how new tools are being used in a real-world setting. Please:

1. Select a community of interest to you. This will likely need to have some level of local government associated with it (e.g., a town, city, or county).
2. Conduct research about how one or more of the six tools learned in this unit is in use in this community. This could be a summary of multiple efforts or a deep dive on a specific use case. The examples could include policies, public programs or services, products or services from private corporations, or otherwise.
3. Conduct additional research about the population living in this community and any challenges or opportunities present there that might influence the potential value of the technological tools of interest.
4. Write a 3-5 page memo integrating (1) your findings on the technologically-driven activities in this community and (2) the features of the local population and the opportunities and challenges pertaining to technology. Be sure to evaluate how you think the efforts are progressing and how you anticipate them proceeding. The memo should include images from your exploration.

Post-Unit Assignment: Proposing a Technological Intervention

This unit of the book has focused on learning about some of the new analytic techniques and technologies that are shaping the practice of urban informatics. This assignment will connect this content with the previous units of the semester, assuming that you have completed Unit III's post-unit assignment. If not, you are welcome to modify as needed.

You will write a policy proposal for the implementation of a technological intervention (or, alternatively, a proposal for why a certain technology is not a good idea or at least not a good idea right now). The technological intervention can be based on any of the six tools learned in this unit. It should be informed by original analysis, preferably that which was presented in your Unit III post-unit assignment.

The proposal should consist of:

- A brief *Executive Summary* that details the main case you are making and the justification for it. It will probably need to briefly define the technology that is the focus of the paper. Think of the audience for this being someone who would benefit from the insights but might not have the time to read the whole paper or the desire to wade through methods.
- A brief *Introduction* that presents the technological intervention of interest and sets up the evaluation of why you believe it would (or would not) be a positive addition to a given community.
- A *Background* that describes the technological tool of interest, including its strengths and weaknesses, and situations when it is most likely to be useful.
- A *Recommendation* that describes the basis of your reasoning for why this would be a good idea (or not a good idea) for a given community. This is where you will incorporate your own original analysis as evidence and justification.
- *Conclusion* that summarizes the argument and makes final recommendations.

Suggested Rubric (Total 10 pts.)

Executive Summary: 2 pts.

Introduction: 1 pt.

Background: 2 pts.

Recommendation: 3 pts.

Conclusion: .5 pts.

Details: 1.5 pts.

Bibliography

Abouk, R. and Heydari, B. (2021). The immediate effect of COVID-19 policies on social distancing behavior in the United States. *Public Health Reports*, 136(2):245–252.

Anderson, C. (2008). The end of theory: The data deluge makes the scientific method obsolete. *WIRED*.

Auguie, B. (2017). *gridExtra: Miscellaneous Functions for "Grid" Graphics*. R package version 2.3.

Badr, H. S., Du, H., Marshall, M., Dong, E., Squire, M. M., and Gardner, L. M. (2020). Association between mobility patterns and COVID-19 transmission in the USA: a mathematical modelling study. *Lancet Infectious Diseases*, 20(2):1247–1254.

Booth, C. (1889). *Life and Labor of the People of London*. MacMillan, London.

Browning, C. R., Calder, C. A., Soller, B., Jackson, A. L., and Dirham, J. (2017). Ecological networks and neighborhood social organization. *American Journal of Sociology*, 122(6):1939–1988.

Cheng, J., Karambelkar, B., and Xie, Y. (2021). *leaflet: Create Interactive Web Maps with the JavaScript Leaflet Library*. R package version 2.0.4.1.

Cruz, P., Wibhey, J., Ghael, A., Shibuya, F., and Costa, S. (2019). Dendrochronology of U.S. immigration. *Information Design Journal*, 25(1):6–20.

Devlin, H., Schenk, Jr., T., Leynes, G., Lucius, N., Malc, J., Silverberg, M., and Schmeideskamp, P. (2021). *RSocrata: Download or Upload Socrata Data Sets*. R package version 1.7.11-2.

DuBois, W. (1899). *The Philadelphia Negro: A Social Study*. Published for the University, Philadelphia.

Eagle, N., Macy, M., and Claxton, R. (2010). Network diversity and economic development. *Science*, 328:1029–1031.

Eagle, N., Pentland, A. S., and Lazer, D. (2009). Inferring friendship network structure by using mobile phone data. *Proceedings of the National Academy of Sciences*, 106(36):15274–15278.

Feinerer, I. and Hornik, K. (2020). *tm: Text Mining Package*. R package version 0.7-8.

Fellows, I. (2018). *wordcloud: Word Clouds*. R package version 2.6.

Fletcher, T. D. (2012). *QuantPsyc: Quantitative Psychology Tools*. R package version 1.5.

Gately, C. K., Hutyra, L. R., Peterson, S., and Wing, I. S. (2017). Urban emissions hotspots: quantifying vehicle congestion and air pollution using mobile phone gps data. *Environmental Pollution*, 229(36):496–504.

Gelfand, S. (2020). *opendatatoronto: Access the City of Toronto Open Data Portal*. R package version 0.1.4.

Grauwin, S., Sobolevsky, S., Moritz, S., Godor, E., and Ratti, C. (2014). Towards a comparative science of cities: using mobile traffic records in new york, london and hong kong. *arXiv*, 229:1406.440 [physics.soc–ph].

Grothendieck, G. (2017). *sqldf: Manipulate R Data Frames Using SQL*. R package version 0.4-11.

Harrell, Jr., F. E. (2021). *Hmisc: Harrell Miscellaneous*. R package version 4.6-0.

Hoffman, J. S., Shandas, V., and Pendleton, N. (2020). The effects of historical housing policies on resident exposure to intra-urban heat: a study of 108 US urban areas. *Climate*, 8(1):12.

Kassambara, A. (2019). *ggcorrplot: Visualization of a Correlation Matrix using ggplot2*. R package version 0.1.3.

Kitchin, R. and McArdle, G. (2016). What makes big data, big data? exploring the ontological characteristics of 26 datasets. *Big Data & Society*, 3(1).

Klare, B. F., Burge, M. J., Klontz, J. C., Bruegge, R. W. V., and Jain, A. K. (2012). Face recognition performance: role of demographic information. *IEEE Transactions on information forensics and security*, 7(6):1789–1801.

Kurgan, L., Cadora, E., Reinfurt, D., Williams, S., and Meisterlin, L. (2012). Million dollar blocks. Available at http://spatialinformationdesignlab.org/projects.php%3Fid%3D16.

Masucci, A. P., Stanilov, K., and Batty, M. (2013). Limited urban growth: London's street network dynamics since the 18th century. *PLoS One*, 8(8):e69469.

O'Brien, D. T. (2018). *The Urban Commons*. Harvard University Press, Cambridge, MA.

O'Brien, D. T., Gridley, B., Trlica, A., Wang, J., and Shrivastava, A. (2020). Urban heat islets: Street segments, land surface temperatures and medical emergencies during heat advisories. *American Journal of Public Health*, 110(7):994–1001.

Ooms, J. (2021). *gifski: Highest Quality GIF Encoder*. R package version 1.4.3-1.

Park, R. E., Burgess, E. W., and McKenzie, R. D. (1925/1984). *The City*. University of Chicago Press, Chicago.

Pebesma, E. (2021). *sf: Simple Features for R*. R package version 1.0-3.

Pedersen, T. L. and Robinson, D. (2020). *gganimate: A Grammar of Animated Graphics*. R package version 1.0.7.

Pigliucci, M. (2009). The end of theory in science? *EMBO Reports*, 10(6):534.

Ratcliffe, J. H., Taylor, R. B., Askey, A. P., Thomas, K., Grasso, J., Bethel, K. J., Fisher, R., and Josh Koehnlein, J. (2021). The philadelphia predictive policing experiment. *Journal of Experimental Criminology*, 17:15–41.

Ratti, C., Sobolevsky, S., Calbrese, F., Andris, C., Reades, J., Martino, M., Claxton, R., and Strogatz, S. H. (2010). Redrawing the map of Great Britain from a network of human interactions. *PLoS One*, 5:e14248.

Rosenthal, J. K., Kinney, P. L., and Metzger, K. B. (2014). Intra-urban vulnerability to heat-related mortality in New York City, 1997-2006. *Health & Place*, 30:45–60.

Rudis, B. (2019). *streamgraph: streamgraph is an htmlwidget for building streamgraph visualizations*. R package version 0.9.0.

Schloerke, B., Cook, D., Larmarange, J., Briatte, F., Marbach, M., Thoen, E., Elberg, A., and Crowley, J. (2021). *GGally: Extension to ggplot2*. R package version 2.1.2.

Spinu, V., Grolemund, G., and Wickham, H. (2021). *lubridate: Make Dealing with Dates a Little Easier.* R package version 1.7.10.

Townsend, A. M. (2013). *Smart Cities.* W. W. Norton & Company, New York.

von Bergmann, J. (2021). *VancouvR: Access the City of Vancouver Open Data API.* R package version 0.1.7.

Wang, Q., Phillips, N. E., Small, M. L., and Sampson, R. J. (2018). Urban mobility and neighborhood isolation in America's 50 largest cities. *Proceedings of the National Academy of Sciences*, 115(30):7735–7740.

Wickham, H. (2019). *stringr: Simple, Consistent Wrappers for Common String Operations.* R package version 1.4.0.

Wickham, H. (2020). *reshape2: Flexibly Reshape Data: A Reboot of the Reshape Package.* R package version 1.4.4.

Wickham, H. (2021). *tidyverse: Easily Install and Load the Tidyverse.* R package version 1.3.1.

Wickham, H., Chang, W., Henry, L., Pedersen, T. L., Takahashi, K., Wilke, C., Woo, K., Yutani, H., and Dunnington, D. (2021). *ggplot2: Create Elegant Data Visualisations Using the Grammar of Graphics.* R package version 3.3.5.

Wilkinson, L., editor (2005). *The Grammar of Graphics.* Springer, Medford, MA, 2nd edition.

Index

For Product Safety Concerns and Information please contact our EU
representative GPSR@taylorandfrancis.com
Taylor & Francis Verlag GmbH, Kaufingerstraße 24, 80331 München, Germany

www.ingramcontent.com/pod-product-compliance
Ingram Content Group UK Ltd.
Pitfield, Milton Keynes, MK11 3LW, UK
UKHW050954280425
457818UK00037B/351

* 9 7 8 1 0 3 2 2 6 4 5 9 2 *